Picturing Technology in China

Picturing Technology in China

From Earliest Times to the Nineteenth Century

Peter J. Golas

HKU PRESS
香港大學出版社

Hong Kong University Press
The University of Hong Kong
Pok Fu Lam Road
Hong Kong
https://hkupress.hku.hk

ISBN 978-988-8208-15-9 (*Hardback*)

British Library Cataloguing-in-Publication Data
A catalogue record for this book is available from the British Library.

Digitally printed

Contents

Illustrations

Color Plates Section (after p. 18)

Color Plates Section (after p. 130)

Preface

Chapter 1

Chapter 2

Chapter 3

Chapter 5

Chapter 6

Closing Comments

Preface

The disparagement that once pervaded much of Western thinking and writing about China's record in science and technology has now long given way to a general recognition that achievements in these realms constitute one of the great triumphs of the Chinese people and their civilization. In the area of technology, more than two generations of scholarly effort, much of it inspired by Joseph Needham and the volumes of his *Science and Civilisation in China*, have fleshed out a remarkable story of Chinese inventive genius and a broad Chinese talent for technological innovation that could hardly have been guessed at before the middle of the last century.

Parts of that story, however, remain to be fleshed out. For our purposes, none is more important than the question of how Chinese approaches to portraying technology influenced the overall development of technology in China. Until relatively recently, much of the work on surviving images of premodern technology analyzed these images, often with considerable skill, to elucidate how a particular technology or piece of equipment worked. From these studies have emerged many interesting insights into how the visual depiction of technology, as well as the technology itself, did or did not change over time.

One of our purposes in this volume will be to pull together many of these insights into an overview of what we presently understand about the illustrations of technology in China up to about the nineteenth century. At the same time, we shall pay special attention to newer approaches that are now beginning to deepen our understanding of the place of both technology and portrayals of technology in the larger Chinese sociopolitical, economic, and cultural order.

It will rapidly become clear that much of what I have to say has been inspired by that vast body of work on the portrayal of technology in European culture, especially in the late Middle Ages and the Renaissance. But despite the fact that the images produced by the artists of the Renaissance generally conform much better than most Chinese images to what we, with our twenty-first-century perspective, tend to expect from technological illustrations. Despite this, I shun the all-too-common approach that sets up the European achievements as some kind of universal standard and then seeks explanations for why the Chinese were unable to meet that standard. I am above all interested in understanding the complex of motivations that led the Chinese to produce the images they did, not why they "failed" to produce images more like those in Europe. When I reference the European experience, it is mainly to help us identify what was particularly distinctive in the Chinese experience, and perhaps suggest roots of that distinctiveness.

Moreover, it deserves to be noted that, seen in a broad historical perspective, the Chinese illustrations by no means infallibly failed to measure up to contemporary European illustrations even when judged by criteria that were much less emphasized by Chinese than by Westerners. As early as the Song dynasty (960–1278), Chinese artists on occasion produced illustrations of technological subjects that displayed a degree of accuracy, intelligibility and even realism generally not found in European images until much later. To take just one example, we can contrast the copy of a thirteenth-century Chinese illustration of the reeling or unwinding of silk fibers from cocoons in Fig. 0.1 with the spinning and weaving scene from a fourteenth-century Flemish manuscript in Fig. 0.2.

To be sure, most Chinese illustrations of technology in this period fell far short of the realism and precision of the remarkable series of painted illustrations of farming and sericulture of which Fig. 0.1 is only one example. But one will search in vain for any European portrayals of technology from these centuries that could match them.[1] After about the thirteenth century, however, one seems to see rather less advance in Chinese portrayals of technological subjects than earlier achievements might have portended. Here, too, attempting to explain why this was the case will require us to carry further our examination of the endlessly fascinating question of the interrelations between technology and broader Chinese culture.

1. For two other similarly limited European illustrations of spinning (from the early and mid-fourteenth century), see Dieter Kuhn, "Chemistry and Chemical Technology: Textile Technology," in *Science and Civilisation in China*, ed. Joseph Needham, vol. 5, part 9 (Cambridge: Cambridge University Press, 1985), 167, Fig. 107 and 168, Fig. 108.

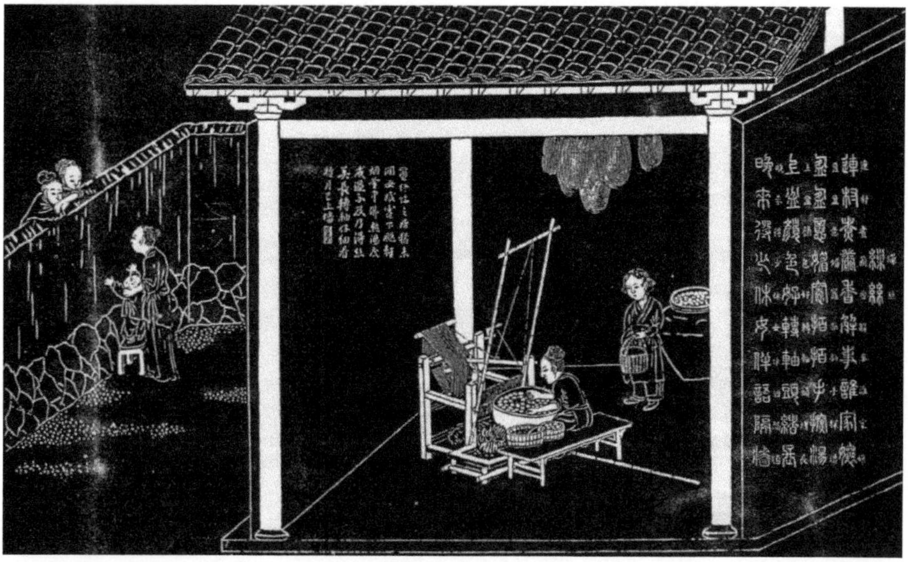

0.1 Reeling silk fibers from cocoons in the Song

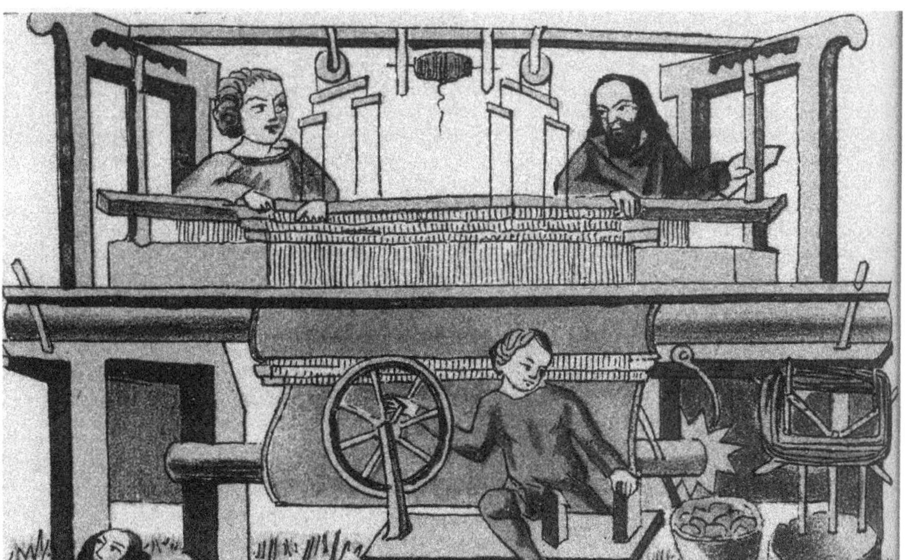

0.2 Illustration from a fourteenth-century Flemish manuscript showing a man and a woman weaving cloth on a horizontal loom powered by an overshot waterwheel while a child spins thread on a spinning wheel

Abbreviations

In the hope that this book will be of use to readers of varying backgrounds, including those with little or no familiarity with the Chinese language, the Chinese works that play an especially prominent role in this story will normally be referred to by an English translation of their titles. For convenience, these works are listed below along with a few other abbreviated titles for works that appear with some frequency. Full citations will be found in the Bibliography.

Agricultural Treatise: Wang Zhen 王禎, *Nongshu* 農書 (The agricultural treatise)
Building Standards: Li Jie 李誡, *Yingzao fashi* 營造法式 (Building standards)
CHC: *The Cambridge History of China*
ECCP: *Eminent Chinese of the Ch'ing Period*
Exploitation: Song Yingxing 宋應星, *Tian gong kai wu* 天公開物 (The exploitation of the works of nature)
HC: Joseph Needham, Wang Ling and Derek J. de Solla Price, *Heavenly Clockwork; the Great Astronomical Clocks of Medieval China*
Military Techniques: Imperial Compilation, *Wujing zongyao* 武經總要 (Collection of the most important military techniques)
Pictures of Agriculture and Sericulture: Lou Shu 樓璹, *Geng zhi tu* 耕織圖 (Pictures of agriculture and sericulture)
SCC: *Science and Civilisation in China*
System Essentials: Su Song 蘇頌, *Xinyixiang fayao* 信儀像法要 (New Armillary Sphere and Celestial Globe System essentials)
Wonderful Machines: Wang Zheng 王徵 and Deng Yuhan 鄧玉函 (Johann Schreck), *Yuanxi qiqi tushuo* 遠西奇器圖說 (Illustrated explanation of wonderful machines from the Far West)

Introduction

A useful starting point for examining how technology was portrayed in traditional China is the recognition of the place of technology in the Chinese mental landscape — or, rather, the absence of such a place. For although virtually all of the technologies that we might imagine in a traditional society were to be found in China, there was no word for and therefore no concept of "technology" in premodern Chinese. Those images that we would tend to see as portraying "technology" are part of a much larger category of images, usually referred to by the Chinese as *tu* 圖.[1] Much of the extremely varied subject matter of *tu* was of a kind that we could comfortably accept as technological, but *tu* were also used to portray an abundance of "non-technological" subject matter such as mathematics, cosmology, geography, music, ethics and magic. All of these images — and they included not only representational images but also diagrams, figures, schemata and the like — were considered to be *tu* or what Francesca Bray has called "templates for action." That is, they were seen as capable of conveying a broad range of specialist knowledge and skills that were "technological" only in the sense that they were meant to be of some kind of "practical" use. The broadness of this category *tu*, together with the absence of any overall concept of "technology," helps to explain why, although the Chinese produced a great many illustrations of technology, they never developed on any basis the concept of a special category of "technical drawings."[2]

Another reason why Chinese never came to think in terms of "technical drawings" derives from the motivations of those who produced technological images. Michael S. Mahoney, writing of Renaissance "practitioners" in Europe who did drawings of machines, provides an interesting and useful checklist of the motives that drove them: "to advertise their craft, to impress their patrons, to communicate with one another, to gain social and

1. For an excellent discussion of *tu*, to which these remarks are greatly indebted, see Francesca Bray, "Introduction," in Francesca Bray, Vera Dorofeeva-Lichtmann and George Métailié (eds.), *Graphics and Text in the Production of Technical Knowledge in China; The Warp and the Weft* (Leiden and Boston: Brill, 2007), 1–6. See also in the same volume her "Chinese Literati and the Transmission of Technological Knowledge," especially p. 299.
2. Even in Europe, as Wolfgang Lefèvre points out, technical drawings in the sense of "drawings traced by technicians for professional purposes or those derived from them" first made their appearance at the end of the Middle Ages and came to flourish only during the Renaissance; Wolfgang Lefèvre (ed.), *Picturing Machines* (Cambridge, MA: MIT Press, 2004), 2.

intellectual standing for their practice, to analyze existing machines and design new ones, and perhaps to explore the underlying principles by which machines worked, both in particular and in general."[3] It is remarkable how poorly this listing serves to suggest the motives of those who illustrated technology in China. Mahoney does not elaborate on exactly what he means by "practitioners," but we shall have occasion more than once to note the dearth of surviving Chinese illustrations produced by "practitioners" of technology, however one may choose to define the term.[4] Rather, images of technology were typically the product of painters or designers or book illustrators who had been trained as such and who did various kinds of pictures and paintings, including sometimes what we would regard as technological subjects. For them, depicting technological subjects was not intrinsically different from portraying people, trees, mountains or the other subjects on which they spent most of their time. The quality of their portrayals was seen to depend on general artistic ability as much or more than any specifically technological knowledge or expertise.

This, then, is a key point in our story: a very large proportion of the portrayals of technology in China were *not* intended, or not mainly intended, to convey technological information. Rather, they were meant to serve other goals, often especially appropriate for the artists who produced them. Aesthetically they might be aimed at arousing the interest and pleasure of viewers, or even generating an emotional response. The fifteenth-century official and writer Xia Shizheng 夏時正 expresses it well in his preface to the *Illustrated Songs on the Salines of Liangzhe*, ". . . these illustrations, which provide a general account of the salines, when taken into one's hands, will both catch the eye and excite the mind."[5] They could also be intended as visual portrayals of ideological, symbolic or moral themes. Paintings of farming practices, for example, could be intended in the first instance to symbolize a well-ordered society under the benevolent rule of the emperor, or to encourage officials to carry out their responsibilities for the welfare of the hardworking people under their jurisdiction, or perhaps to serve as a warning that all was not well in the countryside.[6] For the most part, the viewers of these images would have little or no particular interest in the technology portrayed. Hence what we are tempted to regard as technological illustrations often disappoint us when we approach them for narrowly technological information such as how an implement or machine was constructed or actually worked.[7]

3. Michael S. Mahoney, "Drawing Mechanics," in Lefèvre, *Picturing Machines*, 281. This list could be extended, for example, to include various goals associated with agreements on technological projects (convey a proposal, document an agreement, fix decisions, impose and secure control); Lefèvre, *Picturing Machines*, 4–5.
4. This is in sharp contrast with the situation in ancient and medieval Europe (as well as in Islamic lands) where there existed communities of experts who could and did write for each other while making few or no concessions for the non-expert. Bert S. Hall, "Production et diffusion de certains traités de techniques au moyen âge," in G. H. Allard and S. Lusignan (eds.), *Cahiers d'études médiévals 7; Les arts mechaniques au moyen age* (Montreal: Bellarmin, 1982), 148 ff.
5. Yoshida Tora, *Salt Production Techniques in Ancient China; the "Aobo Tu,"* trans. and rev. Hans Ulrich Vogel (Leiden: E. J. Brill, 1993), 89.
6. For many examples and detailed discussion of this point, see Hammers, *Pictures of Tilling and Weaving*.
7. Peter Jeffrey Booker, *A History of Engineering Drawing* (London: Chatto & Windus, 1963), 213. Most Chinese

For those relatively few cases where elucidation of the technology *was* the prime motive, Chinese authors instinctively relied more on text than on illustrations to convey hard technological information. In exploring why this was the case, we shall often find it necessary to consider a range of aesthetic, economic, social, intellectual and other influences that will take us well beyond any narrow focus on technology itself. Here we might simply note that most of Mahoney's motivations find no counterpart in China: there was no group of "practitioners" using technological writings to communicate with one another; almost by definition, then, no works were written "to gain social and intellectual standing" for the practice of technology; illustrations of technological subjects seem rarely if ever to have been tools for analyzing the functioning of existing machines, still less for inventing new ones;[8] and Chinese in traditional times never felt a need to try to understand the theoretical principles that underlay the functioning of machines.

It is therefore hardly surprising that the Chinese never isolated special qualities that should attach to a good, or at least a satisfactory, drawing of a technical subject per se. It seems only rarely to have occurred even to those with a special interest in technology that portrayals of technological subjects could have standards, for example precision, detail and accuracy, that applied little if at all in other kinds of visual portrayals.

♦ ♦ ♦

What we can learn about technological illustration in China's early history is of course especially constrained by the limits of the evidence at our disposal. Down to the Tang dynasty (618–906), the body of useful surviving images of technologies is particularly fragmentary, and for many periods and technologies even non-existent. Since only a handful of the most important and highly admired paintings or calligraphic works produced during these centuries have survived, almost all as copies,[9] we find ourselves often having to rely on a

---

technological equipment was quite simple and easily constructed by experienced carpenters and other craftsmen. Where more sophisticated knowledge came into play, as with the building of large ships for example, the builders in China as elsewhere typically used no manuals (they were usually illiterate), templates or blueprints, relying instead on a broad know-how, much of which could not be put into words or pictures. Derk Bodde, *Chinese Thought, Society, and Science: The Intellectual and Social Background of Science and Technology in Pre-modern China* (Honolulu: University of Hawai'i Press, 1991), 360; George Basalla, *The Evolution of Technology* (Cambridge: Cambridge University Press, 1988), 83–84; Marshall Sahlins, *Stone Age Economics* (Chicago: Aldine, 1972), 81.

8. For incisive remarks on how the ability to design machines on paper gave the designer a much greater "perceptual span" than the traditional craftsman (e.g. he could manipulate easily all or any of the parts of a machine and even make substantial changes cost-free), see the remarks of J. Christopher Jones cited in Ken Baynes and Francis Pugh, *The Art of the Engineer* (Woodstock, NY: The Overlook Press, 1981), 11.

9. It is tempting, when referring to the period before woodblock printing, to put the word "copy" in quotation marks. Close copies made by tracing seem even after the early and essential invention of paper to have been the rare exception; most often, the "copies" are freehand renderings that can vary greatly in their fidelity to the original.

frustratingly small number of stone carvings, cave paintings, decorations on objects, and other non-manuscript sources.[10]

After the appearance and the spread of woodblock printing in the seventh and following centuries, the situation changed. Much larger numbers of illustrations, especially in the form of outline drawings, were produced and many have survived down to the present. On the other hand, a survey of these illustrations leads quickly to the realization that this abundance is partly illusory. Although constituting the surviving product of something over a thousand years, they equal in quantity far less, it would seem, than what we have just from sixteenth-century Europe with its flood of illustrated books on all kinds of technology.[11] There is also the problem of duplication: a significant proportion of what we have today on the Chinese side consists of more or less close copies of earlier illustrations (though one can of course sometimes learn important things from even small discrepancies). Finally, most of the surviving illustrations come from a relatively limited number of printed works. Though manuscripts continued to be produced in considerable numbers well into the Ming,[12] there is no surviving Chinese parallel to, for example, the manuscripts of a Mariano Taccola in the earlier part of the fifteenth century or the notebooks of Leonardo da Vinci from its end.[13] Nor do *any* of the drawings that have survived seem to have been produced by craftsmen/artisans/technicians in the course of their activities.[14] Consequently, the body of surviving

10. Repeatedly we face the danger of being led astray by odd bits of evidence that may or may not be accurate or representative. The many frescoes in the Buddhist temple-caves of Dunhuang, for example, are hardly rich in technological subjects but they do contain a fair number of illustrations of boats. Nevertheless, these illustrations are as likely to mislead as to provide usable information for the historian of technology. See Joseph Needham, Wang Ling and Lu Gwei-djen, "Physics and Physical Technology: Civil Engineering and Nautics," in Joseph Needham (ed.), *Science and Civilisation in China (SCC),* vol. 4, part 3 (Cambridge: Cambridge University Press, 1971), Fig. 968 (Pl. CDIII) and 455–56.
11. Samuel Y. Edgerton, Jr. "The Renaissance Development of Scientific Illustration," in John W. Shirley and F. David Hoeniger (eds.), *Science and the Arts in the Renaissance* (Washington, DC: Folger Shakespeare Library; London: Associated University Presses, 1985), 184. Estimates suggest that between five and ten thousand European drawings of machines and machine parts survive just from the period 1400–1700; Lefèvre, *Picturing Machines,* 13. Andrea Matthies, for her research on artistic portrayals of building construction in Europe, was able to assemble 339 images of building sites dating from the twelfth to the beginning of the sixteenth century (in contrast to a mere handful of Chinese portrayals pre-1600 and not very many after that either); they contain twenty-nine portrayals of treadmills or windlasses and eighteen pictures of wheelbarrows. Andrea L. Matthies, "Medieval Treadwheels: Artists' Views of Building Construction," *TC* 33.3 (July 1992), 513 and 545.
12. Cynthia J. Brokaw, "On the History of the Book in China," in Cynthia J. Brokaw and Kai-wing Chow (eds.), *Printing and Book Culture in Late Imperial China* (Berkeley: University of California Press, 2005), 16.
13. For a categorization of early modern European sources for machine drawings, nearly all of which have no counterparts in China, see Lefèvre, *Picturing Machines,* 14. Where the illustrations came from inevitably influenced their character. To take just one example, Leonardo's drawing style has been described as "highly personal, idiosyncratic and fundamentally inimitable"; Baynes and Pugh, *The Art of the Engineer,* 29. None of those adjectives apply very well to any Chinese illustrations of technology in traditional times.
14. By one reasonable definition of technical drawings as the drawings, sketches, etc. made by technicians for their professional purposes (Lefèvre, *Picturing Machines,* 2 and 13), we have no "technical drawings" surviving from traditional China. I therefore generally avoid this term. That does not mean, however, that I accept that no technical drawings were made even in early times. Quite the contrary: see, for example, the section in

evidence offers only a very imperfect representation of the totality of Chinese efforts to picture technology in traditional times.

The nature of the surviving evidence thus precludes any attempt here to give equal attention to all kinds of surviving images of technology. Even without the evidential constraints, drawings of mechanical technology would in any case have drawn special attention in our investigation, not least because mechanisms in all their variety and increasing complexity have posed particular challenges for the draftsmen who would portray them accurately and comprehensibly. They thus have a special intrinsic interest. Moreover, drawings of machines and mechanical devices also had a special capacity to provide information that could not be conveyed in words. This helps account for the relatively larger number of drawings of mechanisms or their components that have survived. We will try, however, not to slight other kinds of images that help supplement what we would be able to say solely on the basis of illustrations of mechanisms. Certain kinds of agricultural drawings as well as architectural depictions provide good examples.[15] Especially in earlier periods, they are relatively well represented among the surviving images and related texts. Moreover, they sometimes pioneered interesting advances in drawing techniques, some of which were later taken up in illustrations of other technologies.

♦ ♦ ♦

The history of premodern Chinese portrayals of technology divides rather well into two periods. During the first, from earliest times down to the Song (960–1279) and Yuan (1279–1368) dynasties, a growing ability of Chinese artists to portray technological subjects with greater clarity and accuracy went hand-in-hand with advances in verisimilitude that marked Chinese painting generally. Then, in the following centuries, much of the most prestigious Chinese graphic art turned away from the pursuit of realism. In the Ming (1368–1644) and Qing (1644–1911) periods, Chinese painters commonly eschewed techniques that might have led to the more accurate and effective portrayal of technological subjects. How and why this happened constitutes a major focus of this book.

Since the length and complexity of the story has led us to adopt an essentially chronological approach, this has meant that there were times when discussion of certain topics could not be treated fully in one place and had to be divided among two or more chapters. In other cases, chronological considerations have led to the grouping of rather disparate

---

Chapter 2 on "Models, Automata and Technological Drawings." Nor am I as convinced as Lefèvre that there were no technical drawings in Europe prior to the construction of the Gothic cathedrals; he himself seems to imply otherwise when he writes that *"[a]lmost all of the crafts* performed their professional tasks without drawings." (my italics) Lefèvre, *Picturing Technology*, 2.

15. Interestingly, for Albrecht Dürer, "construction sites were the places where drawing was practiced at its highest level." Cited in Filippo Camerota, "Renaissance Descriptive Geometry: The Codification of Drawing Methods," in Lefèvre, *Picturing Machines*, 200. We shall discuss in Chapter 3 why this was *not* the case at Chinese construction sites.

topics in a given chapter. What follows then will be an attempt to provide a kind of roadmap indicating how many of the most important topics and themes fit into our chronological framework. Readers will also notice a certain imbalance in the story as it is told here, with much more space devoted to the Song and post-Song periods than to what came before. To a great extent, this reflects how much less we know about the earlier period because of the dearth of surviving evidence. But it also results from the fact that one of our overriding questions — why illustrations of technology displayed so little further development in the post-Song centuries — can and must be answered in some detail on the basis of the more extensive evidence for that period and the consequent scholarly attention that has been devoted to it.

Following this introduction, Chapter 1 takes up the appearance and evolution of early graphics in China. We introduce some of the first surviving Chinese efforts to portray technical subjects and the context out of which they arose. In particular, we shall direct our attention to the emergence of certain ideas — a preference for generality over specificity, great emphasis on the moral and persuasive power of pictures — that had an enduring influence on the role illustrations later came to play not only in the graphic arts and in the illustration of technology but also in Chinese culture generally.

In this chapter, we also pay special attention to the unique importance of early illustrations of farming and clothmaking in China. To be sure, the paucity of surviving illustrations from this early period will force us to defer until later our discussion of questions such as the extent to which the kinds of illustrations produced were influenced by the mechanically rather simple technology of agriculture or by the sometimes considerably more complex technology of clothmaking. Here, however, the mainly textual evidence enables us to identify the emergence of clear traditions governing how and why agricultural activities were to be portrayed and to note the social and political basis for a phenomenon unique to China: an extensive body of illustrations depicting agricultural work by peasant farmers but produced by artists from an elite background and reflecting their values.

Chapter 2 deals with the period mainly from the third to the tenth centuries when the continuing paucity of surviving illustrations of technical subjects obliges us to examine general developments in painting in order to tease out what they might be able to add to our knowledge of technical depictions at this time. We shall see that, during these centuries, Chinese painting remained generally committed to realistic narrative representations, with many artists devoted to creating greater verisimilitude in their paintings. But just as the prevailing aesthetic values were encouraging greater realism in painting and drawing, the almost exclusive use of the Chinese brush for visual representations may have impeded the development of certain representational techniques. Moreover, while the invention of woodblock printing toward the end of this period made possible a much wider reproduction and circulation of illustrations, it was a technology that also contained in itself the potential to inhibit advances in illustration techniques.

Chapter 3 brings us to the Song (960–1279) and Yuan (1279–1368) periods, widely regarded as the highpoint in traditional times of both Chinese technological creativity as well as the visual portrayal of technology. Before turning to five landmark works from the eleventh to the fourteenth centuries, each of which represents a new level of achievement in the graphic presentation of technology, we examine one style or technique of Chinese painting, "ruled-line painting" (*jiehua* 界畫), which is the only important non-freehand style of drawing and painting ever developed in China and which, more than any other style, also displayed remarkable potential for the portrayal of technological subjects. We shall see how certain artists realized this potential but also how certain impediments prevented broader use and enhancement of these techniques. We shall also consider the complex role of a government that often supported and promoted technology but also, in its efforts to keep control over certain technologies firmly in its own hands, could serve as a brake on technology itself as well as on its depiction.

The earliest of our landmark works is a military manual, the *Collection of the Most Important Military Techniques*, compiled by imperial order and dating from 1044. As with most early illustrated works on technology, we no longer have its original illustrations and must therefore constantly make judgments, which will always remain tentative, about how closely one or another later copy approaches the original. Nevertheless, it does seem that techniques seen in later versions of the illustrations that would have been "cutting edge" for the early Song period, techniques such as front and rear elevation drawings as well as component parts and assembly drawings, may well have been used in the Song originals. None of them, in any case, were unique to this work.

In addition to technology used for the production of things, some of the most sophisticated technology in China, as in other early civilizations, aimed at the production of information and knowledge. Astronomy and astrology provide the most remarkable examples. The devising of sophisticated astronomical measuring instruments was well underway as early as the Former Han (202 BCE–9 CE). Though we lack confirming evidence, it appears highly likely that something like working drawings must have, from very early times, aided in the construction of these instruments. In any case, such drawings were reaching ever higher levels of sophistication by the Tang (seventh to ninth centuries) and giving birth to a drawing tradition that made possible at the end of the eleventh century the production of one of the greatest masterpieces of traditional Chinese technical illustration, the *New Armillary Sphere and Celestial Globe System Essentials* of Su Song. Its sixty-one illustrations, focusing exclusively on the workings of a single complicated mechanism, display a number of breakthroughs. As in the *Military Techniques*, some were assembly or sub-assembly drawings, but rendered in greater detail than in the earlier work. Many made effective use of labels within the picture, and some even provided answers to the problem of revealing the workings of those hidden parts of a machine that cannot ordinarily be seen.

It is impossible to imagine the *System Essentials* (or Su Song's clock tower itself, which it described) as having been produced without all-out support by the government. One can say much the same for another, almost exactly contemporaneous, masterpiece, the *Building Standards* of Li Jie. Li, like Su, was an official in the central government and was assigned to codify the practices used in the building of various kinds of government edifices. Although it was meant as a handbook for officials who would be supervising government projects and consequently devoted a great deal of discussion to subjects such as the costs of materials and labor as well as how to avoid problems such as pilferage, it was also unique in the detail, accuracy and comprehensiveness with which it depicted actual building practice. It has enabled scholars to reconstruct building techniques in the Song with a thoroughness that would be inconceivable were they confined to examining the very few buildings that survive from that period or later writings on building techniques, none of which ever came close to matching the thoroughness of the *Building Standards*.

Given the importance of agriculture in Chinese life and technology, it should not surprise us that not one but two of our masterworks from the Song-Yuan period focus on agriculture (and the closely-associated manufacturing of cloth, especially but not exclusively silk). The first, dating from the first half of the thirteenth century, is the *Pictures of Tilling and Weaving* by Lou Shu. In two scrolls, Lou presented some twenty-one scenes dealing with agriculture and another twenty-four dealing with silk production. This is the first example we know of in China that recorded extended processes of production in individual illustrations of their successive steps. Lou also included with each scene a lyric connected (albeit sometimes tenuously) with what was happening in the illustration. Just what importance Lou assigned to the portrayal of farming and silkmaking technology is an extremely complex question, as is his motivation for creating these scrolls. Current scholarly opinion tends to place a great deal of emphasis on moral/ideological points Lou was supposedly attempting to make. However that may be, these works had a powerful influence on how agriculture came to be portrayed for centuries afterwards. They are also typical of most other portrayals of farming activities, where the technology in use seldom occupied center stage.

The second of these two major agricultural works was Wang Zhen's *Agricultural Treatise* which dates from the beginning of the fourteenth century. It not only pioneered a new approach to written descriptions of technology but also displays the creative experimentation that marked the best visual portrayals of technology at this time. The treatise can be seen as a kind of culmination of the agricultural handbooks officials had been compiling on their own initiative since the Song in an effort to promote better farming methods. But contrary to the earlier handbooks, the *Treatise* is the first surviving handbook to contain illustrations. And it provides them in unique abundance with almost three hundred drawings and diagrams relating to the practice of agriculture. Moreover, Wang's primary focus on the actual technology in use reveals itself in how often he presents the tools and machines by themselves without landscape settings or people using them. In other words, some painterly elements that are prominent in Lou Shu's scrolls are more sparingly employed here. When farmers and workers do make an appearance, they frequently are portrayed in such a way as to contribute to a better comprehension of the technology.[16] It is productive activities and not aesthetic or moral values that dominate. Here Wang was truly unique. No one who came after him ever wrote such detailed discussions and provided such profuse illustrations of farming equipment. When these kinds of illustrations appear in later works, chances are that they were borrowed from Wang's *Treatise,* either directly or through later copies.

The Song and Yuan dynasties witnessed not only unprecedented achievements in technological illustration but also a major reorientation of much of Chinese life and culture, the subject we turn to in Chapter 4. Many of these developments are reflected in altered thinking about just what the aims of painting should be. By the end of the eleventh century, a new scholar-official elite (often referred to in English as the "literati") now dominated the political and cultural scene in China. To a degree unique among ruling elites anywhere, many of them took up painting as one expression of their cultural superiority. In so doing, they frequently subscribed to a view that rejected the importance of realistic representation in favor of painting that supposedly enabled highly cultivated artists of superior character to give expression to their moral and intellectual pre-eminence. In place of technical skill, many literati critics and connoisseurs even came to admire a certain clumsiness or awkwardness which they held to be harder to achieve than mere skill. These ideas, which influenced not only painting but also other kinds of graphic art, ran directly counter to the kind of exactness and accuracy that make portrayals of technology informative. At the same time, a general consensus on a relatively limited repertoire of subjects suitable for this kind of painting, above all subjects that were easy to paint and could carry significant symbolic baggage, also served to discourage realistic portrayals of technology.

The new approach to painting was further inspired by a vision of the world and the cosmos that was perhaps the most original component of resurgent Confucianism in the Song. It too undoubtedly inhibited to at least some extent both the further development of technology itself as well as efforts to get it accurately onto paper or silk. Its focus on the supposed principles or ideas (*li* 理) of which all things were seen to be manifestations, together with a belief that all objects of a given class shared the same principle, encouraged the scholar-painter to seek to capture the commonalities that united objects rather than to portray those objects in all their detailed individuality. It was only, in the famous words of Su Shi (1037–1101), "a superior man of outstanding talent" who could achieve this in his painting. Mere accurate representations were the work of craftsmen and professionals, who might be highly skilled but who worked according to lower standards and did not deserve mention in the same breath with scholar-painters.

This last point alerts us that there can be a danger of overstating the negative effects of developments in one kind of painting, that by scholar-amateurs, on technological portrayals in general. Beyond the rarefied circles in which most scholarly painting was done, images of technology survive in paintings by professional artists who may have been influenced by the prevailing standards of literati painting, but who gave them less than unconditional

16. This is especially true for the illustrations of sericulture.

adherence. Unfortunately, these works have survived in relatively small numbers compared to the literati paintings prized, collected and commented on by critics and connoisseurs over the centuries since. Moreover, in those that have survived, one detects little inclination and certainly not any general trend toward more effective portrayals of technology.

Because of a great upsurge in book printing during the sixteenth century, however, by far most of the technological depictions we have from the Ming and Qing dynasties are not paintings but rather outline drawings in the form of book illustrations. Inevitably, the demands and the potential of these drawings differed from those of painting. Some of these differences could impact negatively on the portrayal of technology. For example, it was intrinsically more difficult to include fine details in a woodcut than in a painting,[17] all the more because the Chinese regularly used softer plankwood for their blocks instead of harder end-grain wood that would have permitted finer lines and hence finer detail.[18] On the other hand, given a growing and broader readership and the consequent production of many books that focused on providing practical information and entertainment for ordinary readers, realistic representation played a greater role in book illustrations than in paintings. At the same time, in the often highly competitive world of Ming and Qing book publishing, the quality of illustrations often suffered from publishers' efforts to hold down costs. Very crude illustrations of technology such as those found in surviving household manuals that aimed at the lower end of the market provide abundant examples of this process at work. More generally, it is probably fair to say that most of the technological subjects appearing in book illustrations, like most of those in paintings, were treated cursorily and conventionally. As long as enough information was included to make the subjects recognizable, few readers seem to have minded.

Nonetheless, as we discuss in Chapter 5, it was in this environment that there appeared, in 1637, a remarkable compendium entitled *The Exploitation of the Works of Nature* whose author, Song Yingxing, attempted for the first time in Chinese history to provide an overview of most of Chinese productive technology. Its textual descriptions as well as its profuse illustrations — 122 line drawings in the original edition — are our best single source for information on traditional Chinese technology. This is especially true both because traditional Chinese technology had largely reached its maturity by the early seventeenth century and because no other author before the twentieth century ever again attempted to present so comprehensive an overview of that technology.

17. Paintings intended to be reproduced by woodblock printing quite likely were often intentionally simplified or schematized to make the carving of the blocks easier; Liu Heping, "*The Water Mill* and Northern Song Imperial Patronage of Art, Commerce, and Science," *The Art Bulletin* 84.4 (December 2002), 574.

18. Lucille Chia, "Text and *tu* in Context; Reading the Illustrated Page in Chinese Blockprinted Books," in Jean-Pierre Drège (ed.), *Dossier: Texte et image dans le livre illustré chinois*. Special section in *Bulletin de l'École française d'Extreme-Orient* 89 (2002), 243. In Europe, the limits on the ability of woodblock to reproduce minute details were reached already in the mid-sixteenth century; William M. Ivins, *Prints and Visual Communication* (Cambridge, MA: Harvard University Press), 47; Michela Bussotti, "Woodcut Illustration: A General Outline," in Francesca Bray, Vera Dorofeeva-Lichtmann and George Métailié (eds.), *Graphics and Text in the Production of Technical Knowledge in China: The Warp and the Weft* (Leiden and Boston: Brill, 2007), 469.

The richness of this work together with the availability by the late Ming of extensive materials on a great number of topics that relate to its composition enable us to explore in greater depth than possible for earlier works certain questions crucial to our understanding of how the illustrating of a work such as Song's was carried out and why the illustrations took the form they did. Moreover, the illustrations are even more useful because, for every one of the illustrations of the original edition, we have one or more later versions, typically copies, of the same subject. Examining the similarities and differences in the various versions can tell us a great deal about the capabilities and the limitations of Chinese illustration of technology from the seventeenth to the nineteenth centuries.

Our final chapter takes up developments in the Qing period, above all the fascinating story of the Jesuit efforts to introduce into China the most up-to-date painting and drawing techniques from Europe (along with much else from Renaissance culture), and Chinese responses to those efforts. Two outstanding figures will help flesh out this story: Wang Zheng, the first Chinese to attempt to engage seriously with Renaissance mechanics, and the painter Jiao Bingzhen who produced under imperial sponsorship a remarkable new series of illustrations for the *Pictures of Tilling and Weaving* (Chapter 3) that were very much influenced by the new Western techniques, especially linear perspective. Our examination of certain developments in Renaissance Europe that found little or no resonance in China will help us to understand the difficulties Wang encountered in trying to get an intellectual grip on both the ideas and the illustrations of the new science of mechanics as well as the limited influence Jiao's and similar efforts had on Chinese painting practice. In the latter case, the new techniques were met with considerable skepticism as to their value or usefulness even when they enjoyed a certain admiration for their cleverness.

Finally, we shall also examine a kind of last flowering of realism in traditional Chinese painting, the great revival of *jiehua* painting during the Qing. *Jiehua*, as discussed in Chapter 3, was the style of painting that had most in common with the outline techniques of book illustration. Though it displayed many qualities such as precision and attention to details that were essential to good technical illustration, it ultimately failed, probably because of its almost exclusive concern with architectural images, to exert any broad influence on Chinese illustration techniques at the end of the imperial period.

The larger picture that will emerge from this and the previous chapter is that illustrations, sometimes devoted to or including technical subjects, were a vibrant and pervasive segment of China's flourishing book publishing industry from the fifteenth to the nineteenth centuries. Nevertheless, that vitality did not generally express itself in the exploration of new directions in the techniques of technical drawing. This will lead us to a brief closing discussion of possible causal links — in both directions — between technological illustrations and the advance or stagnation of technology.

One

Early Graphics in China

Graphic representations have a very long history in the area encompassed in or surrounding today's China though few are to be found from the specific areas where Chinese civilization first developed. Among the earliest of these representations are rock paintings and engravings (petroglyphs), drawings on the floors of Neolithic huts,[1] and decorations on pottery. Abbreviated as they are, these very early drawings can occasionally provide even fairly specific information relevant to early technologies in East Asia.[2]

Pictorial Writing

Much richer in quantity and technological content are the signs or graphs made up by the Chinese as they began to develop a writing system, probably around the middle of the second millennium BCE.[3] Indeed, on the basis of the early forms of written characters for approximately eighty words "important in scientific thinking" in China, Joseph Needham found twenty-eight to have some kind of technological reference, more than any other category.[4] Many of the early Chinese graphs were pictorial representations of objects used in daily activities including plowing, weaving, sailing, basket making, and calculating. They often convey essential technological details. For example, early characters for "bow" (Fig. 1.1 (a)) leave no doubt that we are looking at a reflex composite bow.[5] A whole series of slightly varied early characters for "boat" (Fig. 1.1 (b)), are consistent in their depictions of straight prows and sterns, giving the boat a punt shape typical of many boats still to be seen in China today (Fig. 1.2); they also often depict the transverse walls that divided the boat

1. Yang Xin, "Approaches to Chinese Painting," in Yang, Barnhart et al., *Three Thousand Years of Chinese Painting*, 1.
2. Wu Hung, "The Origins of Chinese Painting," in Yang, Barnhart et al., *Three Thousand Years of Chinese Painting*, 17 Fig. 3.
3. William G. Boltz presents an excellent discussion of the earliest writing in China, including its dating; see *The Origin and Early Development of the Chinese Writing System* (New Haven: American Oriental Society, 1994), Chap. 2, especially p. 39.
4. Joseph Needham and Wang Ling, "Physics and Physical Technology: Mechanical Engineering," in Joseph Needham (ed.), *Science and Civilisation in China (SCC)*, vol. 4, part 2, 218–31, esp. 231.
5. Tsuen-hsuin Tsien, *Written on Bamboo and Silk; The Beginnings of Chinese Books and Inscriptions* (Chicago: University of Chicago Press, 1962); 2nd ed., 2004, 26; Needham and Wang, *SCC* 4:2, xliii.

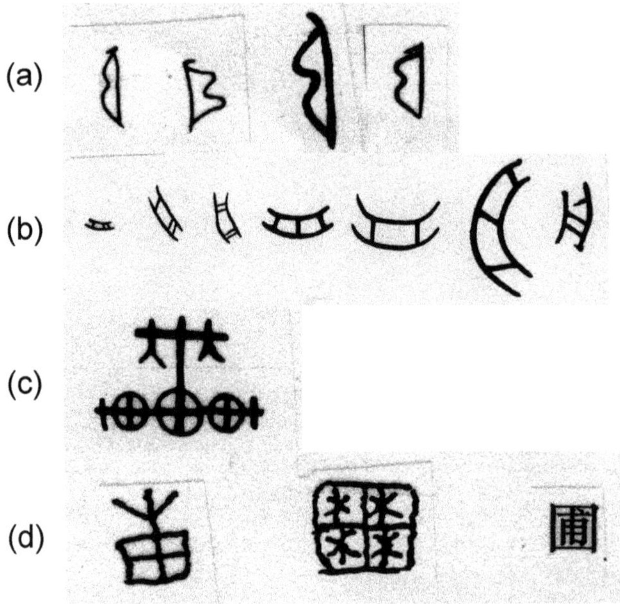

1.1 Early forms of Chinese written characters: (a) bow; (b) boat; (c) cart or chariot; and (d) plants grown in an enclosure and the modern character for vegetable garden.

into sections.[6] Early forms for "cart/chariot" varied more than those for "boat" and the most complex of them (Fig. 1.1 (c)) indicate the basket, the wheels and the inverted V-shaped collars for the normally two horses that pulled the chariot.[7] Sometimes, these early forms of characters go beyond just picturing objects and provide hints about technological practices. For example, the two characters from which parts were later merged to make up the character *pu*, "vegetable garden" or "orchard" (Fig. 1.1 (d)), both suggest plants grown in an enclosure, and thus perhaps indicate that horticulture was a significant part of Chinese agriculture by the early Zhou.[8]

Fascinating and informative as these and other early characters are to the historian of technology, they had virtually no influence on how technology came to be portrayed in other contexts in the early period. And over time, unlike Egyptian hieroglyphics, which strongly maintained their pictorial character, the Chinese writing system developed toward simplification and abstraction and lost in the process most of its pictorial elements, including those related to technology. In later times, a much more interesting question became the relative emphasis on writing and on images in the portrayal and explanation of technology.

6. Cecilia Lindqvist, *China: Empire of Living Symbols*, trans. Joan Tate (Reading, MA: Addison-Wesley, 1991), 150, 375.
7. Lindqvist, *China*, 139–41.
8. H. T. Huang, *SCC* 6, part 5 (Cambridge: Cambridge University Press, 2000), 32.

Geometrical Designs

While only two of the characters assembled and analyzed by Needham could be seen as "pure geometrical symbolism," geometric forms nevertheless played a large role in early Chinese graphic art. The petroglyphs, paintings on pottery, jade carvings and designs on bronzes are typically highly stylized and non-representational, serving mainly as pure ornament. Only occasionally does one find pictorial representations of faces, human and animal profiles, or definable activities.[9] Realistic representation seems to have held little interest for Chinese artists in this period.[10] Rather, as Jerome Silbergeld has noted, there seems to have been an artistic urge to generate forms from a limited vocabulary of abstract shapes."[11] Even in those portrayals where one finds something approaching realism (see Plate 1(a)), the dominant impression is one of timelessness and immobility.[12] The absence of details or attributes often produces portrayals that seem more like diagrams than pictures or representations.[13]

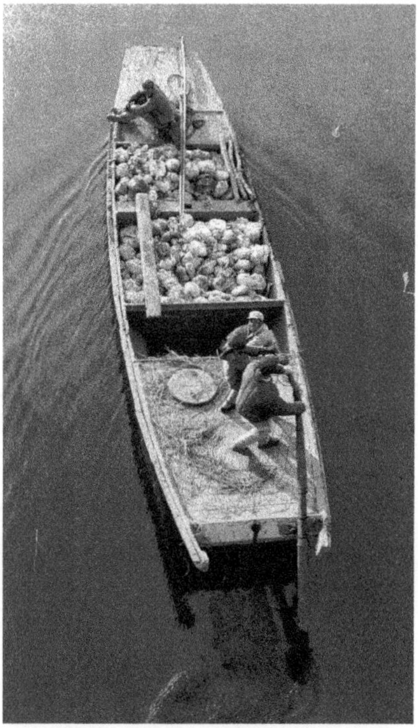

1.2 Boat carrying vegetables on a Suzhou canal. The distinctive construction is reflected in the earliest written characters for boat.

9. Lindqvist, *China*, 30–31.
10. Ladislav Kesner, "Likeness of No One: (Re)presenting the First Emperor's Army," *Art Bulletin* 77.1 (March 1995), 130; Lawrence Sickman, "Chinese Painting," in Wai-Kam Ho et al., *Eight Dynasties of Chinese Painting* (Cleveland: Cleveland Museum of Art, 1980), xiii; Cordell D. K. Yee, "Cartography in China: Chinese Cartography Among the Arts," in J. B. Harley and David Woodward, *A History of Cartography*, Vol. 2, Book 2, *Cartography in the Traditional East and Southeast Asian Societies* (Chicago: Chicago University Press, 1994), 129. Our impressions may, however, be somewhat skewed by the fact that it is mainly ritual objects made of particularly durable materials that have survived to be discovered by archaeologists. It could be that representation was deemed inappropriate for these more or less sacred objects but that it was employed on objects made of less durable materials such as textiles and wood that were created for non-ritual purposes. Robert L. Thorp and Richard Ellis Vinograd, *Chinese Art and Culture* (New York: Harry N. Abrams, Inc., 2001), 111–12.
11. Jerome Silbergeld, *Chinese Painting Style; Media, Methods, and Principles of Form* (Seattle and London: University of Washington Press, 1982), 47 and Fig. 9.
12. Wu Hung, "Origins," 15–19; Silbergeld, *Chinese Painting Style*, 7; Jessica Rawson (ed.), *Mysteries of Ancient China; New Discoveries from the Early Dynasties* (New York: George Braziller, 1996), 34–36.
13. Loehr, "Fundamental Issues," 192; Booker, *History of Engineering Drawing*, 16.

Another striking characteristic of early decorative art in China is the complexity often achieved in geometrical designs (see Plate 1(b)). The more intricate of these precisely planned and executed designs required practical solutions to a wide range of drawing problems such as division of lines or arcs (including circumferences) into equal parts, drawing parallel lines, diagonals and tangents, and constructing regular geometrical figures such as squares and a variety of polygons, circles and ellipses. To solve such problems, artists needed measuring and drawing tools.[14] Hence it is likely that rulers, squares, and drawing compasses[15] were in fairly wide use in China by the late Neolithic.[16] In later periods, the descendants of these tools came to be used in a broad range of non-artistic activities: construction,[17] illustrations of mathematical problems[18] and astronomical measurement. By the second century BCE, measuring instruments (including also the water-level and the steelyard) had even acquired a powerful symbolic character as metaphors for how Heaven regulated the universe or how the sage arrived at correct rules of proper conduct.[19]

Had the Chinese interest (in this early period we might almost call it an "obsession") in manipulating geometrical shapes continued strong, it could well have served as a significant stimulus to the drawing of technological subjects.[20] For the most part, however, and for reasons that are still not clear, Chinese taste took a different direction by the Han dynasty.

---

14. For the early history (including the Neolithic period) of measurement in China, a superb starting point is David N. Keightley, "A Measure of Man in Early China: In Search of the Neolithic Inch," *Chinese Science* 12 (1995), 18–40. See also Li Yan and Du Shiran, *Chinese Mathematics; A Concise History*, trans. John N. Crossley and Anthony W.-C. Lun. (Oxford: Clarendon Press, 1987), 2–3.
15. Lu Jingyan 路敬嚴 and Hua Jueming 華覺明 (eds.). *Zhongguo kexue jishu shi* 中國科學技術史 [A history of Chinese science and technology] (Beijing: Kexue, 2000), 204–10.
16. Rulers with gradations in *cun* 寸 or *fen* 分 have been found from the Shang (Lu and Hua, *History of Chinese Science and Technology*, 156) and it is possible that protractors or calipers-like instruments were also used in this early period, though the evidence for this is not conclusive. Maeda (Robert J. Maeda, "*Chieh-hua*: Ruled-Line Painting in China," *Ars Orientalis* 10 (1975), 125) suggests that early drawing tools were "borrowed from the carpenter's trade" but, given the vagueness of the evidence bearing on when they came into use for different purposes, this can be only speculative. Actually, Wu Jiming's suggestion that measuring tools were first used in the production of pottery is equally if not more persuasive; Wu Jiming 吳繼明, *Zhongguo tuxue shi* 中國圖學史 [A history of Chinese drawing] (Wuchang: Huazhong ligong daxue, 1988), 7. We must also allow for the likelihood of early use of measurement in the sizing and shaping of jades as well as the abrading of their designs; David N. Keightley, "Archaeology and Mentality: The Making of China," *Representations* 18 (1987), 91–128. Later, measurement was also essential for the founding and the decorating of bronzes.
17. Perhaps their earliest use in construction was in the marking off of the circumferences of round houses such as those of the Neolithic village at Banpo, which were often so regular as to have required some kind of compass-like measuring tool; Wu, *History of Chinese Drawing*, 10. Wu Jiming also suggests (p. 25) that working drawings for the construction of wheels (marking out the circumference and the spokes) were used already in the Shang.
18. Wu, *History of Chinese Drawing*, 8.
19. Joseph Needham, Wang Ling and Kenneth Girdwood Robinson, "Physics and Physical Technology: Physics," in Joseph Needham (ed.), *Science and Civilisation in China (SCC)*, vol. 4, part 1 (Cambridge: Cambridge University Press, 1962), 15–17; Bodde, *Chinese Thought, Society, and Science*, 133–47, esp. 135; 338–41.
20. Even in the portrayal of stylized creatures and vegetation, the artists who decorated Neolithic (Yangshao and Longshan) pots were already experimenting with different projections such as frontal and profile; Lu and Hua, *History of Chinese Science and Technology*, 156.

Only traces of the earlier infatuation with geometric forms survived for a time, as in the use of geometrically regular forms for relatively simple shapes like the eyes of humans or horses[21] or in the decorative patterns on mortuary stones in Han tombs.[22]

It was at this point that images with technological content begin to appear with some regularity in the Chinese graphic productions.

From Ornament to Narrative: Warring States and Han Illustrations

The Warring States period (5th to 3rd centuries BCE) witnessed a dramatic shift in China's decorative arts. On the one hand, as we noted, the previous predominance of geometrical designs in decorative or ornamental contexts slowly gave way to a greater emphasis by artists and craftsmen on pictorial representations of people, animals and, generally, of the real world around them or their imaginary vision of another world.[23] As this occurred, many new subjects, a few related to technology, were added to a previously quite limited repertoire of "semi-concrete motifs"[24] and both the old and the new images came to be represented more realistically than previously. Increasingly we see narrative portrayals of mythological and historical subjects[25] such as scenes of battles, processions, concerts, hunts and banquets, as well as more mundane activities (Fig. 1.3).

This development presumably reflected at least in part some basic changes in the way Chinese were coming to think not only about the world around them but also what might await them in the afterlife. Increasingly, they came to see this world in realistic and objective terms, with less importance attached to fantastic and magical forces. At the same time, their ideas of a life after death made it not all that different from what they experienced while alive. These new conceptions are best exemplified in the actual provisions made for the dead. Already by the fifth century BCE, tombs of the elite increasingly reflected the dwellings the dead occupied when they were alive. In the effort to create a somewhat idealized version of the conditions of ordinary life, the tombs not only were outfitted with more objects of daily use including clay, wood or bronze models but were also decorated with portrayals (such as

21. Martin J. Powers, *Art and Political Expression in Early China* (New Haven: Yale University Press, 1991), 25; Anthony J. Barbieri-Low, *Artisans in Early Imperial China* (Seattle: University of Washington Press, 2007), 91.
22. Perhaps the longest and most important survival of geometric shapes for decorative purposes is the endlessly varied wooden or stone lattice designs so beloved by the Chinese in the windows of their houses, temples and palaces.
23. Lucy Lim, "Form and Content in Early Chinese Pictorial Art," in Lucy Lim (ed.), *Stories from China's Past; Han Dynasty Pictorial Tomb Reliefs and Archaeological Objects from Sichuan Province, People's Republic of China* (San Francisco: The Chinese Culture Foundation of San Francisco, 1987), 51; Dieter Kuhn, "Silk Weaving in Ancient China: From Geometric Figures to Patterns of Pictorial Likeness," *Chinese Science* 12 (1995), esp. 80–83 and 104–10; Cordell D. K. Yee, "Cartography in China," 128–31; Barbieri-Low, *Artisans in Early Imperial China*, 45; Peter J. Golas, "Emergence of Technical Drawing in China: The *Xin Yi Xiang Fa Yao* and Its Antecedents," *History of Technology* 21 (1999), 30.
24. Loehr, "Fundamental Issues," 191; Silbergeld, *Chinese Painting Style*, 47.
25. Wen C. Fong, *Beyond Representation: Chinese Painting and Calligraphy, 8th to 14th Century* (New Haven: Yale University Press, 1992), 2, 13.

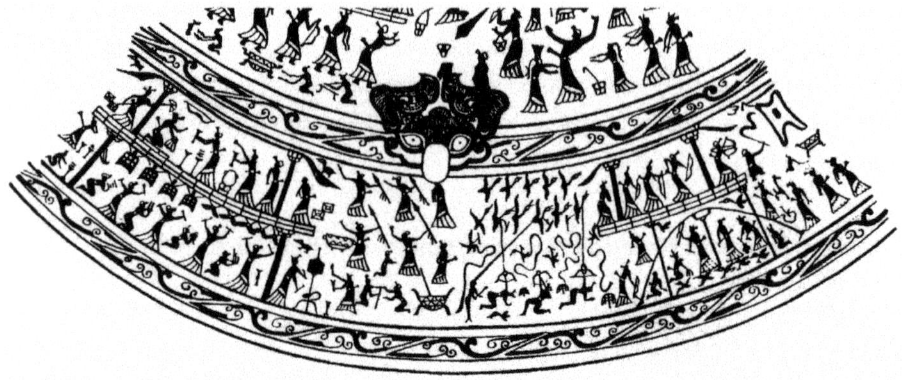

1.3 Designs cast into the surface of a fourth-century BCE bronze bowl. Besides the musical performance and archery, the scene also depicts buildings with extended eaves and elaborate brackets.

stone bas-relief carvings, scenes molded in low relief on ceramic slabs, painted wall murals and funeral banners) of a wide range of daily life activities such as farming, clothmaking, hunting, fishing, cooking and the making of wine.[26]

The information on technology provided by the pictorial reliefs is necessarily limited by the nature of the medium itself, which hardly encouraged the portrayal of specifics and fine details. Thus, in Fig. 1.3, we see enough to know that fairly complex building brackets were in use but, on the basis of these images, can say little more about them. Indeed, as here, the main information we gain is the knowledge that a particular technology had made its appearance by the Han. If the same elements appear in several reliefs, perhaps from different parts of the country, that can provide at least a hint of how widely it was used.

Nevertheless, there are cases where what we learn goes beyond, sometimes considerably beyond, these basics. For example, we have in Fig. 1.4 two looms from a vaulted brick tomb from the vicinity of Chengdu, Sichuan and dated to the Later Han.[27] The first (a) is quite basic, probably used for simple weaving procedures. That is certainly not the case for the second loom (b), which may have been used for the complex weaving of brocades. Richard C. Rudolph provides an excellent description:

> [The brocade loom] has the weaving frame raised twenty-five degrees or more above its horizontal supports, which greatly increases the efficiency of the weaver. The warp beam at the top of the frame has two hourglass-shaped "locknuts" that keep it from rotating after it is set in the proper position . . . At ground level are two foot treadles, an

26. Lothar Lederrose, *Ten Thousand Things: Module and Mass Production in Chinese Art* (Princeton: Princeton University Press, 2000), 66; Rawson, *Mysteries of Ancient China*, 28–29. The pictorial reliefs, by far our richest resource for portrayals of technology in the Han period (and often probably quite similar in style to paintings of the same subjects that have nearly all disappeared), have been extensively studied and have generated a rich bibliography. I have found Lucy Lim, "Form and Content in Early Chinese Pictorial Art," 51–53, to be a particularly illuminating introduction. For a very large selection of these reliefs, see Käte Finsterbusch, *Verzeichnis und Motivindex der Han-Darstellungen* (Wiesbaden: Otto Harrassowitz, 1971).

27. For the complete relief, see Lim, *Stories from China's Past*, 95.

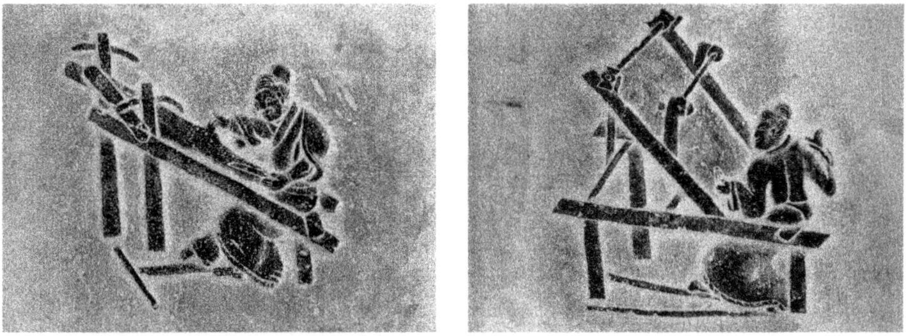

1.4 Later Han stone reliefs showing (a) a woman at a weaving loom for relatively uncomplicated textiles and (b) a woman at a weaving loom for brocades.

> important addition in the evolution and efficiency of the loom. The treadles are connected by cords [not shown] to heddles supported by the "horse head"–the horizontal rod on uprights below the warp beam. When this apparatus is activated by pressing on a treadle, the horse heads swing back and forth, separating the warp threads into upper and lower layers. The angle formed between them is the shed, or passageway, through which the shuttle containing the weft is thrown.[28]

Another extremely valuable piece of evidence, in this case relating to food processing technology, is the stone relief discovered in 1959–60 in Henan province and also dated to the Later Han. It pictures what are almost certainly four steps in the process of making soy bean curd (*tou fu*) and is by far the best surviving visual indication that this process had already been developed by Han times.[29] Finally, the wheelbarrow and its evolution provide another example of the unexpectedly detailed information careful and thorough study can elicit from what might at first seem unpromising sources. Although depictions of wheelbarrows in the stone and clay reliefs from this early period are usually quite simplified, there are relatively so many of them, with sufficiently identifiable variations, that Joseph Needham was able to construct a typology including nine different Chinese wheelbarrow designs![30]

The wealth of evidence on Han technology provided by the stone reliefs makes all the greater our regret that so little of the painting of that period has survived. On the other hand, we should not overestimate how much more we would have learned from access to a greater number of paintings. To be sure, painting must also have been significantly influenced by the growing prominence of the Chinese humanistic view that strongly stressed the importance of human beings and their actions; presumably, this encouraged a greater focus on human activities in all the graphic arts. But it would be a mistake to equate this humanism with individuality. Even in the depiction of people, as well as other subjects, artists

28. Cited in Lim, *Stories from China's Past*, 96. Rudolph continues here with the advantages of this loom, which "makes it possible to produce quickly much more complicated weaves and designs . . . and also leaves the weaver's hands free for other tasks."
29. H. T. Huang, *SCC* 6:5, 305–16, especially 306–7, Figs. 68 and 69.
30. Needham and Wang, *SCC* 4:2, 270, Fig. 510.

often continued to prefer more generalized renderings that avoided the particular and the individual. David Keightley has characterized their work as "an art more of concept than of percept," one in which the inclusion of "vivifying details" for their own sake is far less important than the conveying of general meanings.[31] One particularly interesting example comes not from painting but from sculpture: the seemingly realistic ceramic soldiers who guarded the tomb of the First Emperor turn out, on closer inspection, to have been constructed from a varied but limited repertoire of body part modules that were only slightly modified in the assembling of the completed figures.[32] As Keightley characterizes these sculptures: "The goal was to produce the appearance of individuality without its substance, realism without portraiture."[33]

In the same way, early Chinese paintings display considerable conventionality or generalization in the subjects portrayed, though here and there interesting details are also included. When details are omitted, it could be for a variety of reasons, from the artists' limited command of drawing techniques to aesthetic ideas about the hallmarks of a fine painting. Some degree of generalization may also have been encouraged by reliance on drawing tools as in this Later Han engraving in which the rump, the foreleg and the shoulder of the horse nearest the viewer appear to have been drawn with a compass (Fig. 1.5). In any case, the tendency toward more generalized depictions of subjects at the expense of details and individuality had strong staying power in much of later Chinese art including depictions of technological subjects. Even after a long period down to the Song dynasty during which artists showed considerable ability to create highly naturalistic paintings sometimes replete with realistic details,[34] the preference for more general portrayals reasserted itself once again. As we shall see below, the deep-rooted earlier tendency was in this case further strengthened by conscious and sophisticated aesthetic/philosophical choices.

Early Farming Paintings

Even as Warring States artists were expanding their repertoire of this-worldly subjects, their paintings continued to be invested with "moral and ideological weight" as well.[35] In

31. David N. Keightley, "The Quest for Eternity in Ancient China: The Dead, Their Gifts, Their Names," in George Kuwayama (ed.), *Ancient Mortuary Traditions of China: Papers on Chinese Ceramic Funerary Sculptures* (Los Angeles: Far Eastern Art Council, Los Angeles County Museum of Art, 1991), 13. See also his "Early Civilization in China; Reflections on How It Became Chinese," in Paul S. Ropp (ed.), *Heritage of China: Contemporary Perspectives on Chinese Civilization* (Berkeley: University of California Press, 1990), 37.
32. For example, among the thousands of figures, archaeologists have identified just eight basic types of head! This limited repertoire was effectively disguised by varied combinations of facial components (eyebrows, lips, etc.), which also were standardized into a relatively limited number of types. Lederrose, *Ten Thousand Things,* 72.
33. Keightley, "Quest for Eternity," 23 n. 3; Lederrose, *Ten Thousand Things,* 70–73. This assessment can arguably apply also to the clay slab relief scene pictured in Thorp and Vinograd, *Chinese Art and Culture,* 128–29, Fig. 4.9.
34. See for an example the discussion of the painting *Along the River at the Spring Festival* in Chapter 3.
35. The phrase is James Cahill's; see *Three Alternative Histories of Chinese Painting* (Kansas City, KS: Spencer

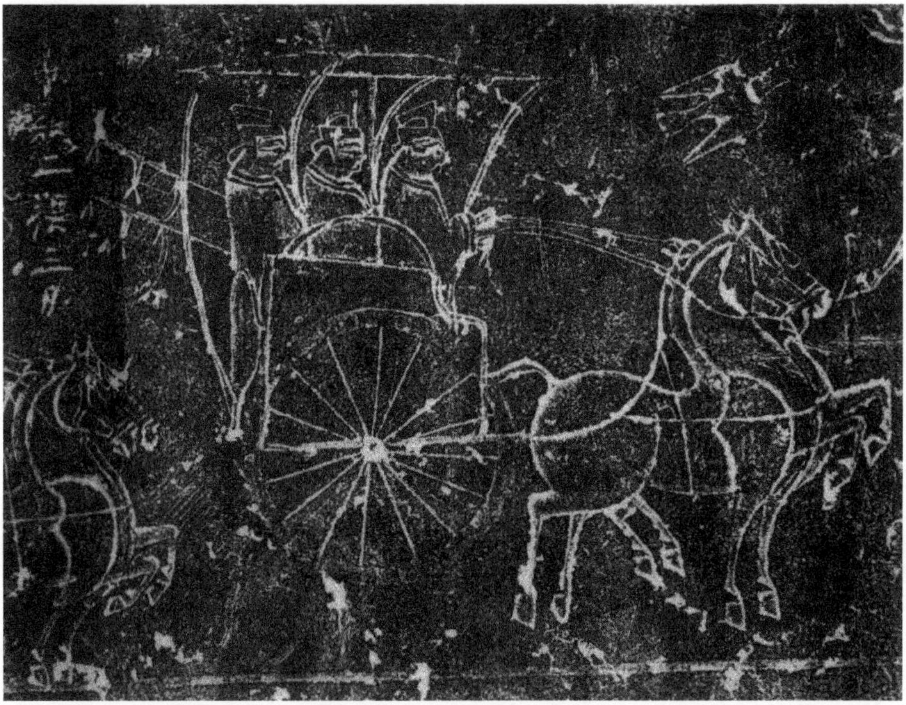

1.5 A first- or second-century CE stone carving of a horse and carriage from a shrine in Shandong

part, this kind of thinking may have derived from the pictographic character of the earliest Chinese writing. This made it easy for Chinese thinkers to conclude that painting and writing shared common origins.[36] From that belief, it was but a short step to conclude that painting and writing also shared common purposes, an important one of which was to encourage morality. In China, more than in other early cultures, much of that morality had to do with political actions. Just as writing could record historical events and personages to serve as warnings and models for those who came later, so too could painting provide visual images of the same subjects, for the same purposes.[37] One influential expression of this attitude by the famous fifth-century CE portrait painter and painting theorist, Xie He 謝赫,

Museum of Art, University of Kansas, 1988), 22. One can compare this tendency with the stress on religious standards for judging paintings in the European Middle Ages, or with the emphasis on aesthetic standards in our own time.

36. Susan Bush and Hsio-yen Shih, *Early Chinese Texts on Painting* (Cambridge, MA: Harvard University Press (Harvard-Yenching Institute), 1985), 50–51. Of course, as in other civilizations, the drawing of pictures in fact long predated the creation of any kind of writing system in the sense of signs used to indicate words with their meaning *and* sound. See Zhu Renfu, *Zhongguo gudai shufa shi* 中國古代書法史 [The history of calligraphy in ancient China] (Beijing: Peking University Press, 1992), 2–3; and Yang Xin. "Approaches to Chinese Painting," in Yang, Barnhart et al., *Three Thousand Years*, 1–4.

37. Though there were some, like Wang Chong (27–c. 100), who rejected such ideas; Bush and Shih, *Early Chinese Texts*, 25.

held that: "All paintings stand for poetic justice; lessons about the rise and fall of ministers over the course of one thousand years can be drawn from paintings."[38]

The lead in this approach to assessing paintings was often taken by officials at the imperial court. From early in the Han, the court was increasingly dominated by an emerging elite of officials who began promoting painting as an "art of persuasion" meant not only to encourage Confucian moral values but also to consolidate and legitimize their power.[39] By the Later Han, a further step in the social history of Chinese graphic arts was also underway. Increasingly, the aristocratic and scholar-official elites took up painting and fine calligraphy themselves as a means to express their high-minded values and thus show themselves as moral exemplars. Being careful to distinguish their painting from the productions of the "artisan" painters who could make a living from their efforts but who were less motivated by moral concerns and therefore by definition incapable of producing paintings of true distinction, they introduced a sharp social divide among painters that would persist in one form or another down to the end of the imperial period.[40]

It is against this background that we can understand the importance that agricultural or farming paintings assumed in the Han dynasty. From early times, China was by far the world's largest agricultural country, where millions upon millions of peasants over the centuries sought to eke out an existence from farming. Assuring that this vast and dense population had at least minimal nourishment was an abiding concern of the Chinese government, based as it was on Confucian teachings that stressed the maintenance of a stable order and the welfare of all the people. The government thus put considerable effort into the promotion of agriculture. A striking side effect of this emphasis on agricultural production was the appearance around the time of the Han of an artistic phenomenon rarely if ever seen in other societies: a significant and quite admired category of paintings, often by members of the highly educated elite from which most officials came or by other painters working under their direction, that not only portrayed rural scenes but even focused on specific farming activities.[41] Such paintings, often seen as symbolizing good government and thereby conferring legitimacy on the whole political order, continued to be painted throughout the imperial period and constitute an important part of the surviving body of

38. Yang, Barnhard et al., *Three Thousand Years*, 2.

39. Lim, *Stories from China's Past*, 51. For examples from the Han to the Tang dynasties of how painting and writing combined to convey moral messages at the imperial court, see Alexander Coburn Soper, *Kuo Jo-hsü's Experiences in Painting* (T'u-hua chien-wen chih): *An Eleventh Century History of Chinese Painting Together with the Chinese Text in Facsimile* (Washington, DC: American Council of Learned Societies, 1951), 7–9.

40. Fong, *Beyond Representation*, 13. For a thorough and incisive discussion of artisans in early imperial China, see Barbieri-Low, *Artisans*.

41. Yonezawa Yoshio, "Chûgoku kaiga ni okeru shomin–toku ni nôminga ni tsuite," [Common people in Chinese paintings, especially peasant paintings], *Tôyô bunka* 2 (1950), 37; Watabe Takeshi, "Chûgoku nôsho *Kôshikido* no ryûden to sono eikyô ni tsuite," [Studies about the spread and influence of the *Pictures of Tilling and Weaving*], *Bulletin of the Faculty of Letters* (Tokai University, 1986), 2.

illustrations of agriculture, sericulture and closely related technologies (e.g. water control) that overwhelmingly dominate surviving Chinese portrayals of technology.[42]

Many of these paintings fall into one of two fairly distinct (at least in retrospect) traditions, inspired in large measure by two literary works that have long been considered in China as fundamental texts on agriculture.[43] The first, with its literary *locus classicus* in the "Wu yi" 無逸 (Against Luxurious Ease) section of the *Book of Documents* (*Shu jing* 書經),[44] stimulated through the centuries the production of works that were explicitly "admonitory" in the sense mentioned above, reminding officials from the emperor down of their responsibilities to the rural population, crucially the promotion of agriculture.[45] These were the paintings that were often to be found in the offices and residences of officials.[46] There even seems to have been a fairly well-established tradition of local officials having paintings of agriculture and sericulture on the east and west walls of their official residences.[47] Yonezawa Yoshio sees a high point for these paintings in the Han.[48] Their decline afterwards may well be related, at least in part, to a decline of Confucian influence in the post-Han centuries.

The second tradition, reflecting an emphasis on the importance of portraying human productive work, derives from one of the best-known poems of the *Book of Songs* (*Shi jing* 詩經), the "Bin feng qiyue" (豳風七月 Odes of Bin, seventh month), which describes at some length the various rural activities through the course of the agricultural year.[49] These paintings, instead of urging or admonishing the rulers, are more like genre paintings having

42. There is a striking contrast here with the situation in other times and other places. For example, in late medieval Europe, illustrations of tools and machines are usually found in connection with surveying, building and warfare. See Eberhard Knoblock, "Technische Zeichnungen," in Uta Lindgren (ed.), *Europäische Technik im Mittelalter 800 bis 1400; Tradition und Innovation* (Berlin: Gebr. Mann Verlag, 1996) 50; Matthies, "Medieval Treadwheels, 510–47.

43. Paul Pelliot, "A propos du *Keng Tche T'ou*," *Mémoires concernant l'Asie Orientale* 1 (1913), 95. A third basic work, the *Yue ling* 月令 from the *Record of Rituals* (*Li ji* 禮記), also inspired paintings closely related to the second category, the genre paintings discussed below. Yonezawa, "Common People," 28; Needham and Wang, *SCC* 4:2, 166.

44. James Legge, *The Chinese Classics*. Volume III: *The Shoo King* (Hong Kong: Lane Crawford, 1865), 464–73; Bernhard Karlgren, *The Book of Documents* (Göteborg: Elanders Boktryckeri Aktiebolag, 1950) (Reprint from *The Museum of Far Eastern Antiquities, Bulletin* 22 (1950), 56–59).

45. Golas, "Emergence of Technical Drawing," 32. Interestingly, the explicit discussion of agriculture actually plays a very small part in this chapter, though it is there right at the beginning.

46. Yonezawa, "Common People," 26.

47. For painting on walls as a long-standing practice for pictures of this kind, see Pelliot, "A propos du *Keng Tche T'ou*," 101, fn. 1 and Otto Franke, "Zur Geschichte des Kêng Tschi T'u: Ein Beitrag zur chinesischen Kunstgeschichte und Kunstkritik," *Ostasiatische Zeitschrift* 3.2 (Juli–Sept. 1914), 187, fn. 2 with the references cited there.

48. Soper agrees; Lawrence Sickman and Alexander Soper, *Art and Architecture of China* (Baltimore: Penguin Books, 1956), 220.

49. Legge, *She King*, 226–33; Arthur Waley (trans.), *The Book of Songs* (New York: Grove Press, 1960), 164–7; Liu Wu-chi and Irving Yucheng Lo (eds.), *Sunflower Splendor: Three Thousand Years of Chinese Poetry* (Garden City, New York: Anchor Press/Doubleday, 1975), 8–11. For the earliest surviving reference to this painting tradition, a painting done by Emperor Ming of the Jin dynasty who ruled from 323 to 325, see Pelliot, "A propos du *Keng Tche T'ou*," 95 and Needham and Wang, *SCC* 4:2, 166.

as their main intent the picturing of rural life, especially farming, in an appealing (and often idealizing) manner.[50] Many of them are found in the tombs of well-to-do landowners and probably were meant to symbolize the family's prosperity.[51] A very interesting related example, but in a religious context, is found in a Buddhist mural from Dunhuang (Plate 2). This portrayal of Maitreya's paradise includes in its upper portion a farmer plowing a plot of land with two oxen as well as other figures engaged in farming activities. The related sutra states that farming is easy and returns are bounteous in Maitreya's paradise.[52] One wonders if this too does not reflect an elite viewpoint: how many peasants after all would look forward to continuing the hard work of farming in "paradise"?

Both of these traditions gave rise to paintings that represented a more-or-less imaginary world against which the real everyday world was measured.[53] For our purposes, it is important to note that the technology portrayed in the farming paintings tended to be there as a by-product; the focus was mainly on the people doing the work and the conditions in which they worked rather than on exactly how the work was done. And given that the paintings often represented the perspective of the elite, there were strong tendencies even in the genre paintings to idealize peasant life while at the same time regarding the peasants as curiosities, sometimes not so far different from the way barbarians were viewed.[54]

On the other hand, as mentioned above, the emergence of an important body of paintings dealing with rural life opened the way to the overwhelming dominance of depictions of rural productive activities in later Chinese portrayals of technology. This in turn may help to account for a major characteristic of so much Chinese illustration of technology, namely, the lack of attention to detail. When a familiarity with the subject can be assumed on the part of most viewers, making the illustration accurate and including all important details becomes less important, especially when the illustration has no particular technological purpose. The imagination of the viewers can be relied on, even if only instinctively, to make the necessary corrections or fill in the missing parts.

---

50. Golas "Emergence of Technical Drawing," 31–32; Thomas Lawton, *Freer Gallery of Art Fiftieth Anniversary Exhibition. II Chinese Figure Painting* (Washington: Smithsonian Institution, 1973), 48–53. The idealization became especially blatant in the drawing of draft animals with human-like happy smiles, as in Song Yingxing's *Exploitation of the Works of Nature*, which we shall discuss in Chapter 5.
51. Watabe, "Spread and Influence," 2.
52. Fraser, *Performing the Visual*, 63, Figs. 6a–6d; 65 and Pl. 3 (between pp. 108–9).
53. Mark Edward Lewis sees Chinese written texts as performing the same function; Lewis, *Writing and Authority in Early China* (Albany: State University of New York Press, 1999), 4.
54. Yonezawa, "Common People," 38. For a much later (thirteenth century) treatment of the "Odes of Bin," see Lawton, *Chinese Figure Painting*, 48–53.

Two

Han to Tang

Realism on the Rise

What we can say about graphic illustrations of technology in the centuries from the end of the Han to the end of the Tang is severely limited by a dearth of surviving drawings or paintings containing elements of technological interest.[1] All too often we can only speculate even about very important topics. For example, the increasing availability of paper in China from at least the fourth century CE[2] must have led to a greater production and circulation not only of texts but also of illustrations, some of which surely included technological subjects. On the other hand, at least by the seventh to the ninth centuries during the Tang, literary evidence indicates that the government was already wary of the spread of information that might be considered a threat to the political authorities. While astronomy and astrology as well as divination especially fit into this category, it would be surprising if the same attitudes did not sometimes apply also to technologies such as the making and use of weapons.[3]

The literary evidence also refers on occasion to drawings that dealt with specialized subjects, some of which may have been technological. The bibliographic chapters of the official histories are a case in point. They include many works now lost that had the word *tu* 圖 or *yang* 樣 (drawing) in their titles, usually indicating that they included visual depictions of some kind (pictures, diagrams, etc.) When writing the *Illustrated Standards for Machines* (*Qi zhun tu* 器準圖) in the early sixth century, Xindu Fang 信都芳 benefitted from being able to draw on a very large library of writings *and* illustrations belonging to his patron, an imperial prince who was intrigued by all kinds of scientific apparatuses.[4] Even earlier, there is considerable evidence for architectural drawings, none of which seem to have survived.[5]

1. Golas "Emergence of Technical Drawing," 32–33.
2. Tsuen-hsuin Tsien, "Chemistry and Chemical Technology: Paper and Printing," *Science and Civilisation in China (SCC)*, ed. Joseph Needham, vol. 5, part 9 (Cambridge: Cambridge University Press, 1985), 43; Frances Wood, *Chinese Illustration* (London: The British Library, 1985), 10.
3. Alexei Volkov, "Geometrical Diagrams," in Bray et al., *Graphics and Text*, 429 and 432–34.
4. Wu Jiming, *History of Chinese Drawing*, 29; Needham and Wang, *SCC* 4:2, 35. The Chinese reference to this library speaks, probably with more than a little exaggeration, of its more than 10,000 items, many of which contained illustrations. By comparison, it is interesting that Cambridge University's library had only 122 volumes in 1424. Norma Levarie, *The Art and History of Books* (New Castle, DE and London: Oak Knoll Press and The British Library, 1995), 67.
5. Wu Jiming, *History of Chinese Drawing*, 65 ff.; Lawrence Sickman and Alexander Soper, *The Art and Architecture of China* (Baltimore: Penguin Books, 1956), 243; Chen Tongbin 陳同濱 et al. (eds.), *Zhongguo gudai jianzhu*

Literary accounts can also suggest developments that may have encouraged the making of drawings that contained what we might regard as technological elements. For example, it appears that the post-Han centuries witnessed considerable interest in mechanical devices or toys, a subject to which we shall return shortly. Given the complexity of these automata, it is likely that some use must have been made of working drawings in their construction. Also, toward the end of this period, we have an important document from the *Jiu Tang shu* 舊唐書 (Old history of the Tang dynasty) relating that the court issued a drawing (*yang* 様) of a standard model of chain pump (*shui che* 水車) and ordered the authorities of the capital area to have a number of the machines built to be used by the people for irrigation.[6] What is especially suggestive here is that it was apparently taken for granted that the illustration was sufficiently accurate and detailed to serve as a construction guide. But in this case as in others, while the texts sometimes provide enough information to enable us to imagine what the devices themselves may have looked like, the lack of a single surviving drawing means we can only speculate about the appearance and techniques of the drawings.

In any case, it seems safe to conclude that the necessity of producing each drawing individually made illustrations of any kind relatively uncommon during those centuries that preceded the invention and spread of woodblock printing. It is true that there appeared at this time a number of techniques for making the process of copying more accurate (always a big problem, greater even for illustrations than for texts)[7] and perhaps somewhat more efficient. But the graphic skills required and copying costs still must have impeded any extensive production and circulation of illustrations.

Aesthetics and Realism to the Fore

During the centuries from the Han to the Tang, moral concerns played an important role in Chinese painting. John Hay would even argue that "almost all painting [down to the seventh century] was essentially narrative in the service of ethical precepts, Confucian and Buddhist alike."[8] Ancient beliefs in the magical potential of paintings also still survived. In the fourth century, one of the first great Chinese painters, Gu Kaizhi 顧愷之 (c. 344–406), was said to have moved a woman by painting her portrait and then pricking the heart area with a needle.[9] As late as the thirteenth century, there was a belief that the replacement of the

---

da tudian 中國古代建築大圖典 [Illustrations of ancient Chinese architecture] (Beijing: Jinri Zhongguo chubanshe, 1996), vol. 2, 1185.

6. Needham and Wang, *SCC* 4:2, 347. I feel quite confident that *yang* here means a drawing rather than a model.
7. Liu Keming 劉克明 and Gong Shixing 龔世星, "*Zhongguo gudai gongcheng tuxue de ruogan chengjiu*" 中國古代工程圖學的若干成就 [Some achievements of ancient Chinese engineering graphics], *Gongcheng tuxue ji jisuanji tuxue* [Journal of Engineering Graphics and Computer Drafting] 1(1992): 79.
8. John Hay, "'Along the River during Winter's First Snow': A Tenth-Century Handscroll and Early Chinese Narrative." *Burlington Magazine* 114 (May 1972), 298.
9. Fong, *Beyond Representation*, 4 and 9n6.

Pictures of Tilling and Weaving[10] on the walls of the imperial palace contributed to political troubles that followed.[11]

After the Han, however, aesthetic criteria loomed increasingly large in the assessment of paintings. This stimulated improvements in painting technique, including techniques that would enable artists to reproduce more exactly what they saw in the world around them.[12] Already in the Han, painters had been experimenting with techniques such as modeling strokes and hatching to convey a better sense of contours and volumes.[13] These experiments continued in the succeeding centuries. In the sixth century, the artist Zhang Sengyao 張僧繇 (502–56), apparently borrowing a technique from India, won praise for the realism of his *trompe l'oeil* flowers painted on a Buddhist temple.[14]

The availability of new techniques, however, did not lead inevitably to their wide adoption. On the contrary, techniques seemingly full of promise for further development were used in a rather halfhearted fashion by only a few artists who seldom showed much interest in exploring their further possibilities. Illusionistic shading provides a good example. Shading has frequently been thought of as an import along with other techniques and styles of Buddhist art. A lacquer painting from the Later Han, however, shows it to have been known in China well before the case can be made for a Buddhist influence (Fig. 2.1).[15]

10. This work is discussed in the next chapter and in Chapter 6.

11. Pelliot, "A propos du *Keng Tche T'ou*," 95. For further examples of supposed magical powers of paintings, see Golas, "Emergence of Technical Drawing," 54n8; Liu Heping, "*Water Mill*," 584; James Cahill, "Confucian Elements in the Theory of Painting," in Wright, *The Confucian Persuasion*, 117.

12. Susan Bush, *The Chinese Literati on Painting: Su Shih (1037–1101) to Tung Ch'i-ch'ang (1555–1636)* (Cambridge, MA: Harvard University Press (Harvard-Yenching Institute), 1971), 13–14 and 17–18.

13. Mary H. Fong, "The Technique of 'Chiaroscuro' in Chinese Painting from Han through T'ang," *Artibus Asiae* XXXVIII 2/3 (1976), 111. For examples of the use of modeling strokes, see Jan Fontein and Wu Tung, *Han and T'ang Murals Discovered in Tombs in the People's Republic of China and Copied by Contemporary Chinese Painters* (Boston: Museum of Fine Arts, 1976), 29, Nos. 7 & 8; 43, No. 20 (ram at lower left); 74–75, Nos. 78 & 82; 113, No. 133; and *passim*; Wu, "Origins of Chinese Painting," 37, Fig. 27. For an example of hatching, see Fontein and Wu, *Han and T'ang Murals*, 99, No. 120. The use of modeling strokes probably developed quite naturally out of the strokes used to depict folds in clothing, a technique already in common use in Han paintings. Hatching, however, may have been more difficult for Chinese artists to accept since, in direct contrast to the importance Chinese painters attached to individual (often calligraphic) lines in painting, hatching was a process that minimized the attention to the individual line.

14. Hubert Delahaye, "Du peu d'effet de la peinture occidentale en Chine aux XVIIe et XVIIIe siècles," in Catherine Jami and Hubert Delahaye, *L'Europe en Chine: Interactions scientifiques, religieuses et culturelles aux XVIIe et XVIIIe siècles* (Paris: College de France: Institut des Hautes Études Chinoises, 1993), 245. For trompe l'oeil in late Northern Song, see Robert J. Maeda, "Spatial Enclosures: The Idea of Interior Space in Chinese Painting," *Oriental Art* 3 (Winter 1985–86), 381 and many other examples from the Baisha tombs in the report of the excavations; Su Bai 宿白, *Baisha Song mu* 白沙宋墓 [The Song tombs at Baisha] (Beijing: Wenwu chubanshe, 1957).

15. Fontein and Wu, *Han and T'ang Murals*, 20 and 28; 29, No. 7; 44, No. 22; 86 and 89, No. 107. This conclusion is further reinforced by the thoroughly Confucian subject matter (paragons of filial piety) of the painting. See also Fong, "Technique of 'Chiaroscuro,'" 115; Sickman and Soper, *Art and Architecture*, 64, 67; Sherman Lee, *A History of Far Eastern Art* (Englewood Cliffs: Prentice Hall and New York: Harry N. Abrams, 1964), 141–42; Cahill, *Chinese Painting*, 15, with a good illustration on p. 17. This calls into question Sullivan's contention that shading was always regarded as a "foreign" technique, appropriate only to foreign subjects.

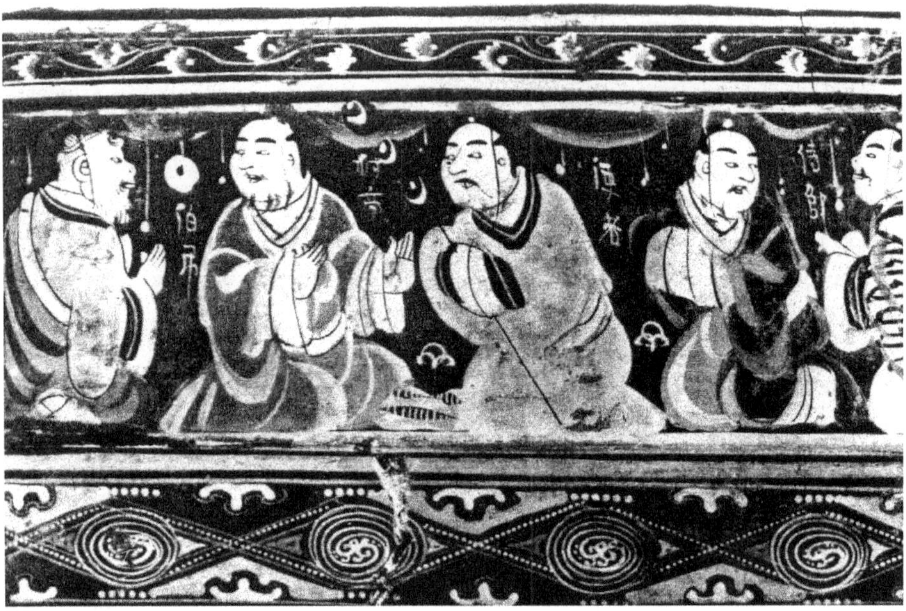

2.1 Detail of a lacquer painting of paragons of filial piety on a basket-work box from Lolang, Korea, c. 100 CE

Shading was also one of the techniques employed in the remarkable paintings of the late sixth century tomb of Lou Rui (婁睿) in Shanxi in which realistic portrayal reached heights not seen in earlier painting.[16] It continued in use into the Tang and is not infrequently to be seen even in later paintings, both colored and monochrome.[17] A most remarkable painting of bamboos (Fig. 2.2) shows what Chinese painters were able to achieve when they pursued the possibilities offered by shading.[18] By and large, however, the Chinese seem never to have done much to develop the technique beyond a very basic level.[19] Of special importance for our story, neither shading nor modeling strokes found their way, except rarely, into the line

Michael Sullivan, *The Meeting of Eastern and Western Art from the Sixteenth Century to the Present Day,* 2nd ed. (Berkeley and Los Angeles: University of California Press, 1989), 54.

16. Yang, Barnhard et al., *Three Thousand Years,* 40–42.
17. See, for example, Zhao Mengjian's *Narcissus* from the thirteenth century (Fong, *Beyond Representation,* 306–7, Pl. 68) and Li Gan's *Bamboo and Rocks* from 1318 (Fong, *Beyond Representation,* 390, Pl. 86).
18. For another example, this one from the Ming, see Dai Jin's 戴進 *Bamboo, Chrysanthemum, and Garden Rock* in Richard M. Barnhart, *Painters of the Great Ming: The Imperial Court and the Che School* (Dallas: Dallas Museum of Art, 1993), 185, cat. 50.
19. Cahill, *Chinese Painting,* 15. It is particularly interesting to realize that "shadowing" in the sense of a shadow actually cast by an object in sunlight seems, with a single exception, to be entirely absent in the Chinese landscape paintings that have survived. See Michael Sullivan, *Symbols of Eternity: the Art of Landscape Painting in China* (Stanford: Stanford University Press, 1979), 8; Anita Chung, *Drawing Boundaries: Architectural Images in Qing China* (Honolulu: University of Hawai'i Press, 2004), 72, Fig. 2.8. Sullivan sums up the case nicely: "Chinese artists, if they used shading at all, used it very sparingly to make an object look more solid, while avoiding cast shadows and a single light source." Sullivan, *Meeting of Eastern and Western Art,* 62, Caption to Fig. 44.

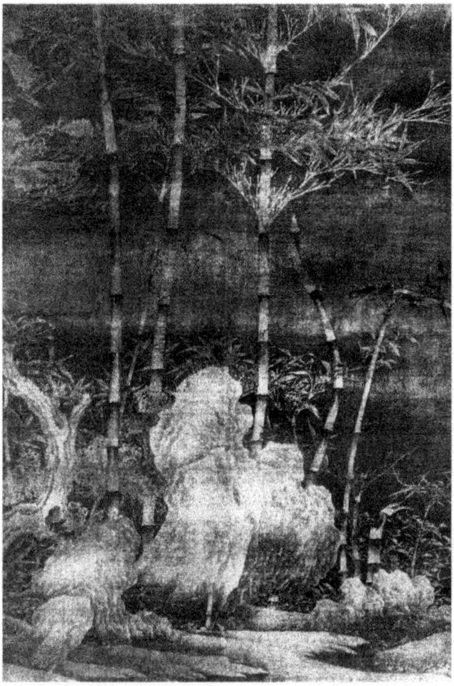

2.2 Anonymous tenth-century(?) painting of *Bamboo and Old Tree Growing by Rocks*

drawings that provide most of our examples of graphic portrayals of technology.

Apart from specific techniques, however, it was the overall Chinese approach to drawing, in these centuries as well as later, that was most important for limiting technological painting. Overwhelmingly, drawing in China was what the Chinese called *baimiao* 白描 (plain drawing) or outline ink monochromatic drawings. Wen Fong's description of *baimiao* draws an important contrast:

> Plain drawing is not drawing at all in the Western sense of the word, since it does not build form with multiple modeling strokes. Rather, it is monochromatic ink painting with a single brushline, a modulated ink outline that captures and reveals the essence of an object spontaneously and completely, without correction.[20]

A moment's reflection suggests how antithetical such a "spontaneous" approach could be to drawings with a technological purpose where one is not interested in the "essence" of a machine but rather in a depiction that reveals accurately and in some detail what the machine looks like or how it is constructed or how it works.

Scale Drawing and Perspective

The Han and post-Han period also witnessed the emergence of drawings to scale, especially in astronomical calculations and the construction of astronomical instruments as well as in the making of maps. The first explicit reference to a scale drawing seems to occur in *The Arithmetical Classic of the Gnomon and the Circular Paths of the Heavens* (*Zhou bi suan jing* 周髀算經) from about 100 BCE.[21] We shall return to this in Chapter 3 when we discuss Su

20. Fong, *Beyond Representation*, 35 and 60–61. Hence, most of the drawings we shall be examining have much more in common with comic books or manga than with fine art drawings as they emerged in the Renaissance Europe.

21. Wu, *History of Chinese Drawing*, 111; Cordell D. K. Yee, "Reinterpreting Traditional Chinese Geographical Maps," in J. B. Harley and David Woodward, *A History of Cartography*, vol.2 book 2 (Chicago: Chicago University Press, 1994), 42. The earliest explicit reference to an architectural scale drawing may date from

Song's clock tower. In the case of mapmaking, which was usually seen in China as closely related to painting, the use of scaled distances (though not a clearly measured surface on which to draw them) may have begun as early as the third or fourth century BCE.[22]

Between the Han and the Tang, the spread of Buddhism in China led to the painting of many large-scale murals. Sarah E. Fraser has examined in detail the kinds of preliminary drawings used in the production of these murals.[23] One might expect that the difficulty of moving from smaller design to large mural would have encouraged the production of scale drawings or at least reduced-scale sketches[24] which, perhaps in conjunction with a grid overlay, could then have been systematically enlarged to provide the patterns for the final mural.[25] Nevertheless, among the surviving materials (admittedly very limited), we have no examples of scale drawings made for this purpose.[26] In the same way, one might have expected scale drawings to have been used in the production of the massive Buddhist icon sculptures so characteristic of this period. Nevertheless, here too we have neither surviving examples or even textual references suggesting the use of scale drawings.

In the area of building technology, by contrast, a piece of evidence from as early as the end of the fourth century BCE might be taken to indicate not only the use of scale drawings but even scale models. It consists of a unique artifact, the *Zhaoyutu* 兆域圖 (Picture of the omen area), a ninety-six by forty-eight centimeter bronze plaque found under the tomb of a feudal ruler of the kingdom of Zhongshan in Hebei that gives a bird's-eye view of the ruler's tomb complex (Fig. 2.3).[27] This roughly scale portrayal in plan of the site and its buildings makes clear that both the concept of a site plan and the notion of scale drawing must have had some currency even by the fourth century BCE. On the other hand, one can question

only the seventh century. Xiao Zhenshi, *A History of Building Technology in China* (Taipei: Po-yuan, 1993), vol. 3, 890.

22. Nancy Steinhardt, "Chinese Cartography and Calligraphy," *Oriental Art* 43.1 (1997), 10, 11. For actual scales of early maps, see Yee, "Reinterpreting," 41–46, and for ideas of Han intellectuals on the practice, ibid., 109–10. Yee nicely makes an important point well kept in mind when considering the use of "scale" in Chinese painting (and therefore also in other kinds of illustration): "In Chinese painting, . . . pictorial scale tends to dominate natural scale; that is, the size of an object depicted was determined by needs of design and not rules of geometric perspective. Foreground features might be diminished to avoid obstruction and overemphasis, while distant objects might be enlarged to act as counterpoint to the middle distance and foreground." Yee, "Cartography in China," 144.

23. Fraser, *Performing the Visual*.

24. Jerome Silbergeld uses the term "reduced-scale sketches" for those preliminary drawings that were not true drawings to scale. Silbergeld, *Chinese Painting Style*, 7.

25. Silbergeld, *Chinese Painting Style*, 7–8; Tseng Yu-ho Ecke, "A Reconsideration of *Ch'uan-mo-i-hsieh*, the Sixth Principle of Hsieh Ho," *Proceedings of the International Symposium on Chinese Painting* (Taipei: National Palace Museum, 1972), 317. In Europe, it is not until the Renaissance that such grid drawings (necessarily drawn to scale) came to be regularly used for painting murals. Francis Ames-Lewis, *Drawing in Early Renaissance Italy* (New Haven and London: Yale University Press, 1981), 26.

26. Fraser, *Performing the Visual*, 75.

27. Yee, "Reinterpreting," 36–37; Wolfgang Behr, "Placed into the Right Position–Etymological Notes on Tu 圖 and Congeners," in Bray et al., *Graphics and Text*, 117–19. Nancy Steinhardt (ed.), *Chinese Architecture* (New Haven and London: Yale University Press, 2002), 29, Fig. 1.19 presents a color photo of the plaque.

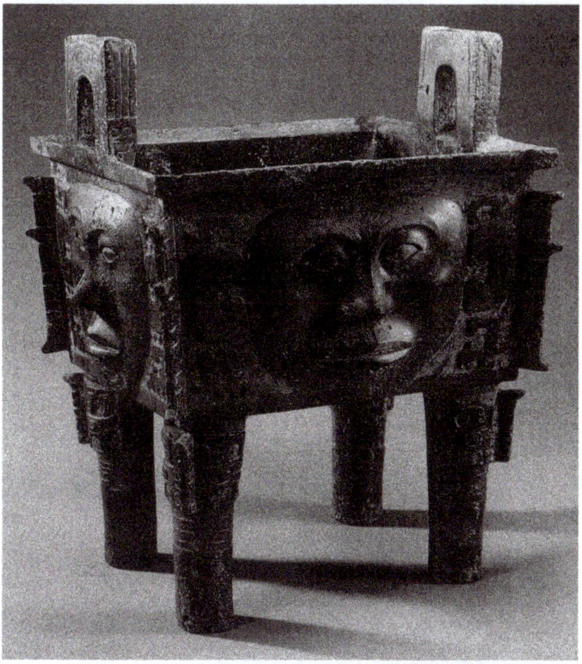

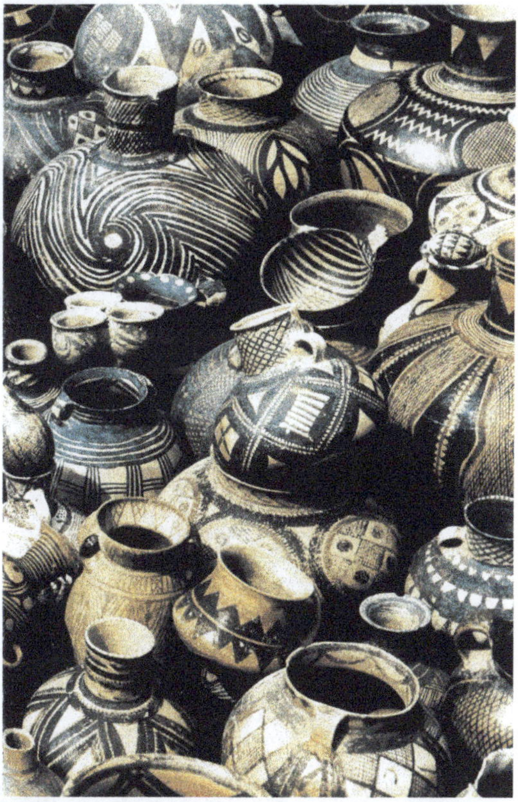

Plate 1 (a) Late Shang (1200–1050 BCE) rectangular cooking vessel with well-modeled face on each side; (b) Geometrical designs on pots of the Yangshao culture (c. 3200–2500 BCE) in north China

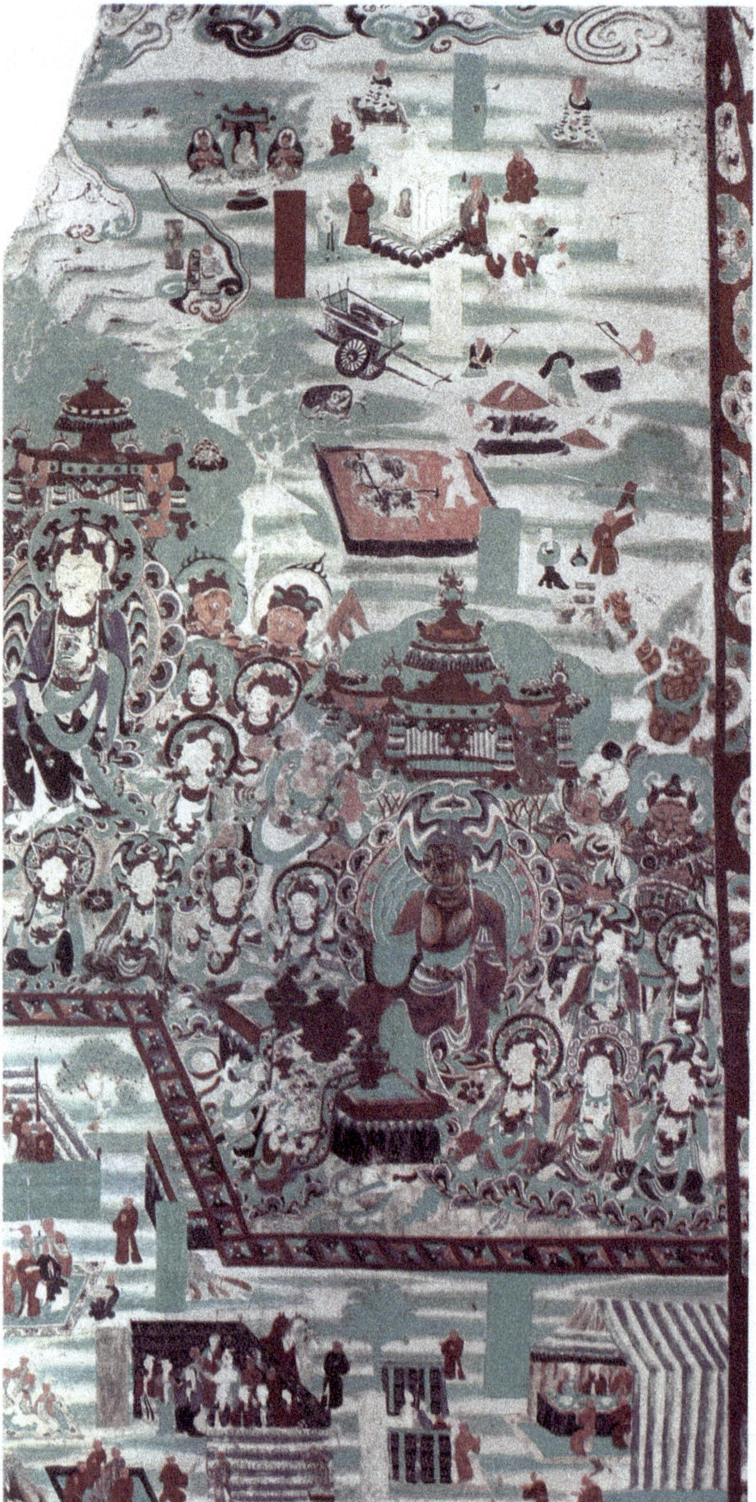

Plate 2 Mural painting portraying Maitreya's Paradise. Cave 196, Dunhuang.

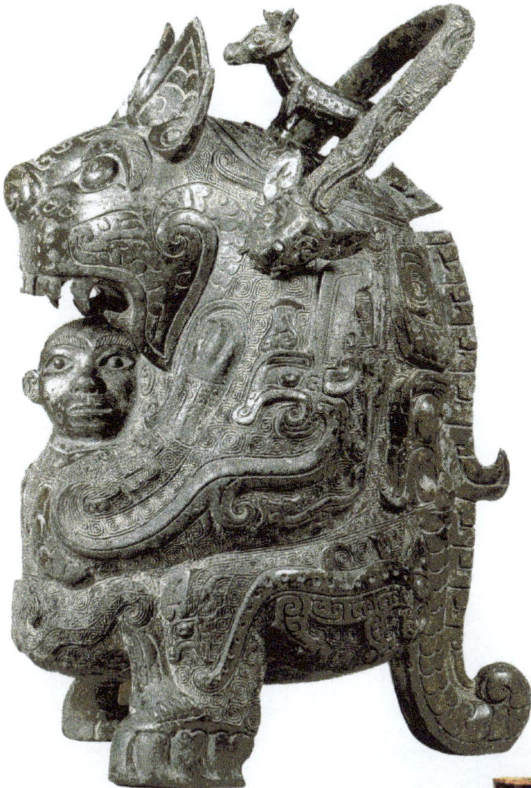

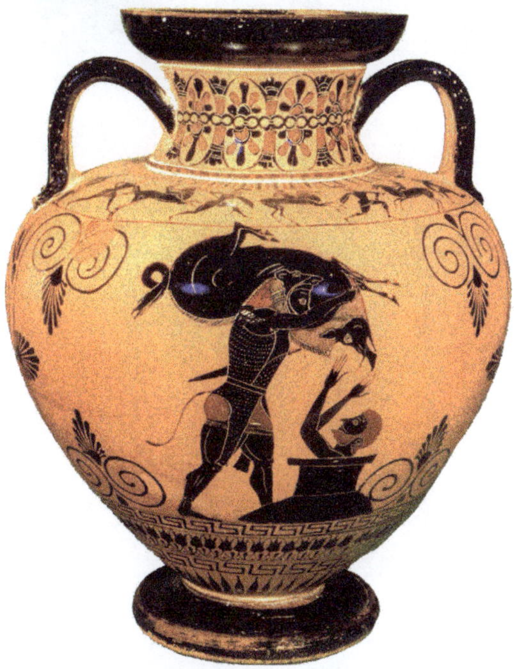

Plate 3 (a) Bronze vessel, perhaps depicting a shaman and his alter ego, from Shang China (c. 1600–1050 BCE); (b) Ancient Greek vase depicting the myth of Hercules, as the fourth of his twelve labors, bringing the Erymanthian Boar back to Eurystheus. The exuberance of the decoration on the bronze contrasts with the rather static placidity of the vase decorations.

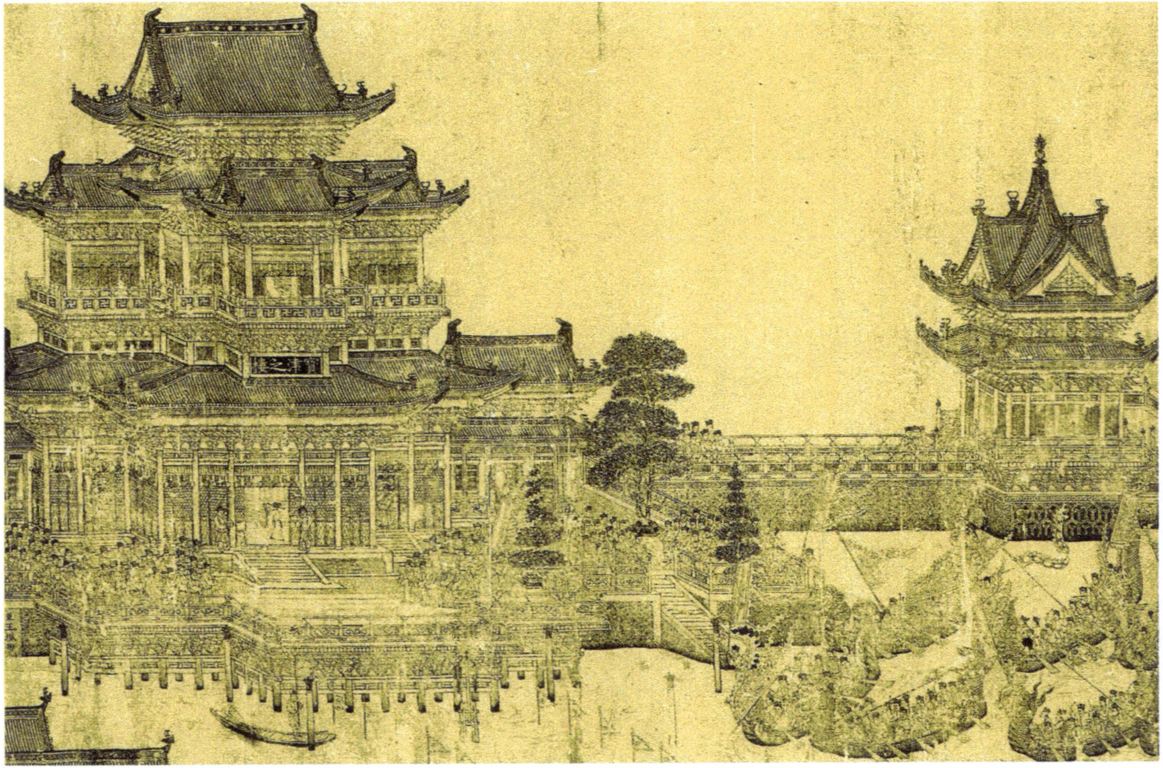

Plate 4 Detail of a fourteenth-century(?) copy of Wang Zhenpeng's *Dragon Boat Regatta on Jinming Lake*

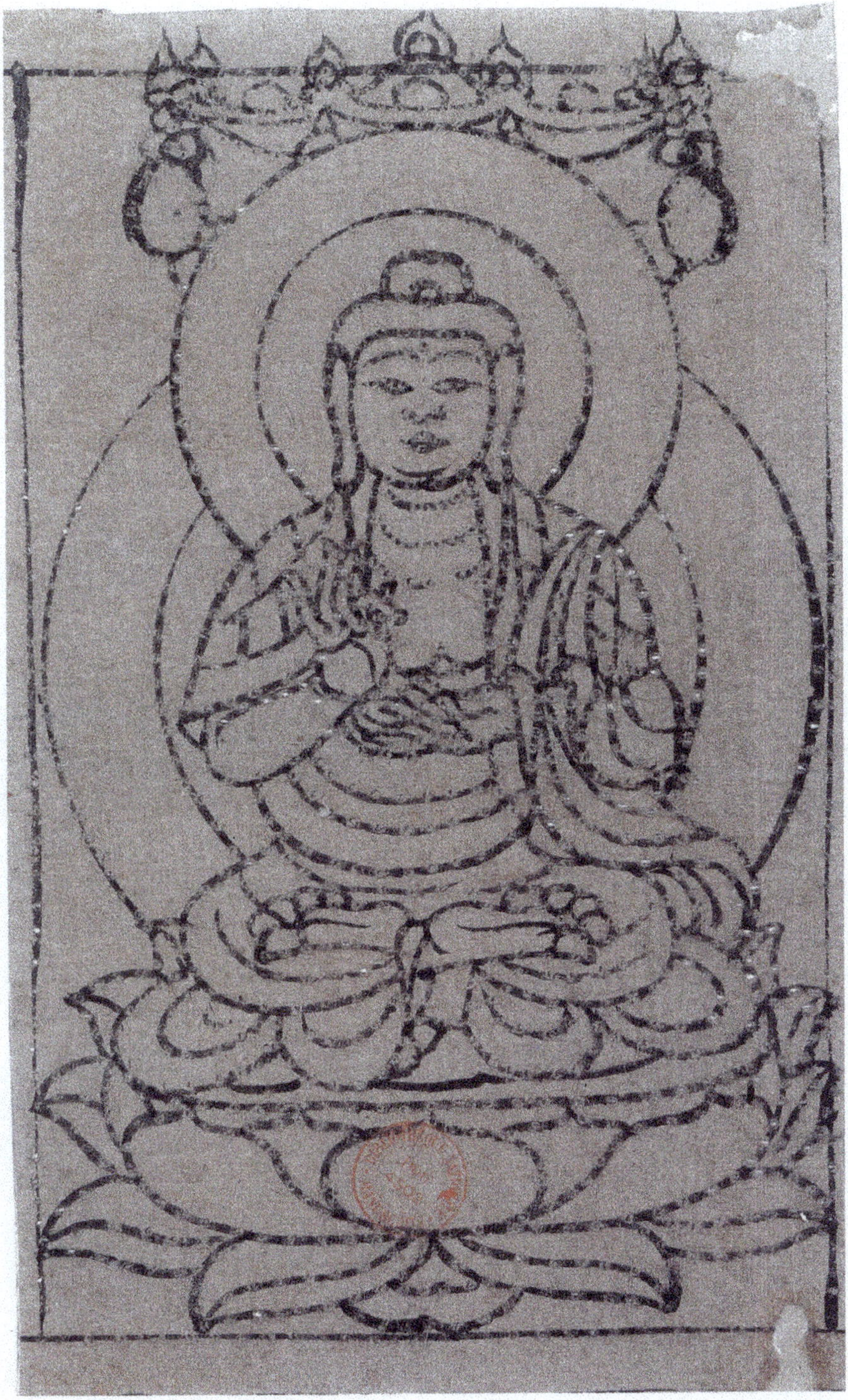

Plate 5 Image of the Buddha with pounce perforations, ninth or tenth century, Dunhuang. Courtesy of the Bibliothèque Nationale de France.

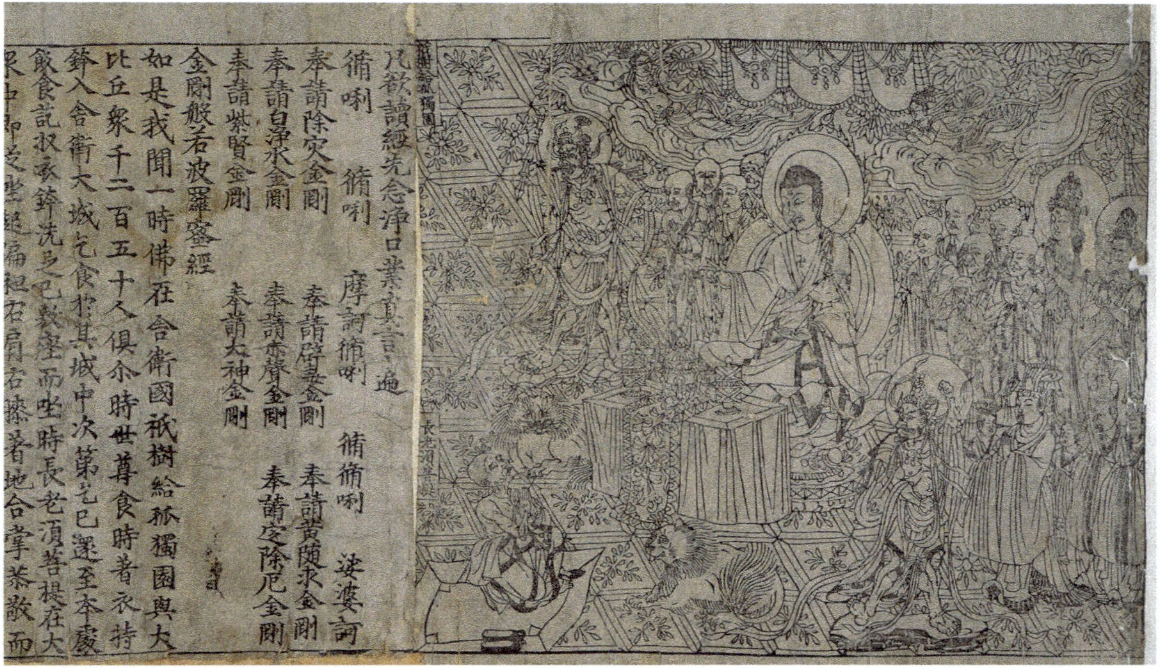

Plate 6 Detail of the ninth-century *Diamond Sutra*, a printed scroll 17 1/2 feet long, discovered in the Dunhuang cave complex in 1907. It portrays the Buddha, his disciple Sabhuti, and numerous other figures. © The British Library Board, OR. 8210/P.2, Frontispiece and text.

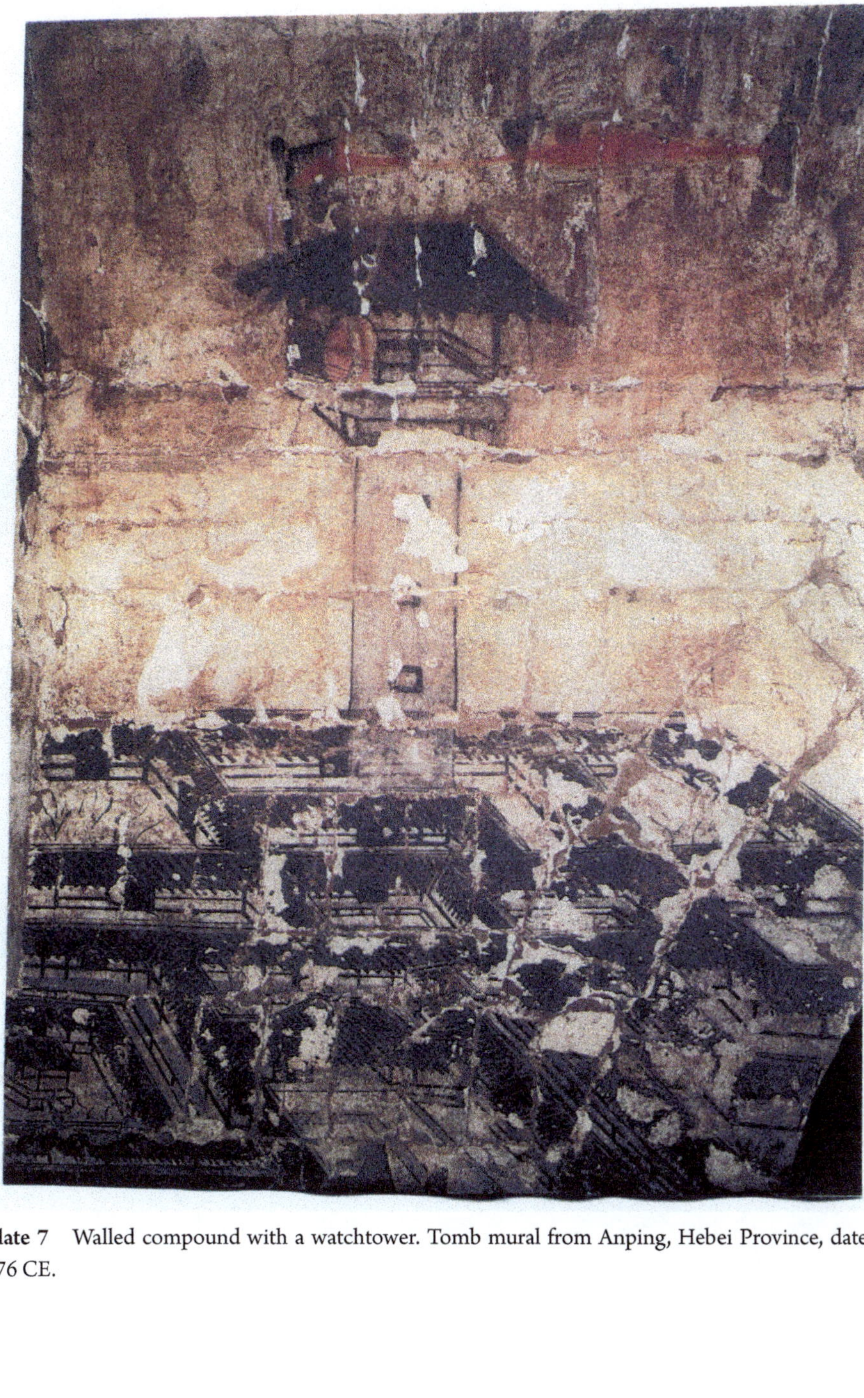

Plate 7 Walled compound with a watchtower. Tomb mural from Anping, Hebei Province, dated 176 CE.

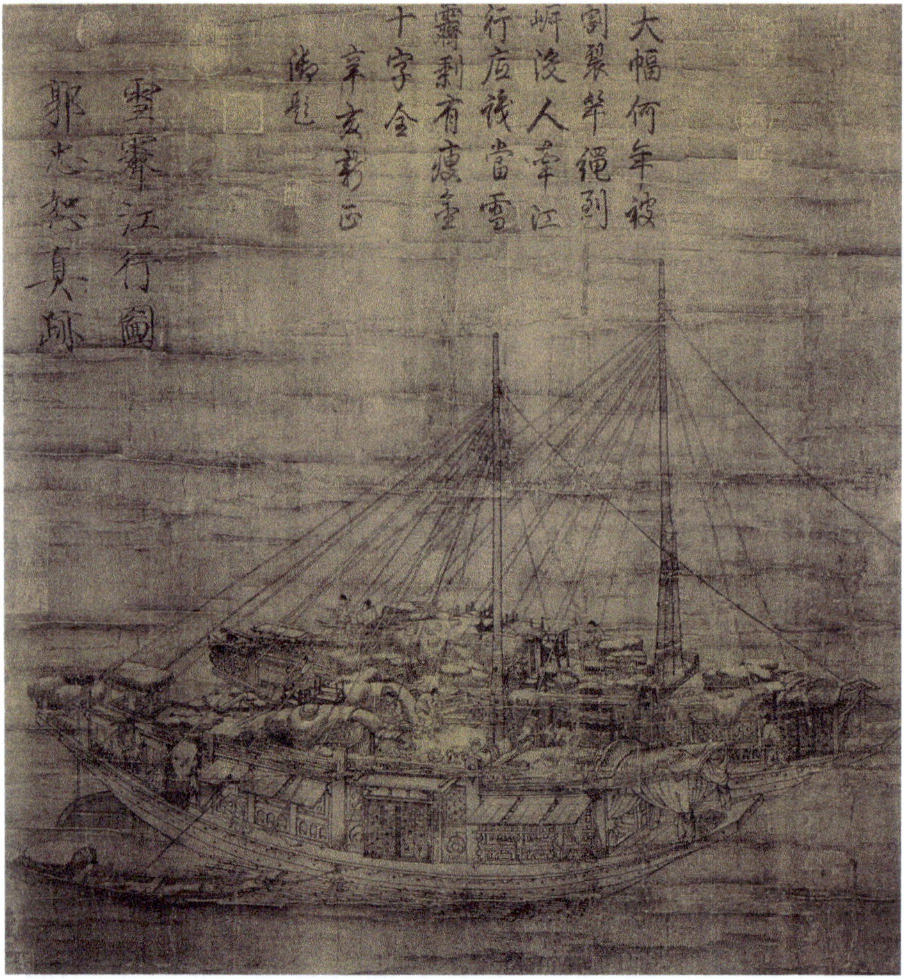

Plate 8 Copy of Guo Zhongshu's, *Traveling on the River in Clearing Snow*.

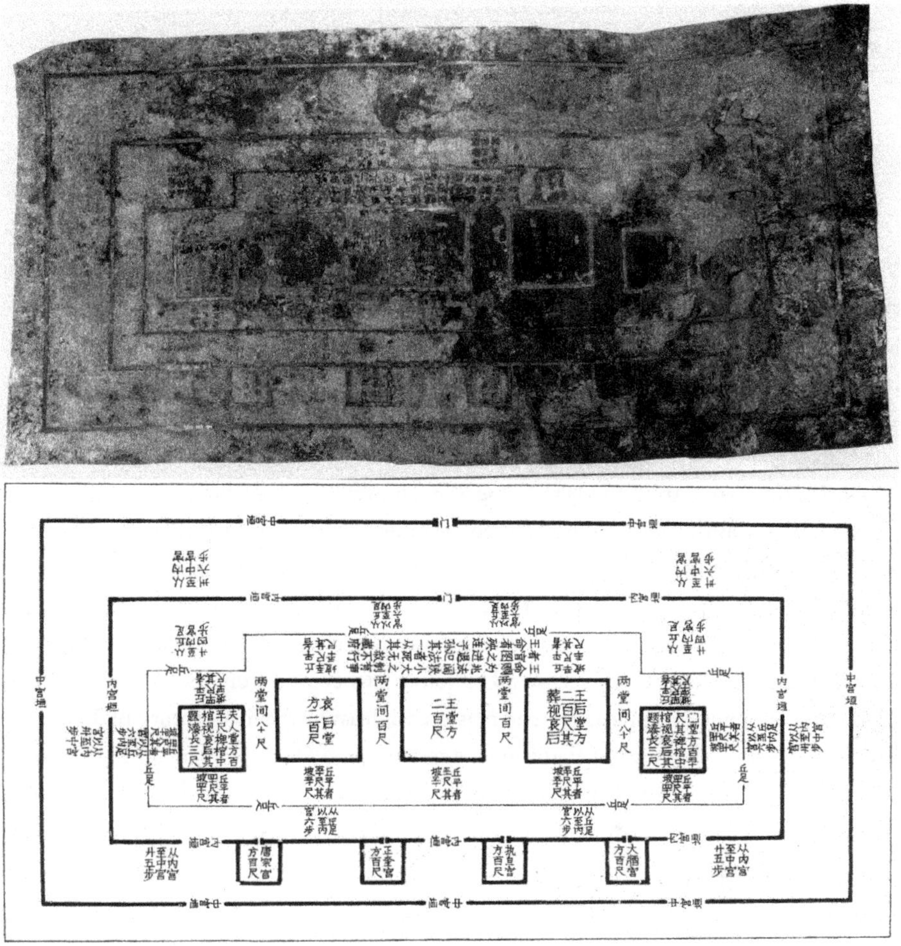

2.3 Bronze plaque showing plan of the Zhongshan mausoleum (late 4th century BCE) together with a reconstruction substituting modern characters for the archaic forms of the original. The inscriptions provide dimensions for the various elements of the complex.

the extent to which the Chinese at this time had techniques sufficiently precise to enable them to avoid inaccuracies in measurement that could lead to mistakes if the drawings were overly relied on for building.[28] This, surely, is also at least part of the reason why the plaque contains inscriptions giving the dimensions of all the architectural features.[29] Interestingly,

28. The ancient Greeks, for example, relied less on drawings in planning and executing their buildings and, instead, aimed for "precision in the widespread repetition of architectural components . . ." See Tom Porter, *How Architects Visualize* (New York: Van Nostrand, 1979), 2. Interestingly, the "repetition of architectural components" nicely describes the modular techniques of construction used in most Chinese buildings, a point to which we shall return below when we discuss the *Building Standards*, traditional China's preeminent manual of building construction.

29. Yee, "Reinterpreting," 36 Fig. 3.1 caption; Wu Hong, "Picturing or Diagramming the Universe," in Bray et al., *Graphics and Text*, 195; Behr, "Placed into the Right Position," 117.

in contrast to the norm for later Chinese portrayals of technological objects where measurements were more often presented not with the images but in an accompanying text, the measurements here are closely related to the elements to which they apply.

How Chinese artists attempted to create a sense of receding distance in their images also influenced their portrayal of technical subjects. Early Chinese painters, like painters in other ancient cultures such as classical Greece, did not create a sense of real space in their paintings (it probably never occurred to them as something they ought to do).[30] Cahill speaks of their constructing a composition by juxtaposing forms without integrating them. "Between and around [the forms] was void; space had no existence except as that which separated one image from another."[31] Actually, this breaking up of space into cells[32] or "compartmentalized units" (paralleled in Chinese views of time)[33] can be seen as another example of the modular thinking to which the Chinese were so prone, what we might call a kind of digital rather than analog way of perceiving.[34]

During the period from the Han to the Tang, however, some Chinese artists did begin to experiment with various methods to convey not only a stronger feeling of real space[35] but also a sense of receding distance in their paintings. Perhaps the most common early means of suggesting distance, at least in pictures presenting broad vistas, was to vary the size and/or height of objects, so that objects drawn relatively smaller or placed higher were understood to be deeper in the scene and therefore farther away from the viewer.[36] Another early development that some artists had mastered at least by the first century BCE, was the use of perspective where parallel, diagonal lines strike off from the plane of the picture.[37] Very effective, especially in landscape paintings, was the use of atmospheric projection in which the manipulation of colors, tints and washes created the sense of distance. Finally, as

30. Our modern consciousness of space influences what we look for in paintings. As Alfred Crosby nicely summarizes it: "For us today, things exist in space like vegetables in an aspic salad. The vegetables may be the chief items of interest, but the aspic is undeniably *there* occupying the area among the items of interest. We do not deny the aspic because it is transparent, and we rarely ignore space even if it is empty." Alfred W. Crosby, *The Measure of Reality: Quantification and Western Society, 1250–1600* (Cambridge: Cambridge University Press, 1997), 170. For the Greeks, see William M. Ivins, Jr., *Art and Geometry: A Study in Space Intuitions* (Cambridge, MA: Harvard University Press, 1946), 40ff.
31. Cahill, *Chinese Painting*, 12. For much the same kind of assessment, referring to Greek painting, see Ivins, *Art and Geometry*, 31.
32. Silbergeld, *Chinese Painting Style*, 50–51.
33. Bodde, *Chinese Thought*, 133.
34. Compare Cordell Yee's characterization of the Chinese view of space as being "particulate." Yee, "Cartography in China," 145.
35. Contrast the two Dunhuang illustrations from the sixth and seventh centuries reproduced in Maeda, "Spatial Enclosures," 375, Fig. 9 and 378, Fig. 12, the second displaying a "much more illusionistically viable space."
36. Benjamin March pointed out many years ago that this technique was *not* normally used for short distances such as a group of buildings, i.e., the distances that would be more typical in technological illustrations; March, "Linear Perspective," 117.
37. Powers, *Art and Political Expression*, 25; see also Wu, "Origins of Chinese Painting," 31. The distinctive characteristic of isometric or parallel projections is that lines parallel in fact are also parallel in the drawing; see Needham, Wang and Lu, *SCC* 4:3, 114, Fig. 778.

early as the sixth century, Chinese painters were even experimenting with vanishing point perspective.[38] A very interesting example is the famous depiction of a Buddhist paradise in a Dunhuang fresco from about 700, a careful examination of which reveals a "modular" approach to perspective in which several separate vanishing points are used.[39]

By the beginning of the Tang, then, Chinese painters were increasingly able to depict a tolerably convincing if not visually accurate world of space and nature.[40] In later centuries, culminating in the Five Dynasties and Song periods, the best artists learned to create, in Cahill's term, a "truly coherent spatial vision."[41] Something of this idea comes through in the comments of Li Zhi 李廌 (late 11th–early 12th centuries) on the architectural painting of the tenth-century painter Guo Zhongshu 郭忠恕 (d. 997): "Roof beams, girders, pillars, and rafters are shown with open spaces between, through which one might move. Railings, lintels, windows and doorways look as if they could really be passed through, or opened and shut."[42] Nevertheless, unlike Western painters since the Renaissance, Chinese painters never developed a notion of space as a measurable geometrical entity with the circle as the ideal form and the triangle as the master of measurements. They therefore never developed mathematical rules to guide the painter in organizing space and in creating more precise spatial relations.[43] As Robert Maeda notes, the Chinese painters were thus able to maintain a much greater sense of spatial dynamism than Western painters, but at the expense of "pictorial orderliness."[44]

Moreover, little of the increasing ability to present a convincing sense of space was applied to the illustrations in the books that, at the end of this period, were being produced in ever greater numbers by means of woodblock printing. Since the majority of our surviving illustrations of technology come from these book illustrations, we shall have occasion to return to this point below as we try to determine what accounted for the particular character

38. Maeda, "Spatial Enclosures," 375–76; Michael Sullivan, *Chinese Landscape Painting*. Vol. 2: *The Sui and T'ang Dynasties* (Berkeley: University of California Press, 1989), 91–94. It is also interesting that the frequently reproduced frontispiece from the earliest surviving woodblock-printed book (the Diamond Sutra, from 868) shows the Buddha in a three-quarter perspective view in place of the flat, full-front view commonly used; see Fig. 2.10.

39. Needham, Wang and Lu, *SCC* 4:3, Pl. CCCIV; Chung, *Drawing Boundaries*, 27–28.

40. For yet another example of the progress that was being made, compare Maeda, "Spatial Enclosures," 375, Fig. 9 and 378, Fig. 12, together with Maeda's accompanying discussion.

41. Cahill, *Chinese Painting*, 29. Nevertheless, Chinese painters even down to the Ming and Qing often continued to exhibit a preference for "pictorial" or "variable" scale where the size of objects depicted was influenced mainly by the requirements of design (avoiding for example overemphasis on certain objects or the obstruction of one object by another) rather than any rules of geometric perspective; Yee, "Cartography in China," 144.

42. Robert J. Maeda, "*Chieh-hua*: Ruled-line Painting," *Ars Orientalis* 10 (1975), 126.

43. This contrast between Western and Chinese approaches to space is nicely elaborated in Delahaye, "Du peu d'effet de la peinture occidentale," 248–49.

44. Maeda, "Spatial Enclosures," 370. March also points out that "there is greater freedom of design possible to a painter who is not bound by the rigid necessities of the single viewing point, vanishing points and other paraphernalia of Western perspective techniques." Benjamin March, "Linear Perspective in Chinese Painting," *Eastern Art* 3 (1931), 137.

of Chinese book illustrations and how this influenced the way the Chinese went about picturing technology.

The Dominance of Brush and Line

We have suggested above that the first Chinese use of basic drawing tools such as the straightedge or ruler, the compass and the square may date back to as early as the Neolithic. By the post-Han centuries, however, there arose a clear prejudice against the use of such tools for painting, at least any painting expected to be taken seriously as "fine" art. At the heart of this prejudice was the strong emphasis, also dating back to very early times, on the ability of paintings to convey a sense of life, movement, vigor, vitality. Even the combining of geometrical ornamentation with totally non-geometric portrayals of vegetation, fauna, and mythological creatures, for example on bronze vessels, served to introduce a remarkable exuberance to what otherwise could have been rather formal and static decorative designs as one sees, for example, in even the anthropomorphic subjects on ancient Greek vases (Plates 3 (a) and 3 (b)).

This strong aesthetic predilection for vitality and spontaneity in paintings does much to explain the importance of the brush in China not only for writing but also for painting.[45] Throughout Chinese history, nearly all writing, drawing and painting has been done with a brush.[46] Evidence especially from early pottery decoration shows that some kind of brush-like writing implement using flexible plant material was in use at least by 2000 BCE but quite possibly much earlier.[47] Given that both animal hair brushes as well as dried ink have been discovered in Warring States tombs, we can be confident that writing and painting with brushes was widely practiced by the middle of the first millennium BCE.[48] The almost universal use of the brush as an instrument for writing and painting was firmly established in the Han dynasty; it was never seriously challenged down to the twentieth century.

Much effort went into producing ever better and more varied writing and painting brushes until they became, in the words of James Cahill, "perhaps the most versatile and

45. Much has been written about the importance of the brush in Chinese painting and calligraphy. Among the works I have found most useful are: Jean François Billeter, *The Chinese Art of Writing* (New York: Skira/Rizzoli, 1990); Chiang Yee, *Chinese Calligraphy: An Introduction to its Aesthetic and Technique,* 2nd ed. (London: Methuen, 1954); Lothar Ledderose, "Chinese Calligraphy: Its Aesthetic Dimension and Social Function," *Orientations* 17.10 (Oct. 1986), 35–50; Yolande Escande, "Perspectives et limites des recherches récentes sur la calligraphie et la peinture," *Revue Bibliographique de Sinologie* 14 (1996), 219–42.

46. Of course, the early Chinese sometimes incised hard materials such as bone and stone; the oracle bones are the most striking example.

47. Sören Edgren et al., *Chinese Rare Books in American Collections* (New York: China Institute in America, 1984), 127; Thorp and Vinograd, *Chinese Art and Culture,* 38, 90. For a Yangshao pot decoration suggesting the use of a brush-like implement, see Wu Jiming, *History of Chinese Drawing,* 14, Fig. 11. Wu would date the appearance of crude brushes much earlier, perhaps even as early as 5000 BCE. However, no surviving brushes predate the Warring States period.

48. Edgren, *Chinese Rare Books,* 127; Billeter, *The Chinese Art of Writing,* 51.

responsive drawing implement devised by man."[49] The Chinese writing and painting brush consisted of a cluster of animal hairs arranged to form roughly a cone shape and fixed into the end of a bamboo tube.[50] Softer hairs suited to absorbing ink or pigment formed the outer layers of the tip while stiffer hairs in the center provided resilience.

Long experience over centuries led to many improvements in the fabrication of brushes. By at least the third or fourth century BCE, brushmakers had discovered how to control the flexibility of brushes by arranging the hair at the tip in a series of concentric layers.[51] Still more flexibility was achieved by either treating the core with wax, which produced a stiff and springy brush whose resilient tip was especially suited to producing lines of varying thickness, or by using softer hairs and leaving them un-waxed, resulting in a brush that could produce lines of more even width.[52] Thick, soft-haired brushes were also produced for applying the "washes" that played such an important role in Chinese painting.[53] Around the end of the Tang or beginning of the Song, a final major advance occurred that set the pattern for brushmaking that has predominated down to the present. Jean François Billeter seems to have been the first to note this advance and to summarize it:

> [T]he improved technique consists in combining . . . longer and shorter hairs of the brush tip in such a way that the ink, instead of collecting in a single large circular reservoir, is retained in thousands of minute pockets distributed throughout a homogeneous tuft of hairs, wherever a shorter hair is packed between longer hairs. In this way the ink is more evenly distributed and its cohesive action is perfectly uniform.[54]

Considerable experimentation also led to the identification of many kinds of animal hair suitable for use in brushes, including rabbit and goat (long the most popular), fox, deer, sheep, and marten. Even mouse whiskers and the hair of human babies were tried. Marten and rabbit hair came to be preferred for fine line work.[55] For marten hair, the best is considered to be that which comes from male wild martens (much better than that of domestic martens) hunted in the autumn.[56]

49. Cahill, *Chinese Painting*, 12.
50. Now essential reading on the construction of Chinese brushes is the revisionist account by Billeter in *The Chinese Art of Writing*, 51–53. Billeter argues for a major advance in the making of Chinese brushes around the end of the Tang or the beginning of the Song.
51. Billeter, *The Chinese Art of Writing*, 51; Tsien Tsuen-hsuin, *Written on Bamboo and Silk: The Beginnings of Chinese Books and Inscriptions* (Chicago: Chicago University Press, 1962), 161.
52. Silbergeld, *Chinese Painting Style*, 5. Such brushes were especially suited for the line drawings that made up the great majority of Chinese book illustrations. See also Chung, *Drawing Boundaries*, 9–10.
53. Silbergeld, *Chinese Painting Style*, 5.
54. Billeter, *The Chinese Art of Writing*, 52. See also the further description that shows why, remarkably, "the quantity of ink released at any given moment is always exactly proportional to the width of the stroke" (pp. 52–53).
55. Chiang Yee, *The Chinese Eye: An Interpretation of Chinese Painting* (Bloomington and London: Indiana University Press, 1970), 193–97. At present, hairs from goats, rabbits and martens (*lang* 狼, an animal related to the weasel, but usually translated "wolf") are the most commonly used.
56. Billeter, *The Chinese Art of Writing*, 53. The extreme variety of hairs used in Chinese brushes is but one more among endless examples of a highly developed Chinese sensitivity to the technological advantages and disadvantages of various natural materials. This probably traces back in large part to the extreme importance in

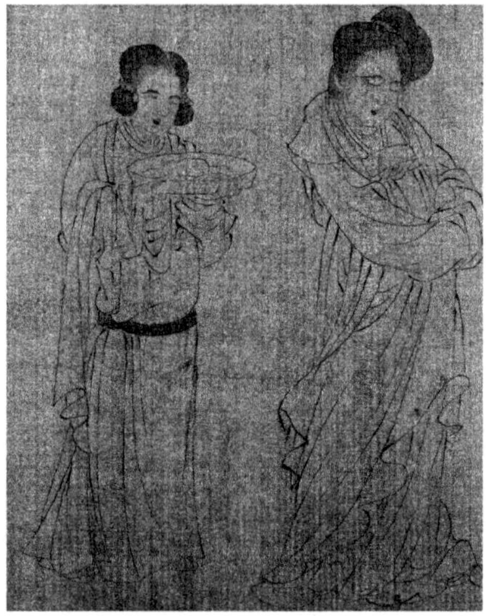
2.4 Detail from Zhou Wenju, *In the Palace,* showing effective use of calligraphy strokes in painting

Of all the standards for evaluating the quality of Chinese drawings and paintings, none was more important than the quality of the brushwork. As neatly summed up by Cahill, "the line drawn by a brush remains the central fact of Chinese painting throughout its history."[57] In contrast to European painters, Chinese painters never downplayed the importance of line in painting in order to concentrate on color or texture or surface contours or creating a sense of mass.[58] Already in the late Zhou, it appears that artisan painters who produced wall and screen paintings for the royal court were divided into two groups: those who delineated shapes using fine lines and those who then filled in the colors. The first group, according to Silbergeld, seems to have enjoyed a higher status.[59] In any case, down to the Tang, this "outline and color mode" predominated in Chinese painting.[60] In Song times and later, the "tendency to work primarily in terms of line" was further reinforced by an ever-closer association of painting with calligraphy and by the rise of scholar-amateur painting as the most highly regarded kind of painting (not only in scholarly circles, but to a considerable extent among all those who took an interest in fine painting).[61] An excellent example is the painting entitled *In the Palace* (Fig. 2.4), a twelfth-century outline copy (*baimiao*) of an earlier painting by Zhou

Chinese technology of agriculture and the efforts over the centuries to maximize agricultural production as well as to the extensive exploitation of a wide variety of minerals in her underappreciated mining industry. See Peter J. Golas, "Chemistry and Chemical Technology: Mining," in *Science and Civilisation in China (SCC)*, ed. Joseph Needham, vol. 5 part 13 (Cambridge: Cambridge University Press, 1999).

57. Cahill, *Chinese Painting*, 11. What is not often stressed when comparing Chinese and Western painting values, however, is that European painters of the late Middle Ages and the Renaissance were also keenly concerned with line. They simply held quite different ideas about what made for a good line in painting and drawing. The Chinese ideas were almost exclusively based on aesthetic concerns focusing on the brushstroke while European artists were much influenced by the importance of geometry at this time. Jean François Billeter points out that Western artists often tended to think of line as a "geometric abstraction which helps us to grasp space in intellectual terms." Lines that did this had to be "thin, uncluttered, free of any variations or apparent irregularities." Billeter, *The Chinese Art of Writing*, 46.
58. A 1603 collection of reproductions of paintings by great painters of the past produced by Gu Bing 顧炳, rested on the assumption that monochrome woodblock prints could provide adequate reproductions even of paintings in which color had figured prominently. Craig Clunas, *Pictures and Visuality in Early Modern China* (Princeton: Princeton University Press, 1997), 138 ff., esp. 143.
59. Silbergeld, *Chinese Painting Style*, 16.
60. James Cahill, "Approaches to Chinese Painting: Part II," in Yang, Barnhart et al., *Three Thousand Years*, 9.
61. Silbergeld, *Chinese Painting Style*, 16–17.

Wenju 周文矩 (active c. 940–97).[62] It provides a particularly good example of how calligraphic brushstrokes could effectively be used in painting.[63]

This example alerts us to the fact that the Chinese brush, with all its advantages, was by no means equally well suited for all writing and drawing tasks. To some degree, the desire to draw different kinds of lines, as discussed above, was met by the creation of different kinds of brushes, each having its particular strength(s). But the lines on which Chinese artists focused in their paintings were not necessarily the same kind of lines that make for a good illustration of a mechanical apparatus. As Chiang Yee has noted:

> Writing lines [i.e., the standard strokes of calligraphy] are used very extensively in Chinese paintings, which indeed are built up of them. Even solid objects usually have their contours outlined . . . [but] they seldom run continuously round the object. Except in delicate or elaborate work we prefer to use a few simple and carefully constructed strokes to convey the essentials of the form.[64]

It is not that the Chinese did not produce some brushes that were better than others for making outline drawings. And a painting master such as Wang Zhenpeng 王振鵬 (active c. 1280–1329) was said to be able to depict buildings in such meticulous detail that they could serve effectively as guides to construct the buildings (Plate 4).[65] The point is that the Chinese writing and painting brush is much better as an instrument for aesthetic expression than for producing images marked by precision and detail. For example, it was always relatively difficult to get a line of consistent width when running a brush along a ruler.[66] Given the general disparagement by the Chinese of painting and sketching that made use of straightedges, this particular limitation of the brush was of little importance to the Chinese artist.[67] For accurate depictions of implements and mechanisms, however, lines just the opposite of those described by Chiang Yee are more appropriate, and here the exclusive use of the brush to the exclusion of any hard-tipped drawing instruments would have been a significant impediment.[68]

62. Fong, *Beyond Representation*, 36–37.

63. The copy also illustrates the absence of color typical in *baimiao* paintings (which we shall discuss further in a moment). The original had actually been in full color. Fong, *Beyond Representation*, 35.

64. Chiang, *Chinese Calligraphy*, 210.

65. Fong, *Beyond Representation*, 397. This is not quite the exaggeration it might appear to be at first reading. A painting could indeed serve this function in major part because of the modular techniques of construction in Chinese architecture that we shall discuss in Chapter 3 when considering the *Building Standards*.

66. Gisbert Combaz, "La Peinture chinoise vue par un peintre occidental," *Mélanges chinois et bouddhiques* 6 (1938–39), 56. The kind of result you are likely to get can be seen in Billeter, *The Chinese Art of Writing*, 45, illustration on lower right. On the same page, Billeter makes the important point that the Chinese did not think of the brush as an instrument to draw evenly thick or thin lines as we think of them but rather to produce "forms of another sort." On the other hand, as we have seen, Jerome Silbergeld points to a distinction between waxed core and unwaxed core brushes with the latter producing a more even line of a kind that was used in ruled-line (*jiehua*) painting. Silbergeld, *Chinese Painting Style*, 5.

67. We shall return to this in our discussion of "ruled-line" painting in Chapter 3.

68. The Chinese also never attempted to use the flexibility of the brush to draw different kinds of lines to distinguish, for example, visible and invisible parts of a machine as Western draftsmen came to use broken and solid lines.

Another characteristic that limited the effectiveness of the Chinese brush for the drawing of technical images was the requirement that its ink supply be constantly replenished. The ink itself was typically prepared at each use by grinding an inkcake made of compressed soot (usually pine) on some kind of suitable stone using a bit of water to make it liquid. This made spur-of-the-moment, on-the-scene sketching inconvenient, especially outdoors.[69] Not for nothing were many Chinese painters in the habit of observing landscapes as intensely as possible and then returning to their studios to paint the scenes from memory. This may have worked well enough for landscapes where absolute fidelity to the original scene was not considered important. But in the case of machines of any complexity, it was hardly likely to be an effective substitute for careful sketching while looking at the machine. Nothing in the artists' background would have trained them to remember accurately technical details.

The nature of the brush also influenced the subjects Chinese artists chose to paint. Besides the aesthetic concern to create paintings full of life, there was also a preference for subjects that lent themselves especially well to depiction by brush and ink, subjects that included certain trees such as pines, bamboos and willows,[70] strange rocks, towering mountains and soft southern landscapes. Even certain seasons: the tenth-century painter, Li Cheng 李成, was partial to wintry landscapes where the contrasts of the black-and-white scenery "served as ideal vehicles for displaying the extraordinary vitality of his brushwork."[71] The painting produced by literati from Song times onward set the standards for what constituted high art in painting. One result was a narrowing repertoire of subject matter that was rich in symbolic significance but that could also be satisfactorily represented by painters well-trained in calligraphy but generally not deeply interested in developing their painterly skills.[72] By the Ming, with the increased availability of painting manuals or copybooks, even the actual way of portraying these subjects tended to become highly standardized.[73] We can see these tendencies fully at play in the illustrations of China's greatest technological

69. This is not to say that no Chinese artists ever made sketches from life. For some examples, see Cahill, *The Painter's Practice*, 88; Sullivan, *Symbols of Eternity*, 8–9; Liu, "*Water Mill*," 593n91. But even in these rare cases, the sketches seem to have been highly abbreviated. Finished paintings, even of landscapes, were virtually always a studio product. Compare Cahill, *The Painter's Practice*, 88 and Liu Heping, "Painting and Commerce in Northern Song Dynasty China," 143n68.

70. Chiang, *The Chinese Eye*, 192.

71. Cahill, *Chinese Painting*, 31.

72. Cahill, "Approaches to Chinese Painting," 8.

73. Craig Clunas, *Art in China* (Oxford and New York: Oxford University Press, 1997), 177, 191; Anne E. McLaren, *Chinese Popular Culture and Ming Chantefables* (Leiden: Brill, 1998), 64. This Chinese form of standardized painting did not, to be sure, completely close the door to creativity. As Lothar Ledderose points out, using Zheng Xie 鄭燮 (1693–1765) as an example: "Painters like Zheng Xie strive to emulate nature in two respects. They produce large, almost limitless quantities of works and are enabled to do so by module systems of compositions, motifs, and brushstrokes. But, they also imbue every single work with its own unique and inimitable shape, as nature does in its prodigious invention of forms. A lifetime devoted to training his aesthetic sensibilities [rather than his artistic skills?] enabled the artist to approximate the power of nature. For the Chinese literati painter, modular systems and individuality are but two sides of the same coin. Its name is *creativity*." Ledderose, *Ten Thousand Things*, 213.

compendium, the 1637 *Tian gong kai wu* 天工開物 of Song Yingxing 宋應星 (discussed in Chapter 5), where the artists, for example, frequently introduced (certainly to no technological purpose) their beloved, highly conventional motifs such as banana trees or Lake Tai rocks to outdoor scenes, and delicate decorative screens to indoor workshops.[74] Inanimate objects, by contrast, offered much less attraction, especially when they were difficult to draw.

Finally, what one might call the Chinese idolatry of the brush made Chinese artists remarkably impervious to the possibilities of other kinds of drawing implements. The Chinese brush may indeed be uniquely versatile but it certainly cannot match the range of possibilities offered by the variety of drawing instruments available to an artist in the time of, say, Leonardo da Vinci: not only the small pointed brush that had been in use since Roman times[75] but also silver-or alloy-tipped rods for metalpoint drawing, quills and reeds, chalks and charcoal.[76] In versatility, no single Western drawing instrument could rival the Chinese brush and ink. But used separately or in combination, the Western tools and materials offered a range of possibilities undreamed of in China.[77] The availability of all of these possibilities was itself a powerful spur to experimentation and creativity in drawing, and led European draftsmen from the late Middle Ages on to develop a whole range of skills appropriate *both* to the requirements of precision and accuracy in technical drawings *as well as* to the aesthetic demands in what have since come to be known as "master drawings."

Models, Automata, and Technological Drawing

Pre-Han and Han burial remains have yielded up many models of everyday objects, including implements, utensils and machines. They provide valuable information on the technology in use when they were made. Most of these were simple models that could undoubtedly have been constructed without any need for design plans simply by scaling down the dimensions of the original.[78] One wonders, however, about more complex examples well represented by the spectacular bronze chariots with their four horses excavated from the tomb of the First Emperor of Qin (Fig. 2.5).[79] Half the size of real chariots, these models were built with extreme precision; one was made up of 3,462 individual parts, surely one of

74. Sun and Sun, *T'ien-kung k'ai-wu*, 68, 108, 150, 158, 166, 194, 240. See also Chapter 5.
75. Henry Petroski, *The Pencil: A History of Design and Circumstances* (New York: Knopf, 1989), 7.
76. Watrous, *Craft of Old-Master Drawings*. The lead pencil had apparently not yet been invented; there is no written or pictorial evidence for its existence before 1564. Petroski, *The Pencil*, 7.
77. To take the goose-quill pen as a single example: it "can be cut in many different ways to produce different shapes, and varying degrees of fineness, of the nib. Because it is an organic material, the quill is flexible and malleable . . . If finely cut, the quill responds to slight changes of pressure or of speed of movement to leave thickened or tonally strengthened lines." In an experienced hand, the quill will glide over the paper surface with smooth fluency. Ames-Lewis, *Drawing in Early Renaissance Italy*, 46.
78. For references to portrayals of many of these models, see Golas, "Emergence of Technical Drawing," 30.
79. Shaanxi sheng Qin yong kaogudui 陝西省秦俑考古隊and Qin Shihuang bingmayong bowuguan 秦始皇兵馬俑博物館 (eds.), *Qin ling er hao tong che ma* 秦陵二號銅車馬 [A bronze chariot and horses from the Necropolis of Qin] (Xi'an: Kaogu yu wenwu, 1983), 1–2, 65; Ledderose, *Ten Thousand Things*, 56, Fig. 3.9.

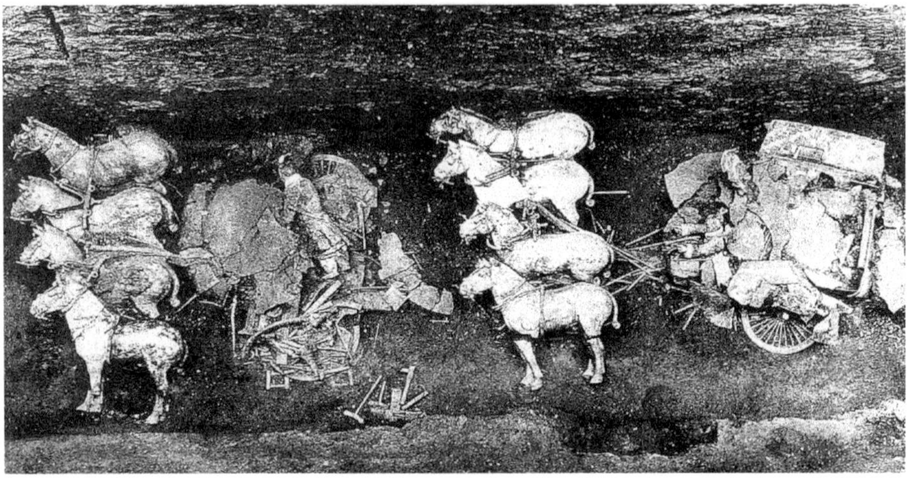

2.5 Two four-horse chariots found in the necropolis of the First Emperor of Qin. Though greatly damaged by the earth that had collapsed on them, they have since been fully restored.

the most complex models ever constructed in the premodern world. Could such a model have been built without any plans or working drawings? Unfortunately, there is no evidence from early mortuary remains or from other sources pointing to the existence of such drawings. One might also ask if it is reasonable to expect, again in the absence of any evidence, that the Chinese were capable at this time of producing working drawings that would have been of real use in planning and constructing such a complex model. Perhaps this model was achieved by systematically scaling down the parts of a full-sized chariot and relying on something in the way of assembly drawings or diagrams.

From early times, the Chinese used models of various kinds and materials (e.g., wood, clay, papier-maché, wax) for designing buildings and other structures or for aiding in their construction.[80] Tomb reliefs and models from the Warring States and Han periods give us a good picture of their appearance, including the complexity they sometimes displayed (Fig. 2.6). Where the making of architectural models was aided by the use of drawings of some kind,[81] they might have served as the predecessors of a rich history of architectural drawing to which we shall return below.[82]

80. Klass Ruitenbeek, *Carpentry and Building In Late Imperial China: a Study of the Fifteenth-Century Carpenter Manual* Lu Ban jing (Leiden: Brill, 1993), 49–50, 169; Needham, Wang and Lu, *SCC* 4:3, 105–6; Xiao, *History of Building Technology*, 3, 890–91; Adam Schall, *Historica Relatio* (Tientsin: Hautes Études, 1942), 90–91. By contrast, architectural drawings and models are remarkably absent in early Europe; Needham, Wang and Lu, *SCC* 4:3, 107, citing L. F. Salzman.

81. The earliest use of drawings to portray buildings could well go back to the beginnings of the Zhou dynasty (Wu Jiming, *History of Chinese Drawing*, 65–6). Wu suggests that "plans" of some sort must have been regularly used to get rulers' approval for the appearance of the palaces they ordered built. We have specific evidence in Sima Qian's *Records of the Historian* showing this was certainly true in the case of the Qin First Emperor. Ibid., 67–68; Xiao, *History of Building Technology*, 3, 888.

82. In our discussion of one of the masterworks of Song technological writing, the *Building Standards* (see Chapter 3), we note Needham's suggestion that it may have been in the drawing of buildings that Chinese

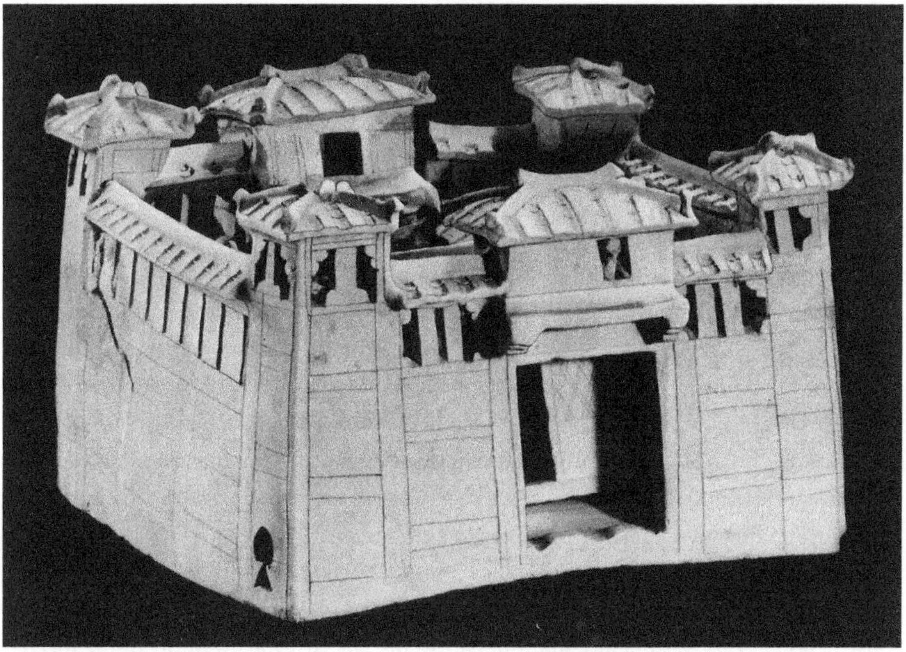

2.6 Fortified manor house dated 76 CE. Within the four towers and two pavilions are two buildings of two rooms each. Inside are also figures engaged in agricultural and domestic activities.

Another use of models may have been in the designing and construction of complex astronomical instruments. It is difficult to believe that a complicated machine such as Zhang Heng's water-powered armillary sphere would have been built in bronze without the use of one or more wooden prototypes.[83] Zhang was reputed to be skillful at painting/drawing (*shan hua* 善畫)[84] and evidence from the titles of books that have not survived suggests strongly that there were also drawings made in connection with the armillary sphere. In the time of Emperor Yang of Sui (r. 605–617), there was at the imperial court a major project to recreate objects from the past that ran a gamut from clothes and many kinds of vehicles to even a seismoscope for detecting earthquakes. The process often seems to have consisted of looking up illustrations in old books, building models on the basis of the illustrations, and then having workers refer to the models when building the full-size versions.[85]

---

technological drawing reached its zenith, at least insofar as precision was concerned. Needham, Wang and Lu, *SCC* 4:3, 106–7.

83. Almost a millennium later, Su Song and his collaborators on the great Northern Song clocktower received government authorization to build the final version only after completing a working wooden model for approval by the emperor. Joseph Needham, Wang Ling and Derek J. de Solla Price, *Heavenly Clockwork: the Great Astronomical Clocks of Medieval China* (Cambridge: Cambridge University Press, 1986), 9, 20, 21.

84. Wu, *History of Chinese Drawing*, 27.

85. Wu, *History of Chinese Drawing*, 30. Only rarely does there seem to be any hint in the Chinese sources, at this time or later, of the complications that could be posed by the transition from even carefully constructed scale models to functioning, full-scale instruments or machines; see for one example, Needham, Wang and de Solla Price, *Heavenly Clockwork*, 128n8. On the limitations of models as indicators of how the full-scale machine

A final and perhaps the greatest stimulus to drawings with technological content in the post-Han centuries were the mechanisms constructed to amuse princes and their courtiers in the many feudal courts of the period.[86] Such toys or "automata" had made their appearance in China before the Han, though the extremely sparse sources hardly support even a guess as to when or to what extent. The somewhat richer evidence from the post-Han centuries, however, reveals considerable ingenuity applied to producing automata of various kinds. Especially popular were doll-like figures that could bow, pour wine, play musical instruments, and dance. The interest in amusing mechanisms flowed over into at least somewhat more functional areas, leading to the invention of "self-moving" carts, doors that could open and shut automatically, several versions of south-pointing carriages (a cart with a standing figure that, by means of gearing, always pointed to the south whatever the turns of the cart),[87] and hodometers (vehicles able to measure distance traveled).[88] The importance attached to these mechanical devices is suggested by the fact that quite a number of their inventors or improvers are known to us by name, an exception to the anonymity so often surrounding inventors and craftsmen in early times.

Unfortunately, our sources almost always limit themselves to describing what the automata could do and tell us nothing about how they did it, i.e., any details of their mechanical construction.[89] This was due no doubt in part to the fact that the inventors of the mechanisms had good reason to keep them secret. Many of them used their secret knowledge and technical skills to win or maintain the favor of emperors or other patrons.[90]

As suggested above, it is difficult to imagine that these mechanisms, some of them quite complex, could have been designed and constructed without at least some recourse to sketches and perhaps even construction drawings of some kind.[91] That nothing of this kind has survived is hardly surprising. While the wider availability of paper encouraged more "writing" of all kinds, texts and illustrations, its fragility assured that virtually all of this writing would fail to survive.[92] With the exception of the considerable number of

will work, see Arnold Pacey, *The Maze of Ingenuity: Ideas and Idealism in the Development of Technology* (Cambridge, MA: MIT Press, 1992), 95; and Friedrich Klemm, *A History of Western Technology*, trans. Dorothea Waley Singer (London: Allen and Unwin, 1959), 162–64.

86. Needham and Wang, *SCC* 2, 53–54; 4:2, 156–65.

87. Needham and Wang, *SCC* 4:2, 286 ff.

88. Needham and Wang, *SCC* 4:2, 281–86.

89. Needham and Wang, *SCC* 4:2, 160, 162. An interesting exception to the silence on mechanisms is the statement that a mechanical puppet theater made by the famous third-century engineer Ma Jun 馬鈞 was powered by a horizontal water-wheel, though we have no further information on how it worked; *SCC* 4:2, 158.

90. Examples include the just-mentioned Ma Jun; Xie Fei 解飛 who devised a large sandalwood cart with animated religious figures for the Emperor Shi Hu 石虎 of Later Zhao in the mid-fourth century; and Huang Gun 黃袞 who constructed one-third life-size boats with animated figures for the Emperor Yang of the Sui dynasty in the early seventh century. Needham and Wang, *SCC* 4:2, 158, 159, 160.

91. For a good idea of the complexity that could be involved, one can turn to just a few of the proposed reconstructions for south-pointing carriages; see Needham and Wang, *SCC* 4:2, Figs. 528, 529, 531, and 533.

92. The earliest piece of paper with writing on it so far found in China dates from c. 110 CE; Tsien, *SCC* 5:1, 42, Fig. 1061. It was during the Eastern Jin (317–420) that paper largely if not completely supplanted bamboo or

documents and books from the cache found in a Dunhuang cave in western China at the beginning of this century (very few of them predating the seventh century),[93] nearly all of the manuscript materials from these centuries have long since perished. To be sure, much of the textual material in these manuscripts has survived either because the texts were printed once woodblock printing came into wide use or because they were cited by later writers who still had access to the originals. But this is much less true for illustrations.

What is more surprising, however, is the silence of the Chinese literature not only in this period but in later periods as well on the subject of, to use Graham Hollister-Short's term, sketches as "vehicles for ideas."[94] Ma Jun 馬鈞, whose puppet theater we have just referred to (see note 89), presents an interesting example. We know a great deal about him, thanks to an essay written by his friend Fu Xuan 傅玄.[95] But it seems that Ma's mechanical genius was matched only by his inability to explain his work. Thus we read in the essay: "[Ma's] powers of exposition fell far behind his mechanical ingenuity, and I doubt if he could express half of what he knew" and "Mr. Ma's gifts are all of the mind and not of the tongue." In connection with his work on a south-pointing carriage, Fu comments: "But again it was almost impossible [for him] to describe (the principle of it) in words."[96] After several statements of this sort, one is left puzzled why the idea of drawings is never raised. It seems that, at least among those relatively rare figures who combined an education in the highly word-oriented culture of traditional China with a serious interest also in technology, using drawings to work out mechanical ideas was seldom if ever attempted.

---

wooden tablets as the material on which books were written. Silk would continue to be used, but it was too expensive for writing on in any but very special circumstances. Tsien, *SCC* 5:1, 43; Wood, *Chinese Illustration*, 10. See also Tsien, *SCC* 5:1, 30 on the measured pace with which newer writing materials replaced older kinds.

93. Tsien, *SCC* 5:1, 45.

94. Graham Hollister-Short, "On the Origins of the Suction Lift Pump," *History of Technology* 15 (1993), 69.

95. The full translation of this essay appears in Needham and Wang, *SCC* 4:2, 39–42 and 158, and Joseph Needham and Robin D. S. Gates, "Chemistry and Chemical Technology: Military Technology: Missiles and Sieges," in *Science and Civilisation in China (SCC)*, ed. Joseph Needham, vol. 5, part 6 (Cambridge: Cambridge University Press, 1971), 200.

96. Needham's take on these statements is not entirely convincing: "The fact that Ma Chün (Ma Jun) could not . . . explain his ideas to the supercilious scholars of the Chin (Jin) court only meant that he was neither a philosopher nor an orator, it did not mean that he could not explain to his own artisans exactly what he wanted made in the world of gear-wheels and link-work." Joseph Needham, *The Grand Titration: Science and Society in East and West* (Toronto: University of Toronto Press, 1969), 39. Though he is making an argument against exaggerating the idea that the Chinese classical written language inhibited the advance of science and technology in China, the question here is whether Ma was simply not very adept in the classical language or, what seems more likely, whether his verbal abilities in general were limited, whether using the classical language or ordinary spoken Chinese. For a more social take on these descriptions of Ma, see Barbieri-Low, *Artisans*, 201.

The Advent of Woodblock Printing

Before the invention of woodblock printing, the Chinese made use of a variety of methods for reproducing manuscript texts and illustrations. These included free hand copies (by far the most common method), tracing, grid techniques, and the use of a kind of stenciling, the pounce method, in which powder made from chalk or some other material was tamped through small perforations made along the major outlines of an original design, leaving behind guide points for redrawing the design (Plate 5).[97]

All of these methods suffered from more or less serious disadvantages. They were slow and tedious, hence relatively costly. Moreover, the very tediousness encouraged frequent errors. In the copying of written text, the need to preserve meaning served as a check against mistakes, although sometimes a very imperfect one. It obviously did not work very well with copyists of limited literacy. The problem could be even greater with illustrations, however. Illustrations of course possessed their own kind of "meaning," but it was of a visual kind that worked much less effectively as a check on accuracy. Add to this the much lesser effectiveness of collating in the reproduction of images as well as the use of conventions in the original that might be unfamiliar to the copyist and we can understand how, as Richard Smith has pointed out, a "copy" of an illustration or a painting could range from a tracing to a free interpretation in the manner of the original.[98] Actually, close copies made by tracing seem to have been the exception; most often, the "copies" were to a greater or lesser extent revised renderings.[99]

Illustrations of technology were also prone to certain factors beyond sheer carelessness that could introduce distortions and mistakes into copies. Common among them was a less than full understanding on the draftsman's part of how a machine of some complexity actually worked.[100] This problem could be exacerbated if the original used conventions unfamiliar to the copyist.[101]

97. Silbergeld, *Chinese Painting Style*, 7; Fong, *Beyond Representation*, 347.

98. Richard J. Smith, *China's Cultural Heritage: the Qing Dynasty 1644–1912*, 2nd ed. (Boulder: Westview Press, 1994), 204 and 209. Indeed, the "original" served often as a stimulus rather than a pattern. Chou Ju-hsi, "Painting Theory in Eighteenth-Century China," in Willard Peterson et al., *The Power of Culture* (Hong Kong: Chinese University Press, 1994), 324.

99. What Ivins notes about copyists in Europe has at least some relevance for China: "When it came to copying a picture, . . . the copyist felt under no obligation to be faithful to either the peculiar forms or the linear syntax of the earlier draughtsman he thought he was copying." Ivins, *Prints and Visual Communication*, 61. See also Arnold Pacey, *Technology in World Civilization: A Thousand-year History* (Cambridge, MA: MIT Press, 1991), 27 and, for some excellent European examples from as late as the sixteenth century, showing how even trained copyists could bungle the portrayal of relatively simple mechanical elements, Eugene S. Ferguson, *Engineering and the Mind's Eye* (Cambridge, MA: MIT Press, 1992), 107–13.

100. One sees this with particular clarity in the efforts of Chinese artists to copy Western technological illustrations as late as the seventeenth and eighteenth centuries. For examples, see Golas, "Technical Representation in China: Tools and Techniques of the Trade," *EASTM* 20 (2003), 18–20, 26–27; Needham and Wang, *SCC* 4:2, 212–14, Figs. 465–68; 548, Fig. 679 and Pl. CCLXVII, Fig. 680.

101. Francesca Bray, "Biology and Biological Technology: Agriculture," in *Science and Civilisation in China (SCC)*, ed. Joseph Needham, vol. 6, part 2 (Cambridge: Cambridge University Press, 1984), 63.

By at least the fifth or sixth centuries, and quite possibly even earlier, the Chinese had devised a technique that greatly facilitated the duplication of texts. It consisted of inscribing texts either in intaglio or in relief and usually on stone. Thereupon, copies (rubbings, dab prints, ink squeezes) could be made by inking the surface, covering it with a piece of paper that was then patted to facilitate transfer of the image to the paper, and then carefully removing the paper.[102] As early as the seventh century, carved blocks of wood were being substituted for the stone carvings and carvers were beginning to discover what kinds of wood[103] as well as what carving tools[104] worked best for different purposes. Now not only text material but also illustrations could be rapidly reproduced in large numbers and at relatively low cost. As the process came into wider use, books became available in unprecedented numbers throughout China. The quality of the printing also gradually improved so that, even by the ninth century, Chinese woodcuts had reached a high level of artistic and technical maturity.[105]

The ability to produce a large number of exact copies by means of woodblock printing seems to have been first exploited on a large scale by Chinese Buddhists.[106] During the centuries from late Han to the Tang, Buddhism provided a two-fold stimulus to woodblock printing. On the one hand, the missionary zeal of the Buddhists encouraged the reproduction of large numbers of Buddhist sacred texts and religious pictures.[107] Secondly, and this is especially important for illustrations, the "religious discipline" of the Buddhists meant considerable emphasis was placed on getting right a complex iconography.[108] These two stimuli came together in the pounce or stencil method. Used in the painting of the famous cave murals at Dunhuang and elsewhere, it helped achieve a high degree of iconographic accuracy while at the same time making possible speedier completion of complex murals since the same design could be used repeatedly.[109]

---

102. The spread of this process as well the following woodblock printing process went hand in hand with the increasing availability in these centuries of suitable kinds of paper.

103. For example, plankwood could produce only limited tonal gradations compared with end-grain wood. Chia, "Text and *Tu* in Context," 243n4.

104. Carvers eventually came to recognize at least sixteen different cutting techniques that could be used with the *quan dao* 拳刀, a standard cutting knife; David Barker, *Traditional Techniques in Contemporary Chinese Printmaking* (Honolulu: University of Hawai'i Press, 2005), 26–27.

105. Tsien, *SCC* 5:1, 253.

106. There is one minor qualification to be made here concerning the word "exact": woodblocks did show the effects of wear and tear over time and use. On occasion, this could result in information being lost. For a striking example, see Golas, "Emergence of Technical Drawing," 45.

107. Lucille Chia, "*Mashaben*: Commercial Publishing in Jianyang from the Song to the Ming," in Smith and von Glahn, *Song-Yuan-Ming Transition*, 286. Personally copying religious works or subsidizing their reproduction was for Buddhists a major means of accumulating spiritual merit. The *Diamond Sutra* to which we shall refer again in a moment was subsidized by a certain Wang Jie as a meritorious act "on behalf of his [presumably dead] parents." Wood, *Chinese Illustration*, 11.

108. Ecke, "Reconsideration," 325.

109. Clunas, *Art in China*, 97.

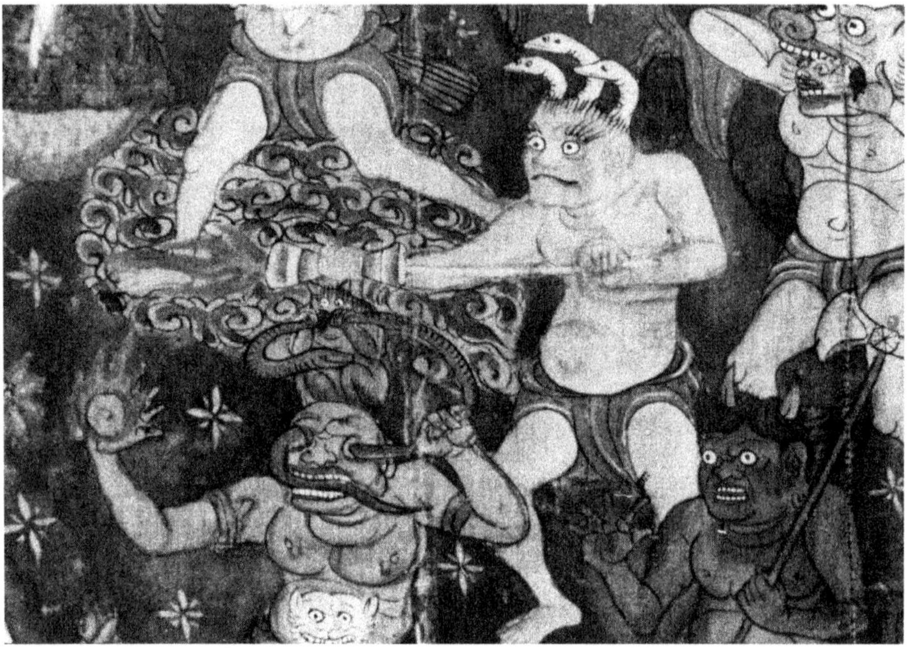

2.7 Earliest portrayal (c. 950) of a "fire lance" making use of a non-explosive gunpowder mix. Buddhist painted banner from Dunhuang.

It is not easy to assess the extent to which the pounce method may have been a stimulus to woodblock printing. Certainly, the two techniques are intrinsically very different. Nevertheless, both had as their goal easier and more accurate reproduction, if in very different media. More important, however, is that both techniques in responding to the desire for iconographic accuracy could get the details right to a degree that was previously rare in any kind of Chinese art and that ran counter to the general neglect of detail that was so typical of much early Chinese art. How far such capabilities had developed in woodblock printing even by the ninth century can be seen in the frontispiece of the world's earliest surviving printed book (in scroll format), the *Diamond Sutra* of 868 (Plate 6).

The extent to which some of this new attention to detail might have appeared also in pictures of technological subjects is impossible to say. The surviving art of this period, most of it Buddhist, rarely portrays technological subjects beyond buildings. Even more rarely do we encounter an image that is significantly informative, such as the tenth-century portrayal of a fire lance (Fig. 2.7).[110] Moreover, there is little either in Buddhist or in non-Buddhist writings during the first three or four centuries of woodblock printing to suggest a significant body of illustrations having to do with technology that may have existed but simply did not survive.[111] It may well have been that this was a period when illustrations simply were

110. Needham, *Science in Traditional China*, 39–40.

111. This contrasts with architectural drawings, none of which have survived but which we know from references in written sources were produced and used with some frequency. Wu Jiming, *History of Chinese Drawing*, 73.

not commonly looked to for this kind of information.[112] But as Chinese technology registered unprecedented advances in the Song, woodblock-printed books with illustrations of technological subjects came to play a significant role in that growth. It was already true then, as it has been since, that much of technological understanding could not be transmitted except by means of illustrations.[113] This makes it all the more ironic that, at a certain point, a kind of inertia to which the technology was particularly susceptible set in so that, instead of a vehicle for spreading new knowledge, woodblock printing came often to embalm technology that was obsolescent if not obsolete. We shall return to this point later, especially in connection with Wang Zhen's *Agricultural Treatise*.

112. Cf. Ivins, *Prints and Visual Communication*, 28–29.

113. Often enough, even illustrations were able to convey technological knowledge only imperfectly (Basalla, *Evolution of Technology*, 83–84). We shall see shortly one very good example in the weeding illustrations of the *Pictures of Tilling and Weaving* which, by themselves, give us virtually no information on how this task is performed, mostly under water.

Three

Song and Yuan

A Golden Age

Ruled-line Painting and Its Critics

As early as the Han dynasty, a distinctive kind of painting emerged in China that, more than any other painting style developed by the Chinese, invited its use for the portrayal of technological subjects.[1] At the heart of this approach to painting, unique in its reliance on tools beyond the brush, was the drawing of straight lines with the aid of a straightedge, a calibrated ruler or a plumb line.[2] In practice, this meant that the technique was originally applied mainly in paintings that included architectural structures[3] since there are few straight lines to be seen in landscapes or in portraits, the two main subjects of Chinese painting in early times.

This style of painting was so well adapted to architectural renderings that the resulting paintings could sometimes be used even as designs for constructing actual buildings.[4] An early example of this kind of painting is the mural on the wall of a tomb in Hebei dating from 176 CE that depicts a walled compound dominated by a watchtower (Plate 7).[5] The painting is especially remarkable for its precocious use of foreshortening which contributes to a strong three-dimensional effect.[6]

Over the centuries, these techniques were further developed and applied to non-architectural subjects, especially carpentered objects such as boats, wheeled vehicles and large weapons. Increasing emphasis on correct proportions and scale relationships even led to some experiments with vanishing points and other kinds of perspective, and with *trompe l'oeil* realism.[7]

1. Liu Heping's tracing of this painting tradition back to the First Emperor of Qin is reasonable but he presents no specific evidence to support it. Liu Heping, "*Water Mill*," 566.
2. Yee, "Cartography in China," 142; Fontein and Wu, *Han and T'ang Murals*, 31 and 32–33, Nos. 11 & 12.
3. Arthur Waley, *An Introduction to the Study of Chinese Painting* (New York: Charles Scribner's Son, 1923), 185.
4. Maeda, "*Chieh-hua*: Ruled-line Painting," 126n18.
5. Yang, Barnhard et al. *Three Thousand Years*, 33; Nancy Steinhardt (ed.), *Chinese Architecture* (New Haven and London: Yale University Press, 2002), 54.
6. Despite such early experiments (see also Sickman, "Chinese Painting before 1100," xiv), and occasional later uses (Sullivan, *Meeting of Eastern and Western Art*, 64 (Fig. 45) and 65), the Chinese would never go on to develop any fixed rules to govern foreshortening.
7. Maeda, "Spatial Enclosures," 370–91, esp. 375 and 376, Fig. 10; 377, Fig. 11; 381 and 388, Figs. 26 and 27.

In the later tenth century, Guo Zhongshu, whose unruly lifestyle contrasted dramatically with a high degree of painterly discipline, took this style of painting to a new level.[8] In works marked by spare, finely inked lines of even strength[9] as well as fastidiously correct proportions and scale relationships, Guo showed himself to be as much a draftsman as a painter. An excellent example is provided by his ships in the painting *Traveling on the River in Clearing Snow* (Plate 8).[10] Paintings like this were the forerunners of the outline (*baimiao*) painting style that became the norm in book illustrations.

In the eleventh century this style of painting eventually came to be christened *jiehua* ("ruled", or "straight-line", or "sharp-edge" painting).[11] It is the only category of Chinese painting that relied not only on instruments such as the straightedge for measuring and drawing straight lines but also made use of other instruments such as the compass and the square for measuring and spacing, for drawing circles, and the like.[12] It was also unique in the Chinese painting tradition in its stress on the importance of technique. In its purest form, it consisted of "a multitude of straight [ruled] lines skillfully interwoven with lines drawn in freehand."[13] *Jiehua* painters also typically paid close attention to portraying details correctly.

Surprisingly, and with something less than complete consistency, Guo Zhongshu himself criticized much of ruled-line painting: "Painters of the present day usually make use of the straightedge, and lay out the bracketing [of towers and pavilions] by ruled lines. The brushwork then becomes complicated, and fails to impart any idea of vigorous beauty or easy

---

Among the experiments in organizing space *not* tried by the Chinese at this time or later was the depiction of space as a measurable entity. Unlike what happened in the European Renaissance, artists in China never even attempted to work out mathematical rules to assist them in the spatial ordering of their paintings; Maeda, "Spatial Enclosures," 370.

8. For two important predecessors, Yin Jizhao 尹繼昭 and Wei Xian 衛賢, see Maeda, "*Chieh-hua*: Ruled-line Painting," 135.
9. Maeda, "*Chieh-hua*: Ruled-line Painting," 130.
10. Maeda, "*Chieh-hua*: Ruled-line Painting," Plate 1; Yang, Barnhard et al. *Three Thousand Years*, 104, Fig. 96. It is difficult to see in this reproduction but Guo also used shading and contour lines (on the masts) to heighten the sense of dimension and mass.
11. Maeda, "*Chieh-hua*: Ruled-line Painting," 123–24, 126; Chung, *Drawing Boundaries*, 10–11 (citing the important passage by Guo Ruoxu 郭若虛 (fl. third quarter of the 11th century) discussing in detail–if not always very clearly–this style of painting and its difficulties). *Jiehua*, widely recognized by the end of the Northern Song as "a difficult field in which to gain skill" (Chung, *Drawing Boundaries*, 11, citing the Song imperial painting catalog) has also posed an ongoing challenge to those who would translate it into English; other renditions include "sharp-edge" painting (Needham, Wang and Lu, *SCC* 4:3, 106), "(fine-line) architectural drawing" (Wu Hong, "Origins of Chinese Painting," 33; Barnhart, *Three Thousand Years*," 105), "measured drawing" (Trousdale, "Architectural Landscapes," 287), or "boundary painting" (Chung, *Drawing Boundaries*; Waley, *Introduction to the Study of Chinese Painting*). Anita Chung interprets the basic sense of the term to be "marking out the boundaries (*jie*) and drawing (*hua*) the outlines"; she also accepts a Northern Song coinage for the term. Chung, *Drawing Boundaries*, 11, 16. Cf. also Maeda, "*Chieh-hua*: Ruled-line Painting," 124.
12. These drawing instruments and techniques had a long history dating back to well before the Han; Chung, *Drawing Boundaries*, 9; Maeda, "*Chieh-hua*: Ruled-line Painting," 125. Recall the discussion in Chapter 1 of the importance in early times of geometrical decoration, which could not have been created without the use of drawing and measuring tools.
13. Maeda, "*Chieh-hua*: Ruled-line Painting," 138.

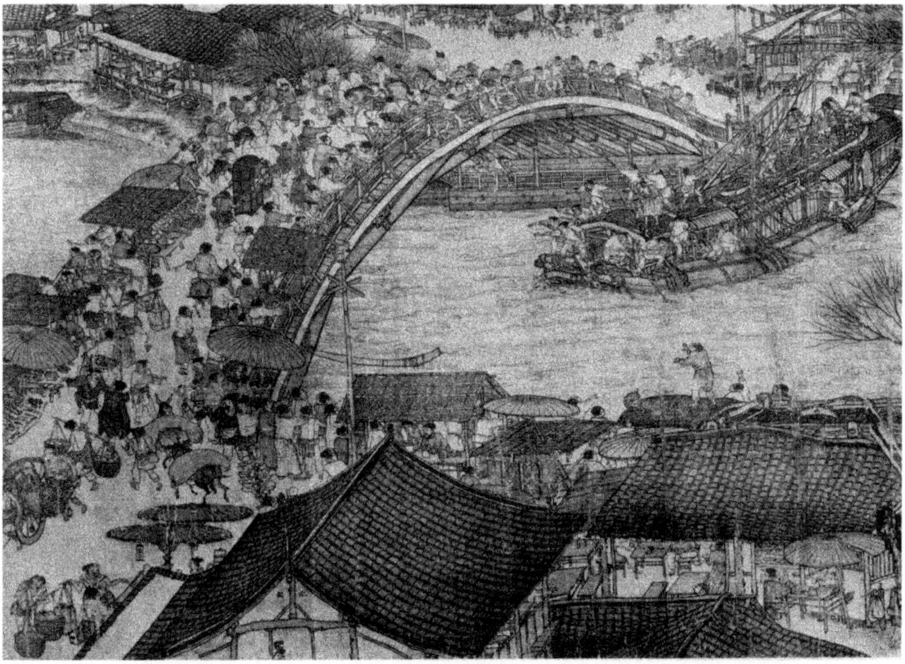

3.1 The famous "Rainbow Bridge" scene from the *Qingming shang he tu* 清明上河圖 (Along the river at the Qingming festival)

elegance."[14] In this, he echoed more general painting values expressed earlier by the famous Tang art critic and art historian Zhang Yanyuan 張彥遠 (fl. 847–874): "Now in painting (any) subject the things which one should especially avoid are a methodical completeness in delineation and coloring, as well as extreme carefulness, extreme detail and the display of skill and finish."[15]

On the other hand, one of the greatest masterpieces of Chinese painting, the *Qingming shang he tu* 清明上河圖 (Along the river at the Qingming Festival) (Fig. 3.1) shows how little such criticisms counted when the technique was used by a master such as Zhang Zeduan 張擇端.[16] Where Guo distinguished himself especially in his ability as a draftsman, Zhang was a consummate painter and certainly one of the finest representational painters in Chinese history. He possessed a remarkable brush technique in which he could vary "the breadth, the weight and the tempo of the line to suit the textural appearance of tile, thatch and wood."[17] At the same time, he had fully mastered a whole series of realistic techniques

14. Soper, *Kuo Jo-Hsü's Experiences*, 15.
15. W. R. B. Acker, *Some T'ang and Pre-T'ang Texts on Chinese Painting* (Leiden: E. J. Brill, 1954, 1974), 1, 185. Similar strictures warned that "A ruled brush makes a dead drawing" (Soper, *Kuo Jo-Hsü's Experiences*, 122n130) and "[I]f one makes use of the line-brush (*jiebi* 界筆) and straightedge (*zhichi* 直尺), the result will be dead painting" (Acker, *Some T'ang and Pre-T'ang Texts*, 1, 182–83).
16. Though the scroll incorporates a broad variety of stylistic elements, Maeda's characterization of it as "mainly an example of [*jiehua*]" seems justified.
17. Maeda, "*Chieh-hua*: Ruled-line Painting," 134. Roderick Whitfield ("Chang Tse-tuan's *Ch'ing-ming shang-ho*

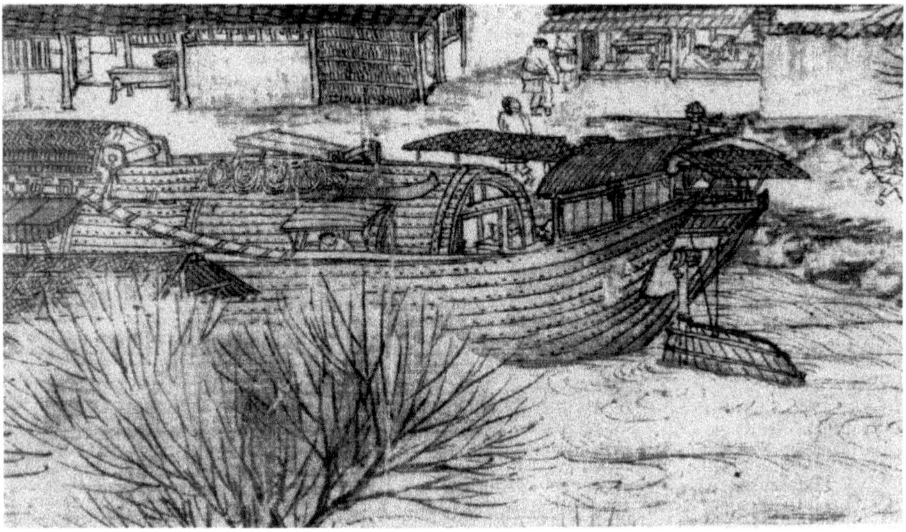

3.2 Detail of a cargo ship from the *Qingming shang he tu*. A comparison of the slung and balanced rudder system as portrayed here with the rudder assembly in the Guo Zhongshu painting above (Fig. 3.2) highlights clearly Zhang's superb ability to present technological details.

such as foreshortening, shading and meticulous handling of details. Zhang's superb combining of the precision of *jiehua* renderings of buildings, bridges, ships (Fig. 3.2), and carts with detailed and differentiated portrayals of hundreds of people (later versions of the scroll are inferior on these as on most other counts) echoes faintly the combination of geometric and non-geometric designs in early Chinese graphic art.

Despite its breathtaking achievement, it is unlikely the Qingming scroll converted many connoisseurs and painting theorists who disparaged paintings done with the help of drawing tools or were impatient with what they regarded as fussy details. Such strictures undoubtedly had their greatest effect among Song and post-Song literati painters, very few of whom would ever deign to paint anything that could be construed as a technological representation. But even the legendary Tang painter Wu Daozi 吳道子 (d. 792), who lived long before these ideals had come to govern what was regarded as the best in Chinese painting, refused in his painting to use a straightedge and was famous for his ability to draw long straight lines freehand.[18]

For our purposes, Zhang's scroll and the other ruled-line paintings that have survived[19] hint at what the *jiehua* tradition might have contributed to technological illustration in

t'u" (Ph.D. Dissertation. Princeton University, 1965), 47) comments on Zhang's ability to depict the infinite variations of moving water: "The rapid movement of the water and its many aspects are rendered . . . by means of fine brush lines, seldom more than a hair's breadth in width and yet thickening and thinning, describing myriad undulations and convolutions in their course, constantly implying the depth of the channel and the movement beneath the surface."

18. John Lust, *Chinese Popular Prints* (Leiden: E. J. Brill, 1996), 253.

19. Such as the untitled depiction of a water mill that Liu Heping incisively compares and contrasts with Zhang's work; Liu, "*Water Mill*," 583–84. See also Zheng Wei 鄭為, "*Zhakou panche tujuan*" 閘口盤車圖卷

China, and not only as a repository of useful techniques for representational drawings and paintings. A sustained improvement in technological representations would also have required artists who, like the artist-engineers of Renaissance Europe, combined artistic talents with keen interest in science and technology and who had a strong desire to convey what they knew and what they had carefully observed, rather than just to entertain their viewers. It is therefore of particular interest that we see in China at this time the emergence of a number of figures who combined technical knowledge, especially of architecture and building practices,[20] with highly developed artistic skills. Guo Zhongshu, for example, is supposed to have "fully mastered the actual practices of masons and carpenters"[21] while the architectural drawings of two other painters, Liu Wentong 劉文通 and Lü Zhuo 呂拙 were used directly as designs for an imperial construction project.[22] We shall see that Li Jie, the author of the *Building Standards* discussed below, was an accomplished architect and builder as well as an admired painter and calligrapher. Wang Zhenpeng 王振鵬, a Yuan master of *jiehua* architectural painting, during his official career was in charge of activities such as transport of grain by water where technological concerns figured prominently.[23] When attempting to understand why technical illustrations failed to register significant advances after the Song, a crucial part of the explanation must consider how the social context of the graphic arts in China mitigated against the appearance of artists with the particular combination of interests and abilities that might have encouraged such advances.[24]

In any case, the achievement of the *Qingming shang he tu*, instead of leading to further experiments and advances in realistic representation, can now be seen to have been a high-point of an approach to painting already falling out of favor. Painters were on the verge of abandoning most of the very techniques that made the best of the *jiehua* paintings so effective.[25] Ruled-line painting, along with realistic painting in general, became a target for many of the theorists and critics who defined the standards of good painting in the Northern Song. As a result, "the development of representational techniques for the visually convincing or 'life-like' rendering of natural objects . . . was scarcely continued as a serious concern of artists afterward."[26] Given the intimate relationship between drawing and painting in China, much closer than in Renaissance Europe, as well as the fact that those who did drawings of technical subjects considered themselves artists first (even if less prestigious professional or specialist artists) and draftsmen second, such strictures also worked against

[Transport carts at the mill]," *Wenwu* (February 1966), 17–25.

20. Needham, Wang and Lu, *SCC* 4:3, 106.
21. Soper, *Experiences*, 156.
22. Soper, *Experiences*, 70–71, 186.
23. Fong, *Beyond Representation*, 396–97. For other examples and discussion, see also Xiao, *History of Building Technology*, 3, 890.
24. See Soper, *Experiences*, 13, on Guo Ruoxu's view of the extensive and complex knowledge painters had to have in order to paint architectural subjects accurately.
25. Sullivan, *Symbols of Eternity*, 70.
26. Cahill, *The Painter's Practice*, 98.

the more precise and accurate drawing of implements, machines, or indeed any inanimate or manufactured objects.

Government, Printing, and Technological Representations

If the sea change in Chinese painting values that began in the eleventh century would prove inimical to advances in visual representations of technology, the growing involvement of the political authorities at this time in promoting and using new technology had distinctly more positive results for technical illustration. Woodblock printing offers an interesting case in point, given the dominant role of printed images in the illustrations of technology produced in late imperial China.

Already in the Tang, the government came to see both good and bad in the emergence and spread of woodblock printing. In one characteristic response, it sought to control the content of printed materials by adopting a two-pronged policy consisting of censorship (fairly limited) and promotion (much more widely practiced). Especially in the early centuries of the new technology, much government effort went into preventing the private printing and widespread distribution of books that were perceived in some way to contain potentially dangerous knowledge: for example, those dealing with astronomy, the calendar, geomancy and occult sciences or others containing certain Buddhist and Taoist religious texts.[27] Such efforts were resorted to from time to time and with varying intensity in later times, culminating in the great "literary inquisition" under the Qianlong emperor in the eighteenth century.[28]

But it did not take the government long to realize printing's potential for furthering its own interests. By the tenth century, the government itself was engaged in major printing projects such as a 130-volume edition of the Confucian classics that appeared in 953 and was offered for sale by the National Academy.[29] On a more material level, government officials interested in encouraging higher levels of agricultural and textile production especially saw the value of making information on advanced techniques more widely available and recognized that printed handbooks, all the more when they were illustrated, could serve this purpose well. Sponsorship by the authorities also led to the printing of major works in other areas of technology in which the state had a keen interest and for which there was no broad audience, including astronomy, architecture and weaponry. This encouragement from court and officials was especially important because of the limited patronage and market for illustrated books dealing with technological subjects. Moreover, because of the financial resources at the government's disposal, the works produced with official and

27. Denis Twitchett, *Printing and Publishing in Medieval China* (New York: Frederic C. Beil, 1983), 30.
28. Joseph McDermott, *A Social History Of the Chinese Book: Printing and Book Culture in Late Imperial China* (Hong Kong: Hong Kong University Press), 62–63; Frederick Mote, *Imperial China* (Cambridge, MA: Harvard University Press, 1999), 925–28.
29. Tsien, *SCC* 5:1, 156.

especially imperial backing often set the standard for high-quality printing of both text and images in the Song.

♦ ♦ ♦

In the remainder of this chapter, we shall examine five outstanding examples of illustrated works on technology from the Song and Yuan periods. Since none of them was produced without direct or indirect stimulus from the government, they illustrate very well the link between government sponsorship and major printing projects. They were all, to a greater or lesser degree, the products of officials writing mainly for an audience of officials. Each work had a direct and strong link to major government concerns: the *Military Techniques* with national defense, the *System Essentials* with cosmic order, the *Building Standards* with the massive constructions that bolstered the legitimacy and power of the government, and the *Pictures of Tilling and Weaving* as well as the *Agricultural Treatise* with the government's responsibility for the economic and social well-being of the countryside.

Strictly for convenience, we discuss these works in the order in which they originally made their appearance.[30]

The *Collection of the Most Important Military Techniques (Wujing zongyao)*

The tenth to twelfth centuries were the last extended period in Chinese history during which heartland China was divided among competing political powers in a kind of multi-state system[31] and when the very survival of the Chinese empire was under threat.[32] One of the results of this highly competitive political environment was significant advances in military technology.[33] Not surprising then was the appearance early in this period of the *Wujing zongyao* 武經總要 (Collection of the most important military techniques), the first printed Chinese military manual as far as we know to have contained illustrations of weapons.[34]

30. We might also have included a discussion of three archaeological catalogs completed in 1092, c.1107 and post-1162. Yves Hervouet, *A Sung Bibliography* (Hong Kong: The Chinese University Press, 1978), 200–201, 203; Liu and Gong, *Ancient Chinese Engineering Graphics*, 171–72. But despite their more than 1200 illustrations, only a very small number present anything of specifically technological interest. Moreover, most of the advanced drawing techniques they sometimes display are represented in one or another of the five works we shall discuss.

31. It was also a period in which China's Song dynasty was forced to adopt realistic policies marked by coexistence and a kind of "diplomatic parity" in its relations with foreign states. Mote, *Imperial China*, 112–14; Rossabi, *China among Equals*, 1–12.

32. It would come to a kind of temporary end, of course, with the Mongol conquest in the thirteenth century.

33. Mark Elvin, "China as a Counterfactual," in Jean Baechler, John A. Hall and Michael Mann (eds.), *Europe and the Rise of Capitalism* (Oxford and New York: Basil Blackwell, 1988), 103; Jacques Gernet, *A History of Chinese Civilization*, 2nd ed. (Cambridge: Cambridge University Press, 1996), 310–12.

34. Its main predecessor, unillustrated, was the *Tai Bo [Bai] Yin Jing* 太白陰經 (Canon of the white and gloomy planet of war) by a certain Daoist named Li Quan 李筌. Dating from 759, it was one of first of China's military

In 1040, spurred on no doubt in large part by the warfare in which the Song was engaged on its northwestern borders with the Xi Xia, Emperor Renzong 仁宗 ordered two eminent officials, Zeng Gongliang 曾公亮 (998–1078) and Ding Du 丁度 (990–1053) to compile a compendium of military knowledge. The finished work, dealing with virtually every aspect of equipping, maintaining and making use of what was a growing Song military force, was presented to the emperor in 1044.[35] Although the emperor himself provided it with a preface, we do not have any clear evidence that it was printed at this time or even later in the Song. Herbert Franke, without going into any details, speaks of its having been "repeatedly" printed in the Yuan and Ming.[36] In any case, no Song or Yuan printed editions have survived.[37] The illustrations in later editions date probably from the Ming but we have no information about when or under what circumstances they were produced and therefore how closely they followed the originals. They are therefore of distinctly limited use for information about illustrational practices in the Song.

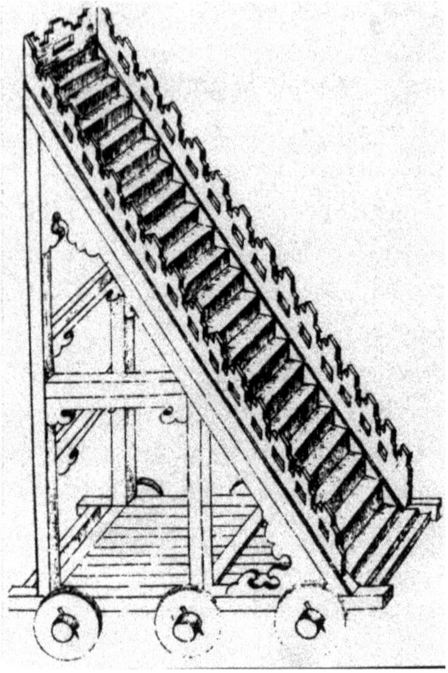

3.3 A "movable sky cart" (*xing tian che* 行天車) from the *Wujing zongyao*

The absence of the original illustrations is all the more regrettable because some of the surviving drawings exhibit considerable sophistication. Many of them, such as the "moveable sky cart" (*xing tian che* 行天車) for scaling city walls, are able to achieve a reasonably effective sense of three-dimensionality through the use of parallel line, equal angle axiometric projections (Fig. 3.3).[38] Others present for what seems to be the first time in surviving

works that, like the *Military Techniques*, gave a detailed description of military technology in use at the time it was written. Needham and Yates, *SCC* 5:6, 25. China of course has a long tradition of works on military theory (tactics and strategy) that date back to the Zhou and which dealt with weapons in a general way. Needham and Yates, *SCC* 5:6, 10–24.

35. Hervouet, *Sung Bibliography*, 235; Herbert Franke, "Siege and Defense of Towns in Medieval China," in Frank A. Kierman, Jr. and John K. Fairbank, *Chinese Ways in Warfare* (Cambridge, MA: Harvard University Press, 1974), 195.
36. Hervouet, *Sung Bibliography*, 235.
37. The oldest printed version now available is sometimes said to date from 1439 but this is probably incorrect. For a discussion of the peculiar edition to which this statement refers, see Needham, Ho, Lu and Wang, *SCC* 5:7, 20–21. The actual earliest surviving printed edition, possibly based on tracings of a 1231 manuscript, dates only from the early 1500s, perhaps 1510. Needham and Yates, *SCC* 5:6, 26; Hervouet, *Sung Bibliography*, 235.
38. Lu and Hua, *History of Chinese Science and Technology*, 160–61; Figs. 4–76 and 4–77.

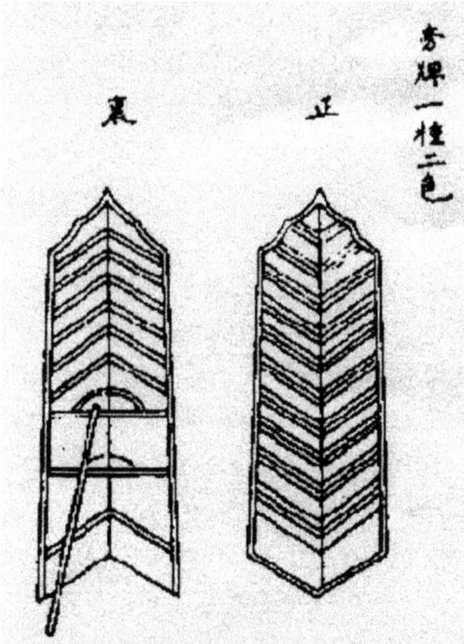

3.4 Front and back view of a shield (*pangpai* 旁牌) from the *Wujing zongyao*

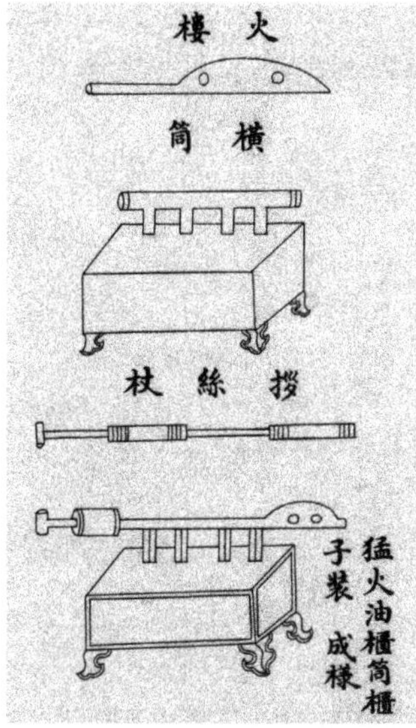

3.5 Flamethrower as pictured in the *Military Techniques*

illustrations both a front and a rear elevation view of an object in order to convey its shape and appearance more completely (Fig. 3.4).[39]

Still others are component parts and assembly drawings, such as the famous portrayal of a flamethrower that provided, along with a detailed accompanying textual description, sufficient information to enable a brilliant reconstruction by Joseph Needham of its double-acting, double piston, single cylinder force-pump (Fig. 3.5).[40] The drawings we now have for the flamethrower could be close to the originals since, as we shall see below, component parts and assembly drawings appear elsewhere in the Song as in the illustrations of the *New Armillary Sphere and Celestial Globe System Essentials.*

Other illustrations from the *Military Techniques* display some remarkable mistakes and omissions suggesting that at least some of the draftsmen who did either the originals or these copies had limited technical understanding of what they were portraying.[41] One can just imagine the kind of landlubber that would have produced a picture of a warship with no masts and sails or other visible means of propulsion (Fig. 3.6)![42]

39. *Wujing zongyao*, qian, 10, 28a.

40. Needham, Ho, Lu and Wang, *SCC* 5:7, 82–85. Needham's Fig. 7 is a composite made up from images in two illustrations of the *Military Techniques* (*qian ji*, *juan* 12, 66a and 66b). For the illustration of Needham's reconstruction, see Needham and Wang *SCC* 4:2), 148, Fig. 434. For the translation of the description of this weapon in the *Military Techniques*, see Needham and Wang, *SCC* 4:2, 145 or Needham, Ho, Lu and Wang, *SCC* 5:7, 83–84 (with the only change in the latter version being the substitution of "naptha" for "petrol" and "oil" in the original). It is a superb example of how adequate the Chinese language *could* be for technological description, particularly when accompanied by an illustration.

41. Hervouet, *Sung Bibliography*, 235.

42. In later centuries, Chinese draftsmen sometimes showed themselves capable of far superior depiction of

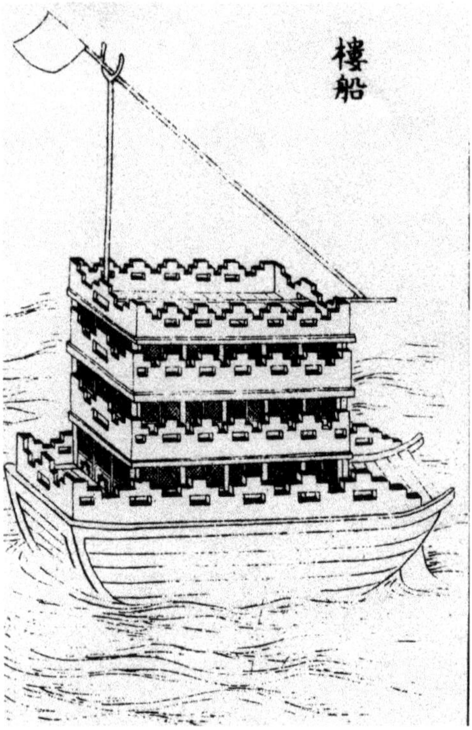

3.6 Multi-decked warship with counterweighted trebuchet but no visible means of propulsion

We shall return to the relationship between advances in military technology and improved technological illustration in Chapter 6 when we discuss certain similarities and differences between developments in Renaissance Europe and in Qing China. Here we can simply note how government sponsorship of technology and written works dealing with it could be, to draw on a military metaphor, a two-edged sword. When the government desired to disseminate a certain message or some kind of information, it had, especially after the spread of woodblock printing, considerable resources (artists, carvers, wood for woodblocks, paper and ink, and a whole workshop apparatus) at its disposal to do so.[43] In the case of weaponry, however, the government was determined to maintain as far as possible a monopoly over weapons in general and "advanced" weapons in particular. It regularly imposed strict controls on the circulation of books dealing with military technology.[44] Only that can explain how the *Military Techniques* came to be the sole major surviving Song work on weapons, out of some 107 known works from the Five Dynasties and the Song.[45] Its circulation was so effectively limited that it may never have been printed in full until five centuries after its compilation.[46]

Thus, in this case, actions of the government almost certainly significantly inhibited advances both in military technology and the portrayal of weapons. Hence Herbert Franke

ships; for examples, see Needham, Wang and Lu, *SCC* 4:3, 405, Fig. 939; 407, Fig. 941; and 410, Fig. 943.

43. In the assessment of Joseph McDermott, the government may have produced close to half of all printed books, at least of those aimed at an elite readership, between the tenth and the second third of the fifteenth centuries; McDermott, *Social History of the Chinese Book*, 68–69.

44. Actually, a surprising number of works on military subjects were compiled throughout the dynasties, as the figures in Needham's table show. Needham and Yates, *SCC* 5:6, 29. Much fewer, however, were printed; often they existed only in small numbers of manuscript copies, as seems to have been the case with the *Military Techniques*. Even when they were printed, it would likely be in quite small runs. A certain amount of material on military technology did make its way into non-official writings such as the *Sancai tu hui* 三才圖會 (Assembled riches from the three realms).

45. Needham, Ho, Lu and Wang, *SCC* 5:7, 19.

46. Needham and Yates, *SCC* 5:6, 26.

finds no basic difference between the *Military Techniques* of the eleventh century and the *Collected Essentials of Military Techniques* (*Wubei jiyao* 武備輯要) of the nineteenth.[47]

The *New Armillary Sphere and Celestial Globe System Essentials (Xinyixiang fayao)*[48]

Two key ideas underlay the importance of astronomy for the Chinese. The first was what Joseph Needham christened the organic view of the universe, namely that we live in a universe where heaven, earth and human beings are closely interrelated so that whatever happens in any one of these realms has consequences in the others. The second idea, so basic that it was rarely made explicit, was that the universe, with all its components, was intelligible to human beings.

From earliest times, the Chinese put a great deal of effort into trying to understand the heavenly component of the universe. Looking at the night sky, they not only perceived that it displayed consistent patterns in both the position and the motion of many of the objects observed but also that not everything in the sky was subject to regularity. Certain events such as comets and supernovas were totally unpredictable. They also gradually discovered that other apparently irregular events such as eclipses or the relative positions of the planets to each other could actually be predicted, though achieving accuracy in those predictions posed huge challenges.

The search for a better comprehension of the heavens, together with the great importance attached to keeping track of time (for practical purposes, for keeping history straight, and for determining auspicious days), led the Chinese to put considerable effort into astronomical observations and calculations. Commonly, just as with the development and production of military weapons, these activities were overwhelmingly within the purview of the government. Since ensuring that the entire universe functioned harmoniously was seen as a key task of the emperor (and a powerful support of his legitimacy), Chinese governments throughout the centuries had a great stake in astronomical calculations necessary to produce the imperially sanctioned calendar and the making of astronomical instruments (armillary spheres, celestial globes, etc.) to aid in those calculations.[49]

The instrument at the heart of astronomical observation in traditional China was the armillary sphere (Fig. 3.7). It apparently developed from a single ring instrument that enabled measurements of declination (distance of a celestial body north or south of the

47. Franke, "Siege and Defense," 198. He would go even further and rate the defense measures described in the Mohist canon as "basically not inconsistent" with what was still being written in the nineteenth century. Franke, "Siege and Defense," 192.

48. This section draws extensively on Golas, "Emergence of Technical Drawing."

49. For the history of astronomical instruments in China, Joseph Needham and Wang Ling, "Mathematics and the Sciences of the Heavens and the Earth," in *Science and Civilisation in China (SCC)*, ed. Joseph Needham, vol. 3. (Cambridge: Cambridge University Press, 1959), 284–390 and Needham, Wang and de Solla Price, *Heavenly Clockwork* are good starting points.

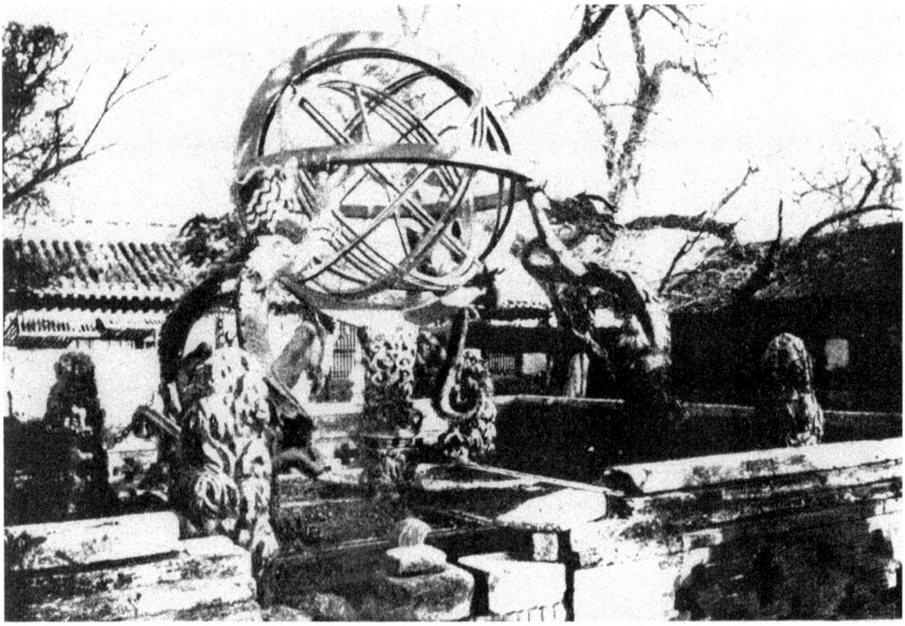

3.7 The ultimate development of the armillary sphere in China: the equatorial armillary sphere of Guo Shoujing (c. 1276)

celestial equator) or right ascension (distance of a celestial body eastward from the vernal equinox) depending on whether it was mounted on the (celestial) equatorial or meridian plane. We know of a model in the first century BCE that had a permanently fixed equatorial ring, while an ecliptic ring was in use by late in the first century CE. The introduction of a horizon ring centered on the observer is associated with the exceptionally ingenious Zhang Heng 張衡 who was active around 125 CE.[50]

Zhang was perhaps the greatest polymath in Chinese history and an accomplished poet and painter as well. He is the first figure we can identify as associated with a second theme in the story of Chinese armillary spheres. Armillary spheres can perform a dual role. Besides performing as a sighting apparatus that could assign positions to heavenly bodies, armillary spheres could also serve as demonstrational instruments to assist calendrical calculations. Zhang was the first to construct a demonstrational armillary sphere, remarkable at such an early period, with a model earth at the center. Zhang also figured out a way to apply waterpower to rotate this demonstrational model, though we know nothing of the details.

We have noted in Chapter 2 that an important advance in the drawing of technological subjects, the emergence of drawings to scale, resulted from the early efforts of astronomers to map the heavens and to track the movements of heavenly bodies. The notion of representations to scale can be seen already in the descriptions of Han armillary spheres which

50. Joseph Needham, "Astronomy in Ancient and Medieval China," *Philosophical Transactions of the Royal Society of London* 276 (1974), 75. For an extended discussion of Zhang Heng's contributions to the early development of astronomical instruments, see *Heavenly Clockwork*, especially 98–112.

were consciously constructed so that between two and four Chinese inches (1 Chinese inch = c. 2.25 centimeters) on the circumference of the outer ring equaled one degree.[51] From an anonymous work called the *Zhou bi suan jing* 周髀算經 (Arithmetical classic of the gnomon and the circular paths of the Heavens), dating from at least as early as the Former Han with some parts likely quite a bit earlier, we have a diagram of the motions of the sun through six zones of the sky at different times of the year. Two sets of scales for two sizes of the diagram work out to 1:18,000,000 and 1:38,000,000.[52] This technique later came to be used in mapping the earth, with scale maps appearing by the third century.[53]

The centuries after the Han witnessed continued improvements in armillary spheres as well as, in the early fifth century, the construction of the first solid celestial globe. In 633, under imperial authorization, the chief court astronomer, Li Chunfeng 李淳風 (602–70), built the first armillary sphere equipped with three layers of rings, which set the standard for later versions. Another major advance occurred in the third decade of the eighth century. The emperor's desire for a new calendar that would more accurately predict eclipses led to the construction by the Buddhist monk Yixing 一行 (682–727), the greatest mathematician and astronomer of his day, in collaboration with the low-ranking but very inventive official Liang Lingzan 梁令瓚 (fl. 720s), of an astronomical apparatus that incorporated in its power train what might have been the world's first "escapement."[54] With its inclusion of two jackwork elements to indicate hours by the striking of a bell and quarter-hours by the beating of a drum, this apparatus has been seen by some, especially Needham and his collaborators, as an escapement clock.[55]

Just as in the case of mechanical toys, it is highly unlikely that these increasingly complex astronomical instruments could have been constructed without recourse to drawings of some kind.[56] Moreover, by at least the Han, books were beginning to be written describ-

51. Liu and Gong, "Ancient Chinese Engineering Graphics," 75–81, esp. 77; *Heavenly Clockwork*, 109, fn. 2.

52. Needham and Wang, *SCC* 3, 256–57; Wu Jiming, *History of Chinese Drawing*, 88, 8; Liu and Gong, "Ancient Chinese Engineering Graphics," 76–77.

53. Around 267, the new Minister of Works of the Jin dynasty, Pei Xiu 裴秀, who has been called by Édouard Chavannes the father of scientific cartography in China, set about making an up-to-date map of the empire. In the preface of the final map, he enumerated six principles he considered essential to good mapmaking. The very first referred to establishing by means of "graduated divisions" the scale to which the map was to be drawn. Without this step, there was "no means of distinguishing between what is near and what is far." Needham and Wang, *SCC* 3, 538–40; Yee, "Cartography in China," 144.

54. In horology, an escapement regulates the driving force of the timekeeping mechanism by dividing the delivery of power into discreet segments.

55. Needham and Wang, *SCC* 3, 74–81. Landes argues very persuasively that this cannot be considered an ancestor of the verge-and-foliot clock escapement as it developed in the West; David Landes, *Revolution in Time: Clocks and the Making of the Modern World* (Cambridge, MA: Belknap Press, 1983), Chap. 1. Even if Landes is right, however, it does not preclude viewing Yixing's and Liang Lingzan's invention as one kind of escapement.

56. This is true even when we allow for the fact that models were also used in the devising of these instruments. To give just one example: Liang Lingzan, despite his low-ranking position in the bureaucracy, became Yixing's collaborator because he was able to demonstrate his knowledge and ingenuity by constructing an accurate wooden model of an advanced-design armillary sphere; *Heavenly Clockwork*, 76. For other examples of the

ing astronomical mechanisms. There is reason to believe, for example, that Zhang Heng may have produced an illustrated account of his armillary sphere.[57] All of these early works, however, have either been entirely lost or survive merely in a few fragments. Only from their titles can we surmise whether they contained drawings of mechanisms, a process that is rendered all the more tenuous since the word *tu* 圖, whose inclusion in the title is often our best clue, can refer not only to illustrations but also to charts, diagrams and the like.[58]

Shortly after the founding of the Song dynasty in 960 there occurred a major milestone in the portrayal of technology in China: the production of an extensive set of sophisticated drawings to illustrate in detail the construction and workings of a complicated astronomical apparatus. This mechanism was an astronomical clocktower built by Zhang Sixun 張思訓 with the support of Emperor Taizong 太宗 (r. 976–97) sometime in the years 976–78. The clocktower itself represented a significant advance over astronomical instruments that preceded it: in rotating a demonstrational armillary sphere for the first time by means of a chain-drive power transmission, it made possible a working model of the heavens that would parallel their movement with a minimum of human intervention.[59] It appears that copies of quite a few of the drawings produced to illustrate the workings of that clocktower as well as further drawings from an armillary sphere or perhaps some kind of clocktower built a century later by Zhou Riyan周日嚴 in the period 1078–85[60] were incorporated in whole or in part into the account of the astronomical clocktower built by Su Song 蘇頌 (1020–1101) and Han Gonglian 韓公廉 in the late 1080s and early 1090s.[61] This was the famous *Xinyixiang fayao* 信儀像法要 or *New Armillary Sphere and Celestial Globe System Essentials*.[62]

The *System Essentials* contains a total of sixty-one illustrations, of which fourteen are star charts and the remaining forty-seven relate to the mechanism of Su and Han.[63] The

use of models, see *Heavenly Clockwork*, 21, 114, 115, 127–28.

57. Wu Jiming, *History of Chinese Drawing*, 28.

58. See the extended discussion of *tu* in Bray et al., *Graphics and Text*.

59. Needham, Wang and de Solla Price, *Heavenly Clockwork*, 114.

60. J. H. Combridge, "The Astronomical Clocktowers of Chang Ssu-Hsun and His Successors," *Antiquarian Horology* 9.3 (June 1975), 295–97.

61. Combridge, "Astronomical Clocktowers," 288 and 291; *Heavenly Clockwork*, 20n4 and 36, esp. n2. Why the Su/Han clocktower was deemed necessary is not entirely clear. Perhaps, as Needham suggests, experimentation in the preceding years had opened the possibility of a significantly improved mechanism. Liu Heping suggests ("*Water Mill*," 577) that the new emperor wanted to demonstrate China's "scientific and natural" superiority to the Liao regime threatening from the north, especially after Su Song had found out on a diplomatic mission during the previous reign of the emperor's father that the Liao had a calendar that was superior in accuracy to the official Chinese calendar; for this episode, see *Heavenly Clockwork*, 6–8.

62. I have used the edition in volume 59 of the *Shou shan ge congshu* 守山閣叢書collectanea, 1922 edition. This work will be referred to hereafter as the *System Essentials*. Combridge, in "Astronomical Clocktowers," reproduces a number of illustrations that were omitted from *Heavenly Clockwork* (208, fn. 6) and attempts, usually quite persuasively, to date all the illustrations in Su's account. I have followed his datings here.

63. I arrive at this number by counting each page on which a drawing appears as a single illustration no matter how may components it may contain. Apart from the 14 star charts, the remaining 47 illustrations all relate to the mechanism of the clocktower: 17 in *juan* 1 (1 overall view, 3 assembly drawings, 13 component parts

abundance of illustrations itself is remarkable for so early a period. But equally remarkable are a number of distinctive features, many seen for the first time in the portrayal of a technological subject in China.[64]

To begin with, there are no human beings pictured in the illustrations. Nothing could reveal more clearly the strong focus on the mechanism and its components, clearly separating these illustrations from the tradition of agricultural/sericultural painting and drawing discussed above. About a dozen of the illustrations provide overall views or assembly and sub-assembly drawings that, together, show the complete mechanism and the relative positions of its different parts (Fig. 3.8). Something over thirty of the illustrations can be viewed as component part(s) drawings. One drawing, for example, pictures the innermost ring of the armillary sphere, showing the sighting tube, cross-struts and polar-mounted declination circle (Fig. 3.9). Another depicts just the parts that made up the sighting alidade assembly (Fig. 3.10). The armillary sphere with its immediate driving mechanism is the best illustrated component, being the subject of no less than sixteen drawings. No other mechanism in China before the twentieth century would receive such detailed illustration.

Another characteristic of the illustrations in the *System Essentials* is a generous use of labels. Indeed, every illustration of the *System Essentials* has at least one overall identifying label. By far the majority also contains one or more labels for component parts.[65] It is true that the use of labels is less than entirely consistent, with the choice of which elements deserve or need a label seemingly quite arbitrary at times.[66] Nevertheless, the labels are often very helpful, even essential, for interpreting the illustrations. This is especially true when the terminology used in the text is imprecise or inconsistent, a problem already well recognized in Su Song's time.[67] Labels (or any writing) also helped to prevent the accidental reversing of an illustration when the almost transparent paper on which it was drawn was pasted on a block of wood for carving. The reversed characters would immediately indicate the mistake.

---

drawings); 3 in *juan* 2 (1 overall view, 2 component parts drawings); 27 in *juan* 3 (4 overall views, 4 assembly drawings, 19 component parts drawings). Some 50 different parts are illustrated. For other totals, see Wu Jiming, *History of Chinese Drawing*, 32; and Liu Keming, "Songdai gongcheng tuxue de chengjiu" 宋代工程圖學的成就 [Achievements of engineering drawing in the Song], *Wenxian* 1991.4, 239. Despite the large number of illustrations and the opinion of the editors of the catalog of rare books in the great eighteenth-century imperial collection *Siku quanshu* 四庫全書 (Complete collection of the four treasuries) that the illustrations with their explanations show "how carefully the author presented the system of construction and method of use of his astronomical equipment" (*Heavenly Clockwork*, 12), text and illustrations together still leave many unresolved questions about the working of the mechanism. This is especially true for the time-keeping water-wheel linkwork; *Heavenly Clockwork*, 206.

64. For a more detailed analysis of the illustrations, on which the following is largely based, see Golas, "Emergence of Technical Drawing," 40–51.

65. By contrast, only one illustration (*System Essentials*, Chap. 2, 4a) has a comment as opposed to a label. (The ten-character comment simply notes that an alternative version of the illustration appears in another copy of the work.)

66. Golas, "Emergence of Technical Drawing," 44.

67. *Heavenly Clockwork*, 17. This problem was made worse by the predilection of the literati to use, for what was perceived as greater elegance, ancient expressions instead of more current terms. Needham, Ho, Lu and Wang, *SCC* 5:7, 284.

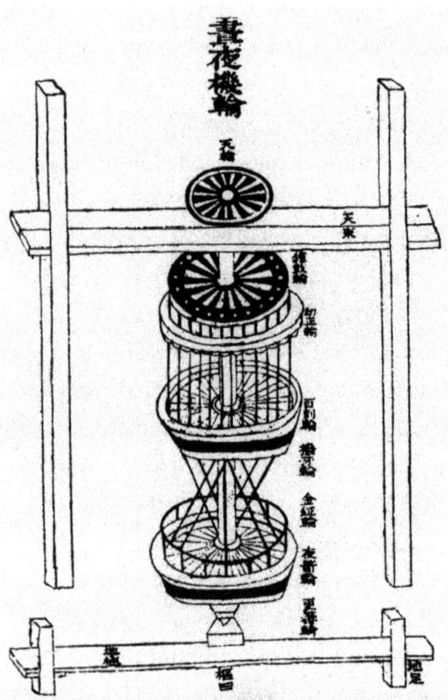

3.8 Assembly drawing of timekeeping shaft and its jackwheels; *System Essentials*, 3, 6b.

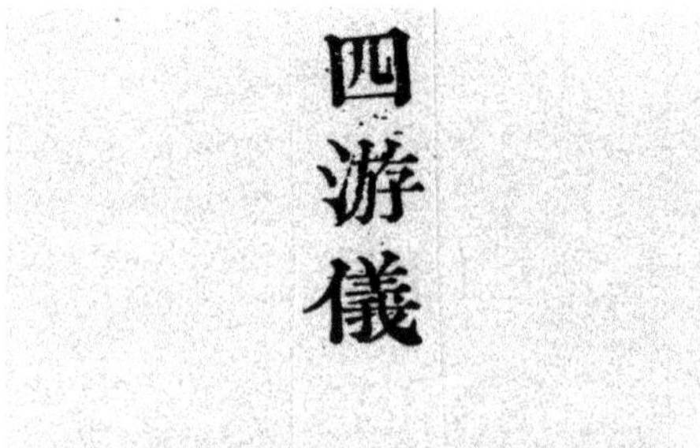

3.9 Component part drawing of the innermost ring of the armillary sphere; *System Essentials*, 1, 9b.

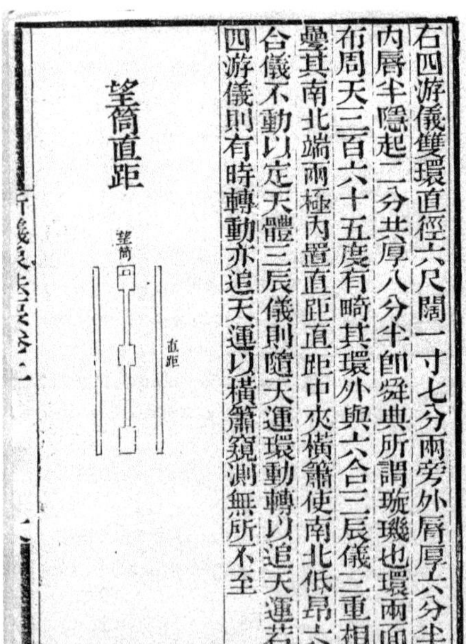
右四游儀雙環直徑六尺闊一寸七分兩旁外脣厚六分半
內脣半隱起二分共厚八分半卽舜典所謂璇璣也環兩面
布周天三百六十五度有畸其環外與六合三辰儀三重相
疊其南北端兩極內置直距直距中夾橫簫使南北低昂六
合儀不動以定天體三辰儀則隨天運環動轉以追天運若
四游儀則有時轉動亦追天運以橫簫窺測無所不至

望筩直距

3.10 Sighting alidade assembly; *System Essentials*, 1, 18a.

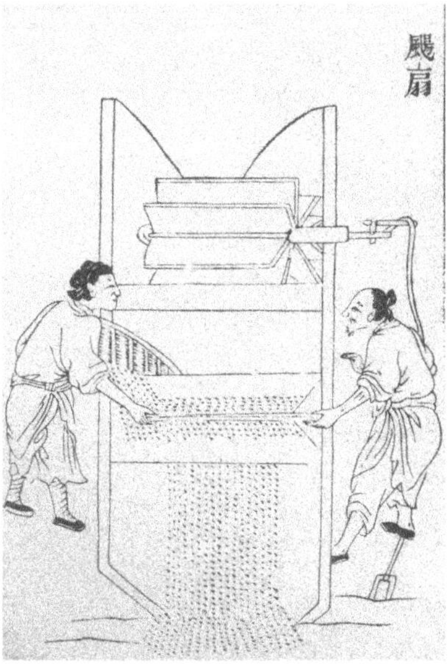

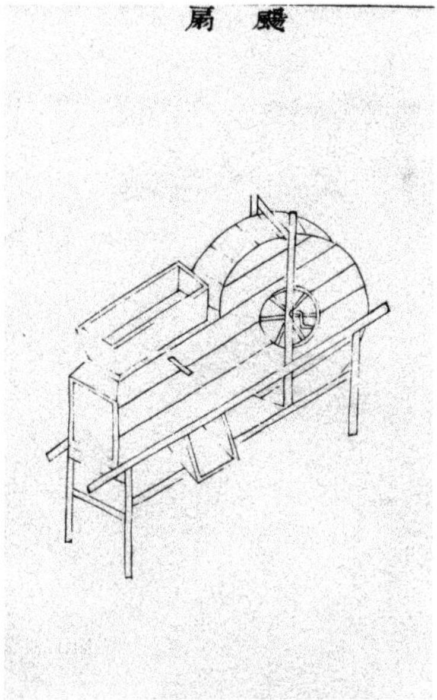

3.11 (a) Plan view illustration of a winnowing fan from the *Complete Treatise on Agricultural Administration*; *Nongzheng quanshu* 23, 11a; (b) Drawing of an enclosed winnowing fan; *Nongshu* 16, 9b. For an early twentieth-century photograph, see Hommel, *China at Work*, 77, Fig. 118 with description, pp. 74–76.

The *System Essentials* also presents for the first time a Chinese answer to one of the perennial problems of technological illustration: how to show the viewer what is happening in those parts of a machine that are enclosed and cannot be seen. One can do this only by creating an artificial condition where all or part of the casing is treated as transparent or is removed. This latter technique has the effect of opening a window as it were to the inside of the machine. It was already in use in Europe in the fifteenth century, being particularly associated with Mariano Taccola and his near contemporary Francesco Martini.[68] The partial cutaway technique seems never to have been devised by Chinese illustrators who even found it difficult to deal with when the Jesuits introduced it in the seventeenth century.[69] The reason seems to be that the Chinese, when they addressed this problem, preferred to treat the entire enclosure or at least one side of it as transparent, as in this famous illustration

68. One of the earliest examples in Europe is a fine fifteenth-century transparent drawing in perspective by Mariano Taccola of a suction pump showing the piston and flap-valve inside; Samuel Y. Edgerton, Jr., "The Renaissance Development of Scientific Illustration," in John W. Shirley and F. David Hoeniger (eds.), *Science and the Arts in the Renaissance* (Washington D.C.: Folger Shakespeare Library; London: Associated University Presses, 1985), 177 and 180; M. T. Wright, "On the Lift Pump," *History of Technology* 18 (1996), 25.

69. For example, Wang Zheng 王徵 who worked so closely with the Jesuit Johann Schreck in producing the *Illustrated Explanations of Wonderful Machines (from the Far West)* (*Qiqi tushuo* 奇器圖說) and was probably unique among Chinese literati in his familiarity with Renaissance mechanics chose, when copying Ramelli's portrayal of a revolving bookcase, to omit the cutaway that so effectively illustrated the gearing enabling successive bookrests to remain at the same angle as the bookcase rotated. Needham and Wang, *SCC* 4:2, 548, Fig. 679 and the facing Plate CCLXVII.

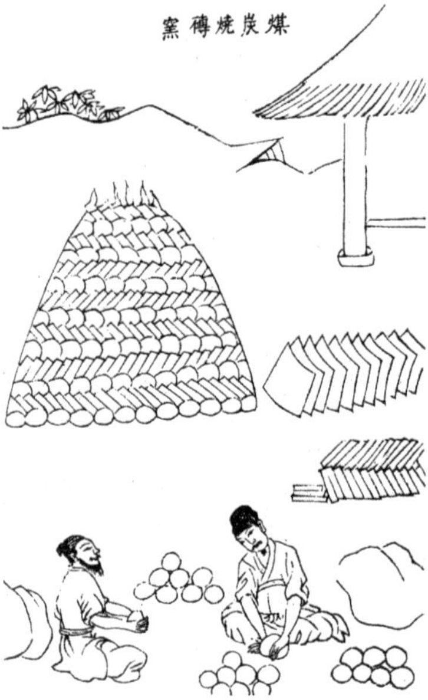

3.12 Coal-fired kiln drawn to show stacking in alternate layers of coal briquettes and the bricks being fired; Sun and Sun, *T'ien-kung k'ai-wu*, 141, Fig.7–4.

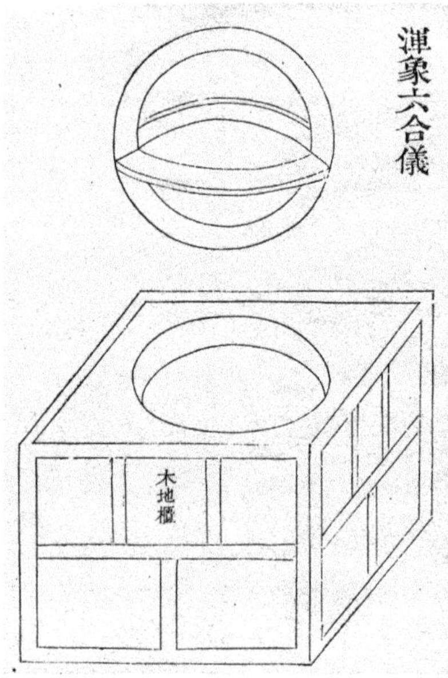

3.14 Subassembly drawing of the mounting system for the celestial globe; *System Essentials*, 2, 3a.

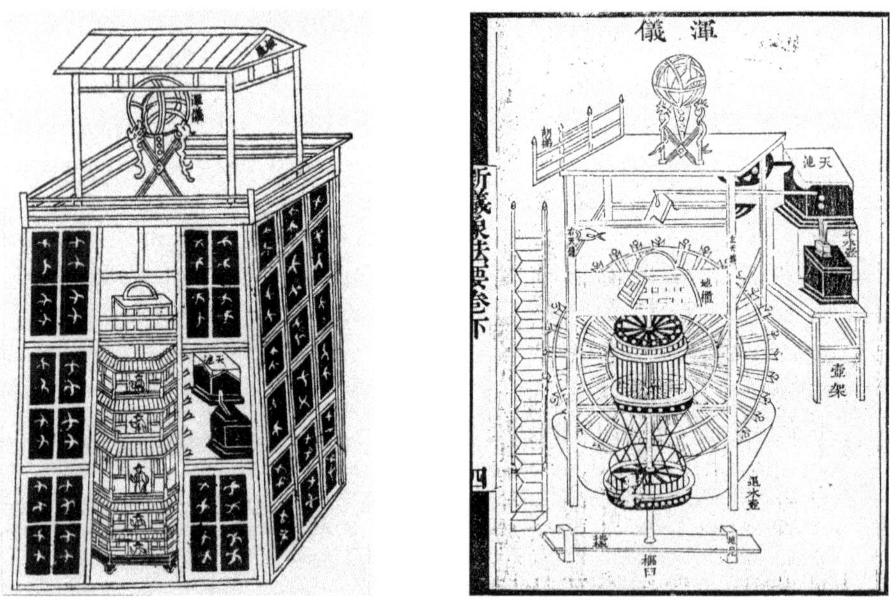

3.13 (a) General view of Su Song's and Han Gonglian's clocktower using the technique of making part of the tower transparent to provide of view of some of the interior components; *System Essentials*, 3, 2a; (b) General view of the interior clockwork made possible by stripping away all of the building except for the main posts; *System Essentials*, 3, 4a.

of a winnowing machine from Xu Guangqi's 徐光啓 *Nongzheng quanshu* 農政全書 (Complete treatise on agricultural administration) (Fig. 3.11 (a))[70] which is also unique in its partial use of a plan view, and which can be compared with the much more traditional rendition from Wang Zhen's *Agricultural Treatise* (Fig. 3.11 (b)), discussed below.

The same solution is used in Song Yingxing's *Exploitation* to portray the contents of a stacked kiln (Fig. 3.12). In the *System Essentials*, however, we are provided with two general views of the entire tower. The first (Fig. 3.13 (a)) makes part of the tower exterior transparent to show from inside the tower the water reservoir and constant-level tank as well as the celestial globe and the box on which it was mounted. The second (Fig. 3.13 (b)) omits everything except the main posts of the framework to provide an unimpeded view of the overall clockwork.[71] The same problem could also be at least partially solved by the component parts illustrations mentioned above and by "exploded" drawings of which, surprisingly, we have a primitive example (in perspective) showing the bronze horizon support and the meridian bearing rings of the celestial globe and the box within which it sat (Fig. 3.14).

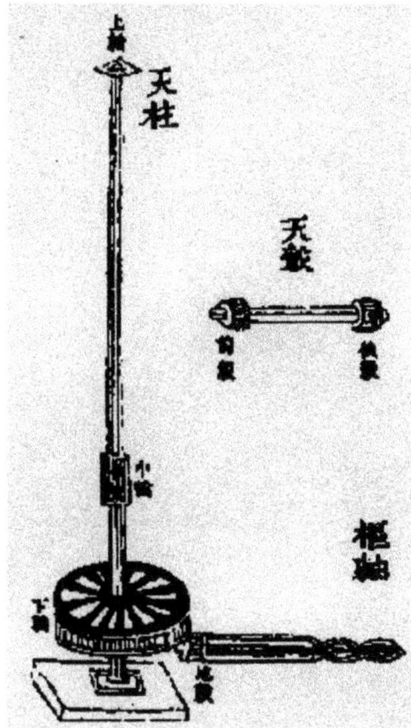

3.15 Assembly drawing of the main vertical transmission shaft; *System Essentials*, 3, 16a.

The illustrations of the *System Essentials* suffered of course from many of the shortcomings common to Chinese illustrations of technical subjects. Some, such as the inclusion of duplicate drawings,[72] seem to have resulted from sheer carelessness. Others, including the relatively poor handling of volume (no striation, hatching, modeling strokes or shading) probably reflected both the tastes and the limitations of the artists. A few were especially egregious, like the inclusion of inappropriate copies of illustrations referring to earlier, less developed mechanisms and the omission of crucial elements even when there is a label referring to them. An example of the latter (Fig. 3.15) is provided by the three gear wheels on the central transmission shaft: not only is the middle "wheel" (*zhong lun* 中輪) drawn as a pinion but the top wheel, despite having a label (*shang lun* 上輪), is entirely missing.[73]

70. Bray, *SCC* 6:2, 372–73.

71. A precedent for this kind of technique can be found as far back as in the "open interior" paintings of the Han "wherein the artist opens up the facade or side of a roofed enclosure in order to show figures inside." Maeda, "Spatial Enclosures," 373 and Fig. 6.

72. Golas, "Emergence of Technical Drawing," 50.

73. Combridge is probably right that these mistakes suggest that this drawing was partially copied from some not very applicable existing drawing; Combridge, "Astronomical Clocktowers," 299. Again, the question arises concerning the familiarity of the artist(s) with the actual clocktower as opposed to more or less relevant

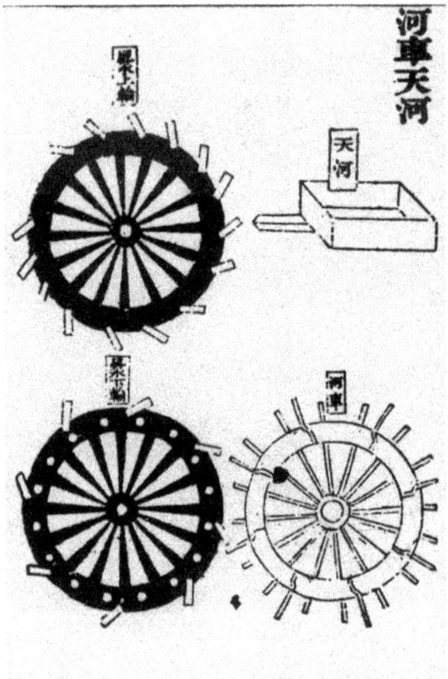

3.16 Component parts drawing showing two norias, the wheel for powering them, and the upper flume; *System Essentials*, 47, Fig. 12

A somewhat surprising aspect of the drawings, especially given that they were unique in the frequency of their focus on component parts and their assembly,[74] is the lack of any consistent or explicit scale, or any dimensions on the drawings.[75] As we have seen, China had a long, if not prolific, history of scale drawing dating back to the Han and earlier.[76] Moreover, this was the period when ruled-line painting flourished: the *Qingming* scroll itself was approximately contemporaneous with the *System Essentials*.[77] Ruled-line painting was known for its extreme precision in the handling of space, scale and proportion. We thus have here an example of a phenomenon that recurs frequently in the history of technical illustration in China: an important drawing technique applied in one kind of painting/drawing was not extended to other kinds where it could have been used to very good effect. For Su Song, it was apparently sufficient that he follow earlier texts of this sort and give very detailed dimensions in the body of the text.[78] Here, Su and his illustrators were no pioneers.

Nevertheless, the drawings of the *System Essentials* represent a major achievement made possible by the fact that Su Song was acting as an "engineer" insofar as he was not only responsible for the designing and planning of the clocktower but also actively supervised

earlier drawings.

74. In most other works, Chinese technical illustrations, together with their accompanying texts, did a better job of providing dimensions than construction details. Golas, "Emergence of Technical Drawing," 61n142. See also Mark Elvin, "The High-Level Equilibrium Trap," in Elvin, *Another History: Essays on China from a European Perspective* (Honolulu: Wild Peony/University of Hawai'i Press, 1996), 61, which notes how Dieter Kuhn, in his reconstruction of Wang Zhen's spinning machine, could and did follow the dimensions provided by Wang.
75. Golas, "Emergence of Technical Drawing," 46–50. In some cases, as in the illustration containing the two norias, the wheel for powering them, and the upper flume (Fig. 3.16), one is tempted to conclude that the size of the components was determined largely by the space on the page available for them.
76. See "Scale Drawing and Perspective" in Chapter 2. Scale drawing does not seem to come into extensive use in Europe until around the beginning of the fifteenth century; Pacey, *Technology in World Civilization*, 95.
77. For the debate on the date of the scroll, see Julia K. Murray, "Water under a Bridge: Further Thoughts on the *Qingming Scroll*," *Bulletin of Sung and Yuan Studies* 27 (1997), 99–107 and Roderick Whitfield, "Material Culture in the Northern Song Dynasty–the World of Zhang Zeduan," in Kai-Yin Lo (ed.) *Bright as Silver, White as Snow* (Hong Kong: Yong Ming Tang, 1998), 61–64.
78. *System Essentials*, Chap. 3, passim; *Heavenly Clockwork*, Chap. IV, passim.

its construction. His drawings constitute for the first time anywhere a *series* of illustrations that truly deserve the name of "technical drawings." They are intended to present accurately what the various mechanisms looked like and how they worked when assembled.[79] Elements that did not contribute to these goals, including human figures, were omitted. With one remarkable exception, the quite charming rendering of the armillary sphere with its two meticulously drawn dragons (Fig. 3.17),[80] almost nothing about the illustrations aims at providing aesthetic pleasure. Rather they have a straightforward, matter-of-fact character that, together with the pronounced concern with details, gives something of the flavor of modern engineering drawings. Indeed, it may well be that many if not most of these illustrations originally constituted working drawings for the construction of the clocktower and were then incorporated into the descriptive account made "for the record."

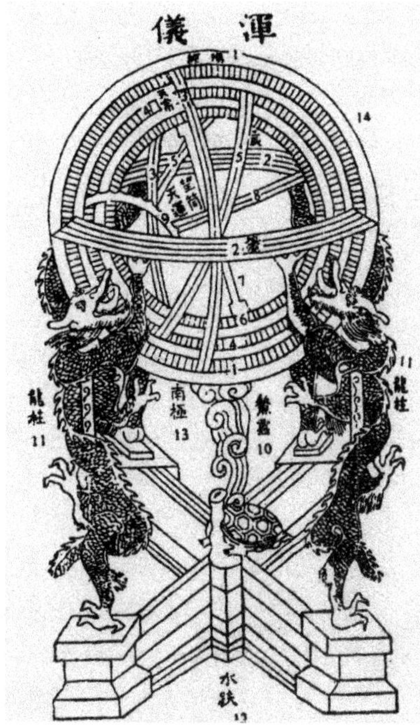

3.17 Armillary sphere with two of its supporting dragons featured prominently; *System Essentials* 1, 6a.

Given the groundbreaking achievement of the *System Essentials*, it is more than a little surprising that this work did nothing to promote the further advance of mechanized instruments for astronomical observations or the keeping of time. Its failure to do so helps us to understand how government sponsorship of technology in China could be very much a two-edged sword. To be sure, there were certain advanced and expensive uses of technology for which only the central government with its considerable resources could provide adequate financial support. Su Song's clocktower as well as its predecessors serve as excellent examples. Moreover, it is unimaginable that Su Song's masterful account would ever have come into being without the full support of the imperial court. Yet, in this area as in others (e.g., weaponry), the government promoted the development and recording of the technology strictly for its own purposes. Indeed what our contemporary zeitgeist might refer to as security concerns often led to efforts by the government to keep knowledge of the technology at least semi-secret.[81] Such concerns seem to have played a major role in limiting the impact of the *System Essentials*.

79. As we have noted elsewhere, how the mechanism was actually constructed or assembled was of lesser importance, at least for the illustrations. Golas, "Emergence of Technical Drawing," 52–53.

80. *System Essentials*, Chap. 3, 6a.

81. This is nicely discussed in *Heavenly Clockwork*, 6n3.

We must remember that all Chinese rulers shared a worldview in which there was a very real link between the legitimacy of the government and its ability to promote a harmoniously functioning universe. Astronomical calculations played a crucial role in ensuring that the government was keeping itself properly aligned with the forces of the universe. Thus, throughout Chinese history, the government has always been inclined to maintain a monopoly on astronomical knowledge as well as the techniques to acquire it.[82] Even within the government itself, access to this knowledge was on a "need to know" basis as a text from the Jin dynasty (265–420) tells us:

> Astronomical instruments have been in use from very ancient days, handed down from one dynasty to another and closely guarded by official astronomers. Scholars therefore have had little opportunity to examine them, and this is the reason why unorthodox cosmological theories have been able to spread and flourish.[83]

Government efforts to monopolize astronomical knowledge, though never entirely abandoned in the imperial period, varied in stringency at different times. While for the Tang we have a very explicit imperial edict from 840 ordering that officials and their subordinates in the imperial observatory keep their knowledge secret,[84] in the Song the study of astronomy seems to have been not at all uncommon in scholarly families, and astronomical questions might even be posed on the civil service exams.[85] By contrast, Ye Mengde 葉夢得 (1077–1148), referring to Su Song's diplomatic mission to the Liao mentioned above (note 61), accounts for the superiority of the Liao calendar with the comment that the "barbarians had no restrictions on astronomical and calendrical study, [and so] their experts in these subjects were generally better . . ."[86] Nevertheless, as an example of the ferocity with which the government might strive to maintain its monopoly on astronomical and calendrical knowledge, we have the remarks by Shen Defu 沈德符 (1578–1642) telling us that, in the early Ming, those who studied calendrical matters in a private capacity were to be exiled to the frontiers while those who had the temerity to actually produce their own calendar were to be punished with execution.[87]

The *System Essentials* provides us with an excellent case study of how, in disruptive times such as those that resulted in the loss of north China by the Chinese in the early twelfth century, the very narrowness of the circles in which certain kinds of knowledge circulated and was understood could lead to eventual loss. A century after the construction of Su's

82. As Needham summarizes the situation, "the majority of the observers who thought and calculated and wrote about astronomical problems were in State service." Needham and Wang, *SCC* 3, 190.
83. Needham and Wang, *SCC* 3, 193.
84. Needham and Wang, *SCC* 3, 193. Indeed, the same edict even ordered that astronomical officials were to segregate themselves not only from common people in general but also from officials in other branches of the government! Ibid.
85. Needham, *Grand Titration*, 33; *Heavenly Clockwork*, 6; Peterson, "Calendar Reform," 50.
86. *Heavenly Clockwork*, 6. For the date, see 7n8.
87. Wu, *History of Chinese Drawing*, 128; see also the discussion in Jacques Gernet, *China and the Christian Impact* (Cambridge: Cambridge University Press, 1985), 60–62.

clocktower, the ability even to reproduce a mechanism of this complexity had disappeared. When the great philosopher, scholar, and amateur astronomer Zhu Xi made a serious effort to understand the workings of Su's clocktower, he was ultimately frustrated.[88] For us, the problem is determining exactly what it was that tripped Zhu up. He himself became convinced that his copy of the *System Essentials* contained "a few mistakes" and that "in a few essential places the discussions cannot be connected together."[89] Rightly or wrongly, but to his mind not unreasonably, he attributed these deficiencies to the secretiveness of the makers of the apparatus who did not want to let people know the whole story. The official Song history, after mentioning Zhu's failure, noted that the *System Essentials* often did not record sizes and dimensions, adding to the difficulty of reconstructing the armillary sphere mechanism. One would like to know what mistakes and what disconnections Zhu Xi thought he had found, as well as what crucial dimensions the compilers of the Song history were referring to. We are not told. In any case, even these incomplete references provide a good example of how government efforts to inhibit the spread of technological information for its own purposes increased the chances that that knowledge might be forgotten, all the more when it represented cutting-edge technology that few people understood well. In this case, the cause of improved technological illustration was dealt an especially serious blow because, as far as we know from surviving works, there was no other manual produced at this time or later that could begin to rival the *System Essentials* for its full set of drawings, using some of the most advanced drawing techniques of the time, to explain a very complicated machine.

The *Building Standards (Yingzao fashi)*

The work known as the *Building Standards* (*Yingzao fashi* 營造法式) is a manual of building methods and techniques for use in the construction of government edifices. Its original version (now lost) dates from 1091, at about the same time Su Song compiled his *System Essentials*. A second version, compiled by an official named Li Jie 李誡 (d. 1110)[90] and presented to the emperor in 1103, survives in only two pages. A further complete revised version, dating from 1145, has served as the basis for all later editions.[91]

88. The following account is based mainly on Kim Yung Sik, "Chu Hsi (1130–1200) on Calendar Specialists and Their Knowledge: A Scholar's Attitude toward Technical Scientific Knowledge in Traditional China," *T'oung Pao* 78 (1992), 109–10.

89. This is an understatement. Zhu's difficulties were undoubtedly compounded by his having available while working on this reconstruction only one *juan* (the second) of the manual's original three *juan*! *Heavenly Clockwork*, 13.

90. On Li Jie and especially how he acquired the expertise to be able to produce his version of the *Building Standards*, see Feng Jiren, *Chinese Architecture and Metaphor* (Honolulu and Hong Kong: University of Hawai'i Press and Hong Kong University Press, 2012), 7–10 and 100–107.

91. For important bibliographical references on the *Building Standards*, see Ledderose, *Ten Thousand Things*, 223n52 and, for the broader background on building techniques, the works cited in the notes below.

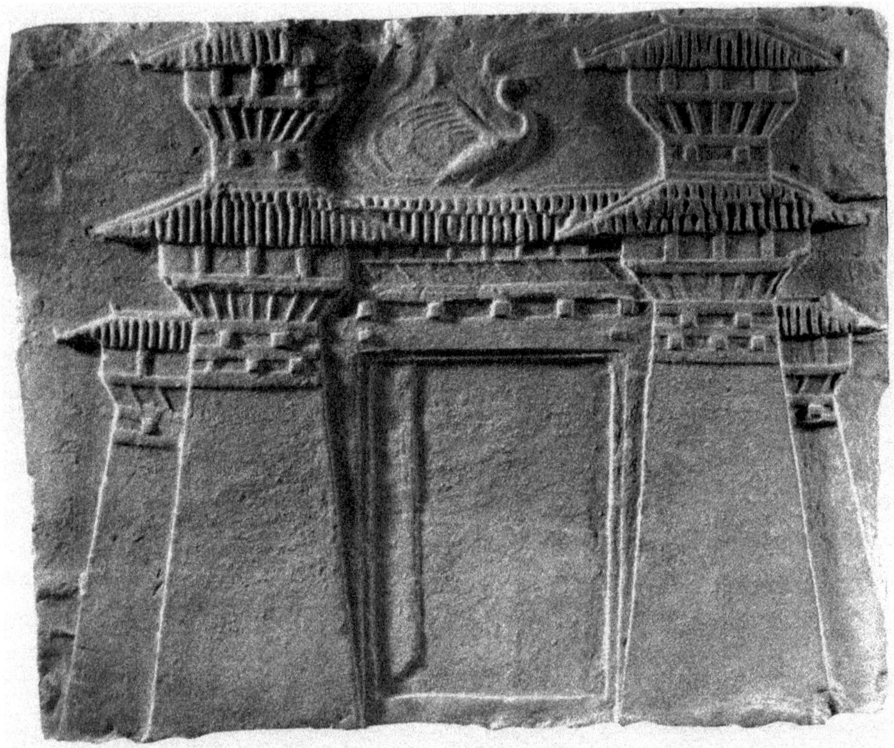

3.18 The "*Que* Pillar Gate," a clay tomb relief from the Later Han

Unlike most other government manuals meant as guides for officials who would be supervising government building projects,[92] the *Building Standards* did not confine itself to bureaucratic concerns such as costs of materials and labor or preventing pilferage but also included an extensive, illustrated discussion of the technology used in the construction of public buildings.[93] No other account and no other illustrations of building practice from traditional China even remotely approach this work in detail, accuracy and comprehensiveness. Insofar as the illustrations we presently have resemble the long-lost originals, they suggest that the originals in certain ways achieved a level of quality as good as any illustrations of technology done in China before the twentieth century.[94]

92. There is a quite large surviving body of building regulations put out not only by the central government but also by government offices in the provinces during the Qing period. See Christine Moll-Murata, Song Jianze and Hans Ulrich Vogel, *Chinese Handicraft Regulations of the Qing Dynasty* (Munich: Iudicium, 2005), especially the "Union List of Handicraft Regulations of the Qing Dynasty," 521–55. For other traditional writings on architecture and building, see Liu, "*Water Mill*," 593–94n96 and Needham, Wang and Lu, *SCC* 4:3, 80–89.

93. There was little expectation in China that even officials supervising building projects would be or would become very conversant with actual building technology. Hence the absence of descriptions or illustrations of the technology in most government manuals dealing with construction.

94. The only other officially-issued comprehensive architectural manual, the *Gongbu gongcheng zuofa zeli* 工部工程做法則例 [Engineering manual for the board of works] did not appear until the eighteenth century and was not originally illustrated. Sun Dazhang, "The Qing Dynasty," in Steinhardt, *Chinese Architecture*, 342–43.

A very long history of using drawings to assist the planning and construction of buildings preceded the compilation of the *Building Standards* in the Song. We noted in Chapter 1, Wu Jiming's suggestion that the regularity of the circumferences of the round houses at the Neolithic village of Banpo may indicate the use of a compass-like measuring tool. Here the "drawing" would have been a marking of the earth to show exactly where the wall was to be built. If this technique was indeed used, it might well be the first example of a connection between drawing and construction in China. The step to using reduced-scale drawings to help imagine or plan the construction of a building may also have been taken as early as the Neolithic.[95] In any case, the growing size and complexity of buildings in the Shang and Zhou periods must have required the use of sketches or drawings of some kind to help in planning constructions and perhaps to work out problems that might arise. Further support for this idea comes, as we have seen in Chapter 2, from the late fourth century BCE *Zhaoyutu* 兆域圖 plaque with its bird's-eye view of a planned tomb complex (Fig. 2.3). Unfortunately, we have scant evidence for the use of construction drawings over the following centuries[96] but we do know that, by Sui and Tang times, drawings of relatively high quality were in wide use, both for public and for private buildings.[97] By the late eleventh century, the careful maintenance of a consistent scale was a crucial element in the architectural painting of Guo Zhongshu.[98] With all these precedents, including the illustrations of the *Building Standards,* it is quite remarkable that no precise scale drawing of a *mechanical apparatus* survives from traditional China.[99]

Advances in painting also contributed to the sophisticated technical drawings of the *Building Standards.* The post-Han centuries saw a growing cultural importance attributed to painting, and this in turn encouraged improved technical competence among painters. No subject, after human beings and the manifestations of nature, offered more attraction for

95. Recall the prolific use of another kind of drawing, the remarkably varied decorations on Neolithic pots, especially those from Yangshao; see Plate 1 (b).

96. We can extrapolate what these early drawings might have looked like by examining the many buildings portrayed on the decorated bricks used in the construction of Han tombs (e.g., Fig. 3.18) as well as on other decorated objects such as the bronze bowl from the fourth century BCE pictured in Chapter 1 (Fig. 1.3).

97. Wu, *History of Chinese Drawing,* 79; Needham, Wang and Lu, *SCC* 4:3, 107.

98. "He has used an infinitesimal to mark off a foot, a foot to mark off ten feet; increasing thus with every multiple, so that when he does a large building everything is to scale and there are no small discrepancies." From the *Huapin* 畫品 of Li Zhi 李廌 (1059–1109), trans. Alexander Soper; cited in Maeda, "*Chieh-hua*: Ruled-line Drawing," 126 and Chung, *Drawing Boundaries,* 17 (who also notes (18) a very similar passage in the *Building Standards.*) Scale drawing likely became an important element in architectural painting in large part because of the role that fixed proportions played in the planning and construction of major buildings in China. Ledderose, *Ten Thousand Things,* 134–35 and Needham, Wang and Lu, *SCC* 4:3, 82–84. In any case, it is not beyond reason that paintings using such techniques might well have served as design drawings in construction projects, as suggested by Anita Chung, *Drawing Boundaries,* 18.

99. In a few cases, one can deduce that the artists were indeed maintaining at least a clear relative scale between the different parts of an object. Liu Keming gives the example of a bronze vessel described in the Southern Song *Xu kaogu tu* 續考古圖 [Continuation of the researches on archaeology with drawings] where the proportions of the dimensions given in the text are maintained in the drawing; Liu, *Chinese Engineering Graphics,* 177.

Chinese painters than the elaborate public and religious buildings, the palaces and temples to be found in both the nation's cities and its countryside. Realistically portrayed buildings had become sufficiently prominent in Chinese painting so that, by the Tang and Song, architectural paintings came to be recognized as a distinct category of paintings.[100] Those produced by skilled *jiehua* painters might even serve directly as designs for buildings.[101] In part, this was because the aesthetic aims of this kind of painting did not preclude attention to details and perhaps even a degree of technological comprehension. An oft-cited statement by Guo Ruoxu 郭若虛 (fl. third quarter of the 11th century) makes the case that architectural painters needed to understand construction details such as beams, columns, brackets, cushion timbers, flying eaves and tortoise-head forms. He lists almost thirty technical terms, some of whose meanings scholars still puzzle over today.[102]

We are probably safe, however, in assuming that most of the painters doing architectural paintings did not have any deep knowledge of building technology. Guo admits as much when he writes: "Few . . . are the painters who have been able to give all this any detailed investigation . . ."[103] Moreover, it is at least questionable how useful such knowledge would have been to the working artist. In the first place, he was portraying essentially surface elements. It would be hard to make the case that understanding the technology that lay behind those surfaces really would lead to a more convincing painting. Moreover, the painter of buildings was often depicting large and highly complex edifices. To what extent could the plethora of details be depicted even if they were understood? Thus, most architectural painters settled for a selection of those visible elements that made possible a portrayal that would *appear* to be simultaneously more or less accurate and, more importantly, aesthetically right.[104]

The ultimate condition without which the *Building Standards* would never have been compiled was its sponsorship by the imperial court. Chinese rulers from very early on understood the role that impressive official edifices could perform as symbols of the government's power and legitimacy. Constructing and maintaining these buildings became an important function of the imperial bureaucracy. By the mid-eleventh century, after a series of major advances, both their basic components as well as their construction methods had become largely standardized in practice. The stage was thus set for what would turn out to be an approximately thirty-year effort to codify those practices. It was the emperor Shenzong 神宗 (r. 1068–85) who, in 1070, issued the original order for the compilation.

100. William Trousdale, "Architectural Landscapes Attributed to Chao Po-chü," 286; Needham, Wang and Lu, *SCC* 4:3, 106–7.

101. Soper, *Experiences*, 70 and 186n578 (Liu Wentong); 71 and 186n583 (Lu Zhuo).

102. Soper, *Experiences*, 13; Trousdale, "Architectural Landscapes," 285; Chung, *Drawing Boundaries*, 10–11. There is a good chance that many of them would have been incomprehensible even to contemporaries. Hence the *Building Standards* opens after an introduction with two chapters dealing with building terminology.

103. Soper, *Experiences*, 13.

104. This was much less the case in the Ming when aesthetic aims often took clear precedence over representational accuracy. Chung, *Drawing Boundaries*, 36–39; Trousdale, "Architectural Landscapes," 312.

The task was entrusted to the Directorate of Construction (*Jiang zuo jian* 將作監).[105] A first version was presented to the throne in 1091, a year before an official by the name of Li Jie 李誡 was appointed to the Directorate. Though nothing survives of this original compilation, we know it was based mainly or even entirely on information in the imperial archives, much of it presumably inaccurate or at least outdated.[106] The result was deemed unsatisfactory. In 1097, Li was ordered to prepare a revised version. With the help especially of a certain Yao Shunren 姚舜仁 who was known for his drawing abilities,[107] the revision was presented to the throne in 1100 and came out in a printed version in 1103.[108]

Li Jie (c. 1065–1110) came from a family of officials[109] and enjoyed considerable esteem not only for his scholarly erudition but also for his talent in painting and calligraphy.[110] In addition, he had considerable ability and experience superintending the construction of buildings: between the years 1100 and 1106, he oversaw the construction of several of the most important buildings in the capital of Kaifeng as well as directing the rebuilding of others.[111] During the same time that he was engaged in compiling the *Building Standards*, he

105. Feng, *Chinese Architecture and Metaphor*, 101 and 257n3; Else Glahn, "On the Transmission of the *Ying-tsao fa-shih*," *T'oung Pao* 61 (1975), 186; K. T. Wu, "Illustrations in Sung Printing," *The Quarterly Journal of the Library of Congress* 28 (July 1971), 186.

106. Not in the least, perhaps, because of the reforms of Wang Anshi. Feng, *Chinese Architecture and Metaphor*, 105–6.

107. K. T. Wu, "Illustrations in Sung Printing," 186; Liu, *Chinese Engineering Graphics*, 179. The two also collaborated in 1105 on plans for a new temple for the imperial sacrifices to heaven (*Ming tang* 名堂). Glahn, "Li Chieh," in Franke, *Sung Biographies*, 2: 528. Needham mentions, in regard to the illustrations in the *Building Standards*, that they were drawn by Li's "drawing-office clerks" (Needham, Wang and Lu, *SCC* 4:3, 107) but I know of no evidence for this and suspect that especially the technical drawings (as opposed to the drawings illustrating decorative motifs) could well have all been drawn by Yao. They certainly display a consistency of approach and technique consonant with the hypothesis that they were the product of a single draftsman. Moreover, given the years devoted to the compilation, even a single draftsman would have had abundant time to complete them.

108. Ledderose, *Ten Thousand Things*, 132; Glahn, "Transmission," 236n3 and 255; Glahn, "Li Chieh," 529. Apart from two pages of the 1103 edition, and a few pages of an 1145 revised version (discussed below), the work has survived only in copies and copies of copies of the 1145 version.

109. Since Li did not get a biography in the official history of the Song dynasty (*Song shi* 宋史), all accounts of his life must rely primarily on his tomb inscription written by Fu Chongyi 傅沖益 and collected by Cheng Ju 程俱 (1078–1144) who had worked under him and presumably knew him well. (Cheng is often mistakenly given as the author of the inscription; see W. Perceval Yetts, "A Chinese Treatise on Architecture," *Bulletin of the School of Oriental and African Studies* 4: 476n3; Liu, "*Water Mill*," 594n99.) The best biography in English, on which I have relied extensively, is Else Glahn, "Li Chieh." For a good Chinese account, see Liu, *Chinese Engineering Graphics*, 179–80.

110. Li once presented to the court a horse painting that was greatly admired by Emperor Huizong 徽宗 (r. 1101–25); Glahn, "Li Chieh," 525.

111. Lothar Ledderose enumerates many of the buildings for which Li supervised the construction, enlargement or restoration: "a residence for five princes, one of them the later Emperor Huizong, various other palace buildings, the ancestral temple of the dynasty, a Buddhist temple in honor of Emperor Huizong's mother, the administrative offices of the metropolitan prefecture, a new office building for the ministry, a campus for the national university comprising 1,872 columns, and the main city gate" (*Ten Thousand Things*, 133). But I do not follow Liu Heping in calling Li an "architect" and speaking of "his profession as an architect" (Liu, "Painting and Commerce," 132). The surviving evidence does not persuade me that he had a "professional"

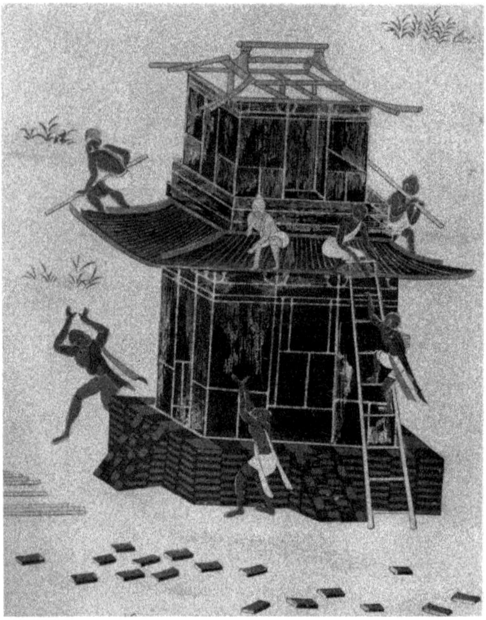

3.19 Building a pavilion or watch tower; eight-century cave-temple fresco from Dunhuang.

also successfully directed the construction of the Wu-wang Di 五王邸, the residence of the five younger brothers of the emperor Zhezong 哲宗 (r. 1086–1100).

Li, a specialist to a degree rare among Chinese officials, emphasizes in his preface to the *Building Standards* that he made a great effort to learn the practices and the orally transmitted rules of master-carpenters and artisans.[112] The *Building Standards* provide abundant evidence to back up his claim. On the other hand, it would be a mistake to assume that it was only the practical side of building that drew Li's interest. Antiquarian curiosity deriving quite naturally from his classical education led Li to comb the classics, histories and other earlier writings for references to the architectural terms presented in the first two terminological chapters of the *Building Standards*. These chapters also undoubtedly gave his book a special aura of erudition that could only have enhanced its appreciation by Li's more rigorously bookish fellow officials. They would also have respected Li's learning in other areas, which led him to write another six books that we know of but which have unfortunately all disappeared. Each of these books was very concise, with none longer than ten chapters (*juan*),[113] thus suggesting that Li in all his writings habitually expressed himself as clearly and succinctly as he did in the *Building Essentials*.

In addition to his general learning and his command of building technology, Li was also highly regarded as a painter and as a calligrapher proficient in all the major styles of Chinese calligraphy.[114] Although there is no evidence that he ever produced architectural paintings, Li's artistic talents must have had a positive influence on the illustrations in the *Building Standards*. Indeed, it could be further argued that Li, in his mastery of art and technology, more than any other Chinese resembles the great artist-engineers of the Renaissance even

---

capacity for *designing* buildings.

112. Needham, Wang and Lu, *SCC* 4:3, 84; Feng, *Chinese Architecture and Metaphor*, 8–9.

113. Their titles and chapter lengths are given in Liu, *Chinese Engineering Graphics*, 180. They dealt with history, geography, the ancient seal script, horses, the Chinese short lute (*pipa* 琵琶) and the board game "six sticks" (*liubo* 六博).

114. Glahn, "Li Chieh," 525, 526.

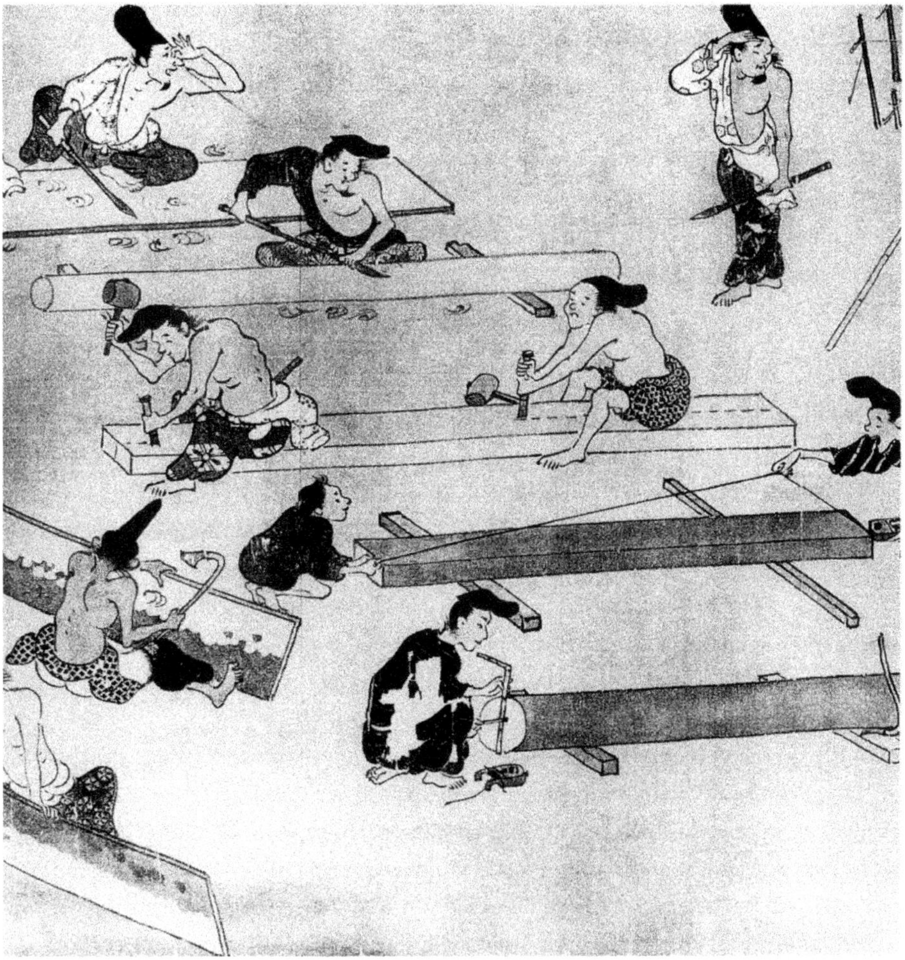

3.20 Medieval Japanese carpenters at work (from a 1309 scroll)

though their focus tended to be on mechanical devices, something in which Li seems never to have taken any particular interest.[115]

Given the unique achievements of Li Jie,[116] it is especially disappointing that we do not have direct access to his original *Building Standards* and its illustrations. We have no idea how many copies of Li's work were actually printed when it was first published in 1103

115. We can only regret that his compendium contains not a single actual building scene and thus provides us with no information on actual building methods and the tools and machines used. Indeed, surviving portrayals of construction scenes from pre-twentieth century China are very rare, an exception being Fig. 3.19. Needham found none important or informative enough to be included in the section on "Building Technology." Needham, Wang and Lu, *SCC* 4:3, 58–144. We can see an excellent example of the kind of information we are consequently lacking for China in a detail Lothar Ledderose presents from an early fourteenth-century Japanese scroll (Fig. 3.20).

116. That he never rose higher than the middle levels of the bureaucracy and did not have his biography included in the official Song history may be attributable to his early death at about age 45.

but, in any case, when Jurchen invaders sacked the capital, Kaifeng, a quarter of a century later and forced the Song court to relocate south of the Yangtze, most copies of the work seem to have been lost.[117] In the 1140s, when the emperor Gaozong 高宗 (r. 1127–62) turned his attention to making Hangzhou an architecturally worthy capital, he ordered a new edition of the *Building Standards* which was duly produced from an "old copy" with some editing by a bibliographer named Wang Huan 王喚.[118] Since Li Jie was long since dead, and since Wang Huan commanded little of the technological expertise Li brought to his task, it is reasonable to assume that the new edition fell short of the original. Nowhere was this more likely to be true than in the very detailed illustrations. In any case, as we noted, only small fragments survive from these first two editions. Although the book was included in the Ming *Yongle dadian* 永樂大典 imperial encyclopedia (completed 1408) and the Qing imperial library collection *Siku quanshu* 四庫全書 (late 18th century), it was largely forgotten in the post-Song period except by those few people, especially collectors, who were able to put their hands on a surviving manuscript copy.[119]

Not until the early twentieth century did the *Building Standards* become widely available in China. A manuscript dating from 1821 and belonging to a certain Ding 丁 family of Hangzhou was found in the Jiangnan Library in Nanjing and was photolithographically reproduced by the Commercial Press in 1919 (reduced size) and 1920 (facsimile).[120] This manuscript provides many examples of errors introduced into the illustrations by one or more copyists unfamiliar with building techniques.[121]

A much finer edition, more carefully edited and quite elaborately produced (e.g., with the colors in the decorative patterns, indicated only by characters in the earlier editions, painted in), appeared in 1925.[122] It is in many ways a masterpiece of Chinese bookmaking. Nevertheless, it poses serious problems for those who would use it for historical purposes, and especially for historians of technology.[123] It cannot, for example, be viewed as truly representing the 1103 edition and is "only partly a true reproduction of the 1145 edition." Part of the difficulty stems from the fact that this version was edited following bibliographical criteria and without sufficient attention to or understanding of the problems of the

117. The entire contents of the imperial libraries were lost at this time; Paul Demiéville, "Review," *Bulletin de l'Ecole Francaise d'Extreme-Orient* (BEFEO) 25 (1925), 229n6.

118. Or Wang Ao 王奥: compare Glahn, "Transmission" 521 and Glahn, "*Ying-tsao fa-shih*," 186; Needham, Wang and Lu, *SCC* 4:3, 84, fn. c.

119. Yetts, "A Chinese Treatise on Architecture," 484. The later history of the text is told in considerable detail in Demiéville, "Review."

120. Yetts, "A Chinese Treatise on Architecture," 474, 485; Needham, Wang and Lu, *SCC* 4:3, 84, fn. c; Chen Mingda, *Studies on Structural Carpentry*, 2, 232; Else Glahn, "Unfolding the Chinese Building Standards: Research in the *Yingzao fashi*," in Steinhardt, *Chinese Traditional Architecture*, 50.

121. Glahn, "Unfolding," 51.

122. For a full discussion of the production of this edition, see Yetts, "A Chinese Treatise on Architecture." Feng Jiren points out that "new drawings made then [1925] were added to it. Feng, *Chinese Architecture and Metaphor*, 111. See Feng 111–12 for further details on the various editions.

123. I rely for the following discussion especially on the work of Else Glahn.

technological descriptions.[124] The drawings posed special problems. In the manuscripts consulted for this edition, they were in particularly bad shape. It was therefore decided to have new drawings made by a master builder who had worked in the imperial palace before the fall of the Qing dynasty in 1911. The result was that the new illustrations failed to take into account many changes in construction methods, shapes, proportions and even decorative colors that had taken place between Song and Qing.[125] To summarize the situation, then, we are simply unable to say with any precision just how close the illustrations we have today are to the Song originals. Nevertheless, the extensive work done by outstanding Chinese scholars such as Liang Sicheng 梁思成, Chen Mingda 陳明達 and, most recently, Jiren Feng[126] suggest that these illustrations do give a generally good idea of the Song originals.[127]

The *Building Standards* in its current form consists of (1) a very general introduction dealing mainly with rules for basic preliminary calculations (building orientation, leveling, walls and roofs);[128] (2) the two chapters previously mentioned that include a glossary of forty-nine construction terms as they appear in a wide variety of works (classics, histories, poetry, dictionaries etc.) dating from the Zhou dynasty up to the Song;[129] (3) thirteen chapters giving a very practical description of building practices including sections on stonework, carpentry, wall building, tile making, and decoration; (4) another thirteen chapters on materials and labor with much emphasis on amounts needed and their costs; and (5) 193 illustrations grouped together in the final six chapters.[130]

Given the book's considerable emphasis on elucidating building technology (thirteen of thirty-four chapters), some scholars have chosen to view it as a kind of technological manual or handbook for imperial construction projects in the sense of enabling the responsible officials to supervise the building process. For example, Li Jie explicitly describes how a 100-to-1 scaled construction drawing should be made on a wall before commencing the building; dimensions could then be measured during construction directly from that drawing.[131]

124. Chen, *Studies on Structural Carpentry*, 2, 232.

125. This became increasingly clear in later years when more original Song buildings were discovered in China and were subjected to careful examination and measurement.

126. Feng Jiren's *Chinese Architecture and Metaphor*, which I came upon only as I was finishing the final revision of this manuscript for publication, now takes its place as another indispensable resource for any serious study of the *Building Standards*.

127. Much of the most persuasive evidence derives from careful comparison of the discussion and illustrations of the *Building Standards* with surviving Tang and Song buildings.

128. Demiéville ("Review," 233–34) suggests that this may have represented a sort of résumé for the officials who would check the work before it was presented to the emperor.

129. Needham and his collaborators, on the basis of the tenuous connection between these chapters and the following more practical discussion of the actual building methods, conclude that "Li Chieh [Jie] never quite succeeded in fusing the scholarly and the technical traditions." Needham, Wang and Lu, *SCC* 4:3, 84.

130. Needham, Wang and Lu, *SCC* 4:3, 84–85; Steinhardt, *Chinese Architecture*, 189. For fuller summaries of the work, see Glahn, "*Ying-tsao fa-shih*," 187; Sickman and Soper, *Art and Architecture*, 258; and, in characteristic detail, Demiéville, "Review," 233ff. Liang Sicheng's *Annotated Translation* (*Yingzao fashi zhushi*營造法式注釋 [An annotated translation of the *Building Standards*]. 2 vols. Beijing: Zhongguo jianzhu gongye chubanshe, 1983) is a masterpiece of twentieth-century Chinese scholarship.

131. Xiao, *History of Building Technology*, 3, 892. As Ledderose emphasizes, a basic principle of Chinese architecture

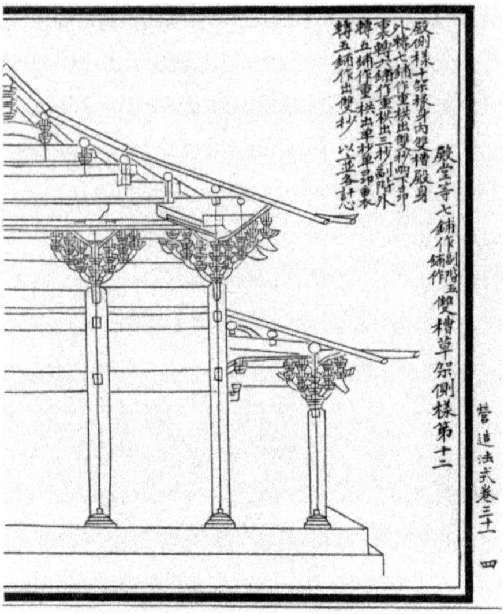

3.21 Cross-section of a hall showing six corbel bracket assemblies; *Yingzao fashi*, 31, 3a–b. (For a partial reproduction of a very similar illustration, see Needham, Wang and Lu, *SCC* 4:3, 96, Fig. 762.)

But this view of the *Building Standards* has been resolutely rejected by the leading Western expert on the work, Else Glahn, who argued that its "aim . . . was to facilitate the accounts of public buildings for which the ministries were responsible. It was not to teach officials to become architects, for the craftsmen knew how to build."[132] In other words, according to this view, the book had a quintessentially bureaucratic purpose: enabling officials to manage and determine the costs of labor and materials (just over one-third of the book deals with labor requirements and materials) so as to prevent the wasting or pilfering of government funds.[133] Given the persuasive evidence that can be marshaled to support either of these views, it seems reasonable to conclude that Li had both goals in mind, though not necessarily to the same degree.[134] Moreover, scholar that he was, Li was keenly concerned to present his material so that it would be comprehensible whether or not the reader wished to or was able to make practical use of it. This indeed was his major justification for the many illustrations he included. As he himself writes: "In the discussions of the construction and processing of components for all systems, if there are things whose principles cannot be understood unless illustrated, I specifically provide drawings for all of these, so that the systems are clarified."[135]

In any case, there are several ways in which the illustrations of the *Building Standards* constitute a milestone in Chinese depictions of technology. In a general sense, one can point

was that the measurements would not be specified in absolute, but rather in relative or proportional, terms; Ledderose, *Ten Thousand Things*, 136–37.

132. Glahn, "Unfolding," 48. Liang Sicheng holds essentially the same view, stressing Emperor Huizong's intent that the manual would help prevent waste and corruption in the building process. Liang, *Annotated Translation*, 6.

133. Feng, *Chinese Architecture and Metaphor*, 2. Lothar Ledderose suggests that bureaucratic control of the production process may have been the primary aim of *all three* of the earliest known manuals of architecture; Ledderose, *Ten Thousand Things*, 133.

134. This is essentially the view presented in Xiao, *History of Building Technology*, 3, 920. See also Liu, *Chinese Engineering Graphics*, 180.

135. Feng, *Chinese Architecture and Metaphor*, 110 (slightly modified). Feng effectively argues that the last decades of the eleventh century marked a major transition in Chinese building practices that created a strongly felt need for a manual that would clarify the new developments. He also provides (112) a very handy list of the most important elements of building knowledge that Li felt required illustration.

to their overall prominence in the book.[136] In a civilization where enormous authority was vested in the written word, illustrations by comparison were almost always undervalued and under-used as a means for conveying information and knowledge.[137] In the *Building Standards*, this was not the case. Its inclusion of profuse illustrations (by one count, 541 of them[138]) to help in explaining the technology surely was due in no small part to the extensive use of drawings on a day-to-day basis in the construction of large and complicated buildings as well as the fact that Li Jie, who had carefully studied building and carpentry practices, as well as the work of earlier writers on these topics, must have been thoroughly familiar with such drawings.

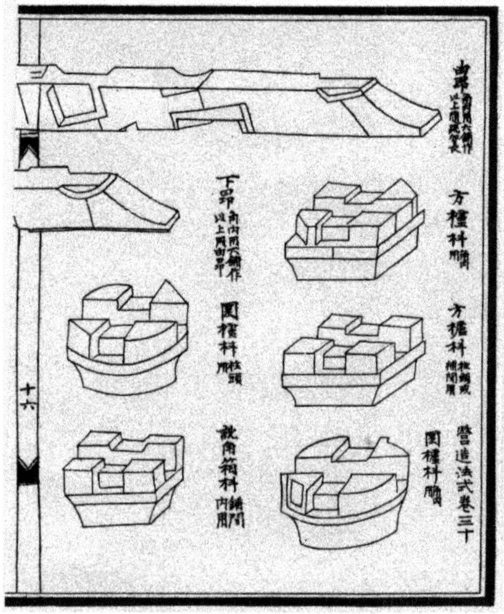

3.22 Two bracket assembly cantilever arms and five bracket-arm bases; *Yingzao fashi* 30, 16a.

The illustrations of the *Building Standards* also incorporate certain drawing techniques that might have been very useful for mechanical and other kinds of technical drawing. The use of (ground) plans (*dipantu* 地盤圖), elevations (frontal and profile) (*zhengyangtu* 正陽圖), and sections (*ceyangtu* 側陽圖) (Fig. 3.21) are not unprecedented but there seems to be here a new appreciation of these techniques.[139] Other techniques include the many drawings of component parts (beams, columns, bracket arms and blocks) (Fig. 3.22) as well as techniques (Fig. 3.23). On the basis of the detailed information provided by these drawings, Needham and his colleagues are prepared to accept them as very close to "working drawings in the modern sense."[140]

136. Wang Qingzheng, "The Arts of Ming Woodblock-printed Images and Decorated Paper Albums," in Chu-Tsing Li and James C. Y. Watt (eds), *The Chinese Scholar's Studio: Artistic Life in the Late Ming Period* (New York: Thames and Hudson, 1987), 56.

137. As opposed, for example, to entertaining.

138. Xiao, *History of Building Technology*, 3, 923.

139. Steinhardt, *Chinese Architecture*, 189; Wu Jiming, *History of Chinese Drawing*, 79–81. Curiously, the only other place one finds the considerable use of plan, elevation, and section drawings at this time is in archaeological catalogs where they were however devoid of any technological content; Liu Keming, *Chinese Engineering Graphics*, 171–77. The sources are silent on whether there was a mutual influence running in either direction. In later times, plan and section views especially and multiple views of the same object seem to have been only rarely used. (An exceptionally informative plan view, at least for terminology, is given in a component drawing from a shipbuilding manuscript, probably from the early nineteenth century; Needham, Wang and Lu, *SCC* 4:3, 408, Fig 942. It is worth noting that the emphasis is mainly on the names and relative locations of the various modules or components.)

140. Needham, Wang and Lu, *SSC* 4:3, 107.

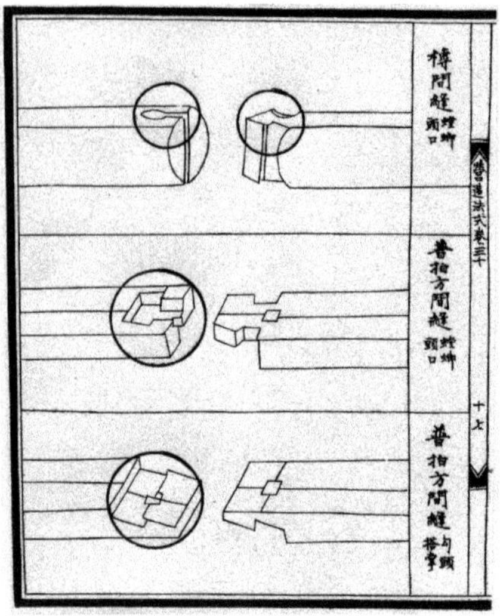

3.23 Joinery techniques illustrating mortise and tenon joints used for tie-beams and cross-beams. (The circles have been added to highlight distortions in the drawings.) *Yingzao fashi* 30, 17b.

The illustrations in the *Building Standards* also rival the *System Essentials* in their prolific use of labels within the illustrations, as seen for example in Fig. 3.23. The insertion of labels, as we have noted, was facilitated by the fact that the great majority of the labels consisted of only a small number (often as few as two or three) Chinese characters which could be inserted in the normal vertical format or in a horizontal format, depending on space available and aesthetic considerations.[141] On the other hand, in contrast to the *System Essentials* where only one illustration incorporates a longer comment as opposed to a label,[142] the *Building Standards* illustrations frequently include comments that supplement or explain in a significant way the information contained in the illustration itself (see Fig. 3.21) as well as encouraging greater consistency between illustration and text.

The result of the care lavished on both text and illustrations, including the unprecedented recourse to the most advanced drawing techniques available, does not mean that the illustrations are without very real limitations. When we view them carefully and with sufficient understanding (always keeping in mind the question of how close the illustrations we have today are to Li Jie's originals), we find that they are sometimes more flawed than appears at first sight. Liang Sicheng has compiled an extensive list of the kinds of shortcomings one finds, including the absence of any clear drawings to scale,[143] no attempt to distinguish components from the overall structure by use of differentiated lines, and mistakes such as lines that are too long or too short, missing, or even fictitious.[144] Not infrequently, the illustrations can be just plain wrong or incomprehensible. Of the three jointing techniques illustrated in Fig. 3.23, the top illustration suggests "trying to fit a square peg into a round hole" while the circled areas of the middle and bottom drawings have elements that just make no sense.

As we have noted, the *Building Standards* was a product of a period when, by and large, traditional Chinese building practice for major edifices had reached its zenith and most of its

141. And of course, the insertion of labels and comments was facilitated by the fact that woodblock printing could reproduce text and illustrations with equal ease.

142. *Heavenly Clockwork*, 52, Fig. 28. It simply notes that an alternative version of the illustration appears in another copy of the work.

143. Despite its inclusion of a clear description of scale drawing! Cheng, *Drawing Boundaries*, 18.

144. Liang Sicheng, *Annotated Translation*, 11–12.

advances were behind it.[145] The *Building Standards* thus represents an elaborate codification of these practices. There would in later periods be modifications but no major transformations in architectural theory or practice that in turn could have been expected to stimulate greater reliance on and, perhaps, advances in architectural drawing. Moreover, the manner of building in China may have been such as to encourage, more than in most technologies, the oral passing on of the knowledge of master builders, thus reducing the need for written manuals.[146] Lothar Ledderose in his recent, brilliant book on the role of modules in Chinese culture devotes a chapter to these building techniques.[147] He describes in great detail how traditional Chinese buildings were constructed by the assembling of standard, generally small, and generally light (i.e., made of wood) components, a kind of Lego style of construction. This method lent itself especially well to hands-on learning where habitual rules were the closest one came to anything resembling theory. As Percival Yetts remarks: "[P]racticing craftsmen [were] the real architects of China."[148]

This module style of building differed in almost every way from the methods for building the great edifices of the later Middle Ages and Renaissance in Europe. Among other things, it minimized in China the need for planning buildings using geometric methods and drawings. In Europe, by contrast, the building of the great cathedrals stimulated perhaps the earliest explicit use of drawing to think through questions and problems of design.[149] Also, with their modular style of construction (and greater reliance on wood rather than stone), the Chinese were much less reliant on great hoisting machines of a sort that gave such stimulus to the mechanical creativity and visual experimentations of great engineers like Leonardo.[150] Nevertheless, there is abundant evidence in the earlier record and above all in the *Building Standards* to support the suggestion that it may have been in the drawings

---

145. As Fu Xinian argues, "every form now associated with traditional Chinese architecture came to fruition during this [Sui-Tang] period and, furthermore, set patterns for future construction principles and practices." Steinhardt, *Chinese Architecture*, 127. Ruitenbeek on the other hand would see the end of major innovations in official architecture and building around 1400. Ruitenbeek, *Carpentry and Building*, 28. Even after this, there was some further evolution in regional styles. Ibid., 7.

146. Ruitenbeek describes how this applies to carpentry: "The actual shaping of the various wooden parts and cutting the joints to fit them together is what may be called the carpenter's work proper. The knowledge and skill needed to accomplish this are not to be found in any book. To describe in detail every step of the carpenter's work, it would be necessary to have worked with Chinese carpenters for many years. Even so it would be impossible to convey every knack of the trade into words." Ruitenbeek, *Carpentry and Building*, 64. See also Needham and Wang, *SCC* 2, 121–27 and Needham and Wang, *SCC* 4:2, 47–50.

147. Ledderose, *Ten Thousand Things*, Chap. 5. Note that Ledderose defines a module not as an interchangeable part but rather a standard unit.

148. Yetts, "A Chinese Treatise on Architecture," 473.

149. Pacey, *Maze of Ingenuity*, 47–48, 50; see also James Ackerman, "The Involvement of Artists in Renaissance Science," in Shirley and Hoeniger, *Science and the Arts in the Renaissance*, 94. This does not contradict what was said above about Li Jie's advocacy of a scaled construction drawing to be completed before commencing the building. Li's drawing seems to have been meant to serve as a tool to check dimensions of the building during construction, not as a means of working out the structure of the building beforehand.

150. Paolo Galluzzi, *Leonardo da Vinci: Engineer and Architect* (Montreal: Montreal Museum of Fine Arts, 1987), 50.

related to building technology that Chinese technological illustration achieved its greatest successes, especially in terms of precision.[151]

Why further advances proved elusive can perhaps be related at least in part to a "deterioration of building techniques" in later centuries, as suggested by William Trousdale.[152] In any case, the contrast with what happened in Europe could hardly be greater. It may have taken the Europeans until the sixteenth and seventeenth centuries to reach the level of architectural drawing achieved by Chinese in the Song.[153] But even while their drawing techniques may have been slow to improve, new developments in architecture and building techniques provided European architectural draftsmen with a continuous flow of new stimuli and challenges. Growing understanding of stresses and strains, a flowering of mechanical ingenuity that led to massive hoists and cranes, along with of course a sophisticated command of geometry, made possible the building of the cathedrals of high medieval Europe.[154] Then came the efforts of people like Alberti to rediscover and recreate the aesthetic achievements of classical architecture, and architecture dramatically changed direction. As a result, when it reached the rough equivalent of what turned out to be the high point of Chinese architectural drafting, European architectural draftsmanship did not even pause. The draftsmen continued to develop their skills, for example, by using a series of scale drawings to construct geometrically accurate perspective views.[155]

Finally, when considering Needham's further suggestion that Chinese drawings of machines never measured up to Li Jie's standards,[156] one might note that, in all Chinese architectural history, only Li Jie measured up to Li Jie's standards and nothing even remotely bearing comparison to his work was ever produced in the centuries that followed. On the other hand, even when mechanical drawings were manifestly flawed, what they succeeded in conveying may sometimes have represented an achievement indeed worthy of comparison with Li Jie's drawings.

The *Pictures of Tilling and Weaving (Geng zhi tu)*

At some time in the early 1130s, not long after the Song armies had lost north China to the Jin and the court had retreated to the lower Yangtze region beginning a period that would come to be known as the Southern Song (1127–1279), a certain Lou Shu[157] 樓璹 (1090–

151. Needham even suggests that one can view the *Building Standards* drawings almost as "working drawings in a modern sense–perhaps the first time in any civilization." Needham, Wang, and Lu, *SSC* 4:3, 107.

152. Trousdale, "Architectural Landscapes," 313.

153. Needham, Wang, and Lu, *SCC* 4:3, 110, fn. c.

154. Ivins, *Prints and Visual Communication*, 8.

155. Pacey, *Maze of Ingenuity*, 63. As Pacey points out, it was these advances that made possible more complete thinking-out of structural designs before construction and thus provided more opportunity for experimenting with new ideas.

156. Needham, Wang and Lu, *SCC* 4:3, 110, fn.c.

157. The pronunciation of the second, extremely rare character that serves as Lou Shu's personal name has also often been given as "Shou"; Wang Chaosheng and Francesca Bray use "Chou." I follow James Cahill and

1162), was serving as the local magistrate of Yuqian 于潛 county (about twenty-five miles west of Hangzhou in an area of highly advanced rice farming). In addition to his regular duties, he produced at some point two remarkable series of paintings, twenty-one dealing with rice agriculture and twenty-four with the production of silk and silk fabric.[158] Each painting was inscribed with a 40-character lyric by Lou relating to the activity depicted.[159] The entire work was titled the *Geng zhi tu* 耕織圖 (Pictures of tilling and weaving).[160]

If we can rely on later accounts written in 1210 by his nephew, Lou Yao (or Yue) 樓鑰 and his grandson, Lou Hong 樓洪, Lou Shu's reputation for erudition eventually brought him an audience with Emperor Gaozong 高宗 (r. 1127–62) at which he presented a copy of the paintings with their lyrics.[161] The emperor responded enthusiastically to the work, explicitly comparing it to two famous classical texts on agriculture, the "Wu yi" (Against luxurious ease) section of the *Book of Documents* and the "Bin feng qiyue" (Odes of Bin, seventh month) poem from the *Book of Songs*.[162] He ordered Lou's paintings to be displayed in the "back palace" or the imperial residential quarters.[163]

Lou Shu, like Li Jie, came from a distinguished family with a tradition of official service: his grandfather, great uncle and father had had successful careers as local officials[164] and his great grandfather had even been invited by Wang Anshi 王安石 (1021–86) to teach in his prefectural school.[165] Given this family tradition of government service and Confucian scholarship, as well as the self-indoctrination resulting from his years of study for the civil service examinations, Lou shared with many other local officials a strong Confucian sense of responsibility for the general welfare of the people entrusted to his care.

Roslyn Hammers in using "Shu."

158. It is unclear whether the paintings were in album or handscroll format. Bray sees them as "albums" in the broad sense of "a sequence of poems and paintings on a single theme." Bray et al., *Graphics and Text*, 525. There are also conflicting views on whether Lou himself did the paintings; compare Cahill, *The Painter's Practice*, 139 with Yoshida, *Salt Production Techniques*, 81n68. If Cahill is right that Lou, who is not otherwise known as a painter, did not himself do the final paintings, he could still have had a role in their design, perhaps providing sketches or rough draft paintings on which the final versions were based.

159. Each lyric *shi* 詩 consisted of eight lines of five characters each, the standard pentasyllabic regulated verse (*wu yan lüshi* 五言律詩) form. Thus the total text consisted of only some 1800 Chinese characters.

160. When I began working on the *Geng zhi tu*, I followed a suggestion by Joseph Needham (Needham and Wang, *SCC* 4:2, 166, fn. e) and used "Pictures of Agriculture and Sericulture" as my working translation of the title. Eventually, I decided to change that to "Pictures of Tilling and Weaving" since I had come to feel that the earlier translation tended to emphasize technological content more than was warranted. My choice has been reinforced by Roslyn Hammers's use of the second translation in her recent *Pictures of Tilling and Weaving: Art, Labor and Technology in Song and Yuan China* (Hong Kong: Hong Kong University Press, 2011) which is now the basic reference for this work.

161. For an up-to-date discussion of the possibility that this happened in 1145, see Hammers, *Pictures of Tilling and Weaving*, 9, 217 and 248n1; Bray et al., *Graphics and Text*, 524–25 prefers 1153 or 1154.

162. We have noted in Chapter 1 (see "Early Farming Paintings") the appearance of agricultural scenes in early Chinese painting, even though the painters had only limited interest in portraying farming technology itself.

163. Bray (*SCC* 6:2, 49) says that "Kao-Tsung and his wife, who were both calligraphers, themselves annotated some of the poems," but gives no source for this point.

164. Yonezawa, "Common People," 58n5.

165. Hammers, *Pictures of Tilling and Weaving*, 42 and 81.

As we have noted earlier, one of the most crucial tasks for local officials was the promotion of agriculture (*quannong* 勸農).[166] In the Song period, this responsibility was recognized throughout the government as never before. Practical actions to promote agriculture, including written efforts to spread more widely knowledge of improved techniques, were undertaken on an unprecedented scale.[167] It was therefore nothing out of the ordinary for Lou Shu to state in the second lyric in the agricultural section of the *Pictures of Tilling and Weaving* "(As an official,) I have taken 'Encourage Agriculture' as my motto."[168] Conditions in China during Lou's early adulthood reinforced that commitment. The insecurity of the Song state just after the loss of north China gave special impetus to the efforts to increase agricultural production in the areas of south China under its control in order to assure sufficient food production not only for the native population but also the large number of refugees from the north.[169] One of Lou's motivations for creating his masterpiece may well have been to encourage the restoration of old patterns of life and work after the wartime disruptions.[170]

Other motivations quite likely also came into play. It is reasonable to ask, for example, whether Lou might not have undertaken this project as a means of advancing his career. James Cahill at any rate seems to believe that the paintings were aimed from the beginning for presentation to the emperor[171] though that idea is somewhat called into question if indeed a decade or even two elapsed between the completion of the paintings and their presentation to the emperor. Nevertheless, there is much to be said for Francesca Bray's contention that "production of social harmony and political order"–something that would surely appeal to the emperor–seems to have been as much at stake as material production.[172] As the accompanying poems make clear, the illustrations can be seen as an iconic or symbolic interpretation of an ideal rural order with "families working together to achieve abundance"; this was certainly the way that they tended to be viewed by most Chinese in the centuries that followed.[173] It was of course the government's responsibility to bring about and to maintain this ideal order. Such a view dovetails nicely with the suggestion that even international considerations may have had a role to play during both the Southern Song and

166. See Chapter 1, "Early Farming Paintings."

167. Bray et al., *Graphics and Text*, 523–24; Peter Golas, "Rural China in the Song," *Journal of Asian Studies* 39.2 (February, 1980), 310.

168. Franke, "Zur Geschichte des Kêng Tschi T'u," 193; Pelliot, "A propos du *Keng Tche T'ou*," Plate XII. For a somewhat different interpretation of this statement, see Hammer, *Pictures of Tilling and Weaving*, 167.

169. Bray et al., *Graphics and Text*, 524.

170. Otto Franke, *Kêng Tschi T'u: Ackerbau and Seidengewinning in China* (Hamburg: L Friederichsen & Co., 1913), 66–67, 70; Dieter Kuhn, "Die Darstellungen des *Keng-chih-t'u und ihre Wiedergabe in populär-enzyklopädischen Werken der Ming Zeit*," *Zeitschrift der Deutschen Morgenländischen Gesellschaft* 126.2 (1976), 338. With the widespread use of woodblock printing by this time, it would not have been unreasonable for Lou to hope that his paintings and lyrics would also be printed and thereby gain a wide audience. Indeed, this did happen, but not as quickly as Lou might have wished.

171. Cahill, *Three Alternative Histories*, 23.

172. Bray et al., *Graphics and Text*, 532.

173. Bray et al., *Graphics and Text*, 534.

the Yuan dynasties in government support for the *Pictures of Tilling and Weaving*, serving as a tool in the competition between Chinese and non-Chinese rulers for the "minds and hearts" of the Chinese people.[174]

Absent any direct evidence in his own words, we can only conjecture the importance Lou attached to portraying the actual technology of rice farming and silk production. Contrary to what many have thought, a good case can be made that this did not rank high in his priorities. For example, again according to his grandson, Lou meant his illustrations to portray exhaustively the "conditions" of the farmers (*tu hui yi jin qi "zhuang"* 圖繪以盡其狀) while the emphasis of the lyrics was on giving full expression to the "feelings" of the people as they carried out their workaday activities (*shi ge yi jin qi "qing"* 詩歌以盡其情).[175] With his keen desire to encourage farming, Lou in all likelihood did take the trouble to observe personally and with some care the techniques and the equipment of rural producers.[176] On the basis of the illustrations and lyrics, however, one is hard pressed to decide how thoroughly he understood the more complex of these activities.

In any case, where Lou seems to stand out as a major innovator, is in his decision (1) to portray in two very long series of paintings at least most of the main steps in the two most important and elaborate production processes in traditional China, and (2) to provide a text to elucidate each scene. What inspired this approach?

Apart from a variety of earlier textual and graphic treatments of farming and cloth production,[177] Lou may also have been inspired by a long-used if by his time somewhat archaic format for scroll paintings that consisted of "sections of text followed by pictures that illustrated them."[178] That no agricultural paintings or illustrations utilizing this format have survived could well be an accident of history, reflecting conditions in the period before the widespread availability of books and copies thanks to woodblock printing. However, the fact that Lou's almost exact contemporary, the Southern Song scholar Zheng Qiao 鄭樵 (1104–62), discussed in a famous essay the roles of both text and image in conveying knowledge, including technical knowledge of various kinds, but made no mention of

174. Cahill, *Three Alternative Histories*, 18–23. Roslyn Hammers in her *Pictures of Tilling and Weaving* (especially Chapters 2 and 3) would go so far as to see this work as a highly political statement following the ideas of the Northern Song reformers Fan Zhongyan 范仲淹 (989–1052) and Wang Anshi. While I have some reservations about how far to take this interpretation, one can only applaud her insistence that a true understanding of the work must rest on thorough analysis of *both* the illustrations and the lyrics. (I am very grateful for the generosity with which she has shared her ideas and work with me.)

175. Franke, *Ackerbau und Seidengewinnung*, 67. Singing while working is explicitly mentioned twice in the scrolls, in the lyrics "Transplanting Seedlings" (*cha yang* 插秧) and "Picking Mulberry Leaves" (*cai sang* 採桑). Hammers, *Pictures of Tilling and Weaving*, 174 and 196.

176. At least once he refers to just such on-the-spot inspection of farming: "Staff in hand, I head to the (cultivated fields in the?) eastern countryside." See "Tilling" (*geng* 耕), the second of the farming lyrics. Hammers, *Pictures of Tilling and Weaving*, 167 and Franke, "Zur Geschichte des *Kêng Tschi T'u*," 193.

177. See, for example, Chapter 1.

178. Cahill, *Three Alternative Histories*, 16–18. Such paintings were much favored by Emperor Gaozong; Julia K. Murray, "The Role of Art in Southern Sung Dynastic Revival," *Bulletin of Sung Yuan Studies* 27(1997), 54–55.

agriculture may perhaps be taken as evidence for an absence of earlier efforts to combine text and image in portraying farming practices.[179]

Another much more likely precedent for Lou's possible innovation were single portrayals of production scenes which included several steps in the production process, many of which date back at least to the Han. An especially remarkable surviving example is what appears to be an inscribed stone slab from a Later Han (25–200 CE) tomb that depicts from right to left three steps in the production of silk cloth: twisting the threads by hand, quilling (combining the threads into yarn) using a wheel, and weaving the cloth on a loom (Fig. 3.24).[180] To be sure, it seems unlikely that Lou had encountered any very early examples of this genre. However, a number of single scene portrayals of production processes from around the time of the Song have survived down to the present and suggest that this may have been a fairly common subject in the genre paintings that constituted an important category in painting at this time.[181] A quite remarkable example is a horizontal handscroll in the Shanghai Museum (Plate 9). It bears no title or name of its creator though there is evidence to suggest it may have been painted by a certain Wei Xian 衛賢 who was active in the second half of the tenth century or another painter much influenced by him.[182] It portrays a skillfully ordered scene[183] that depicted, as the scroll was unwound from the right, the whole production sequence by which grain, almost surely wheat, was transformed into flour at a water-powered mill, sacked and shipped. Step by step one can follow the delivery of the grain to the mill, winnowing of the grain using two large suspended sieves, the grinding of the grain on a waterwheel-powered millstone, sifting by means of a reciprocator powered by the waterwheel on the right, drying of the grain in the sun, sacking and sending on the

---

179. Bray et al., *Graphics and Text*, 522–23.

180. For other multi-step production scenes down to the end of the Han, see Lim, *Stories from China's Past*, 103, Pl. 18 and Yu Weichao (ed.), *A Journey into China's Antiquity* (Beijing: Morning Glory Publishers 1997), vol. 2, 142, Fig. 64 (fermenting wine or, possibly, distilling spirits). See also Rawson, *Mysteries of Ancient China*, 199–200 and Huang, *SCC* 6:5, 217–21; Lim, *Stories from China's Past*, 109, Pl. 24 and Fig. 3 and 189 (kitchen preparations); Lim, *Stories from China's Past*, 26, color plate 14 (hulling and winnowing grain); Huang, *SCC* 6:5, 305–16 and 331–33 (making soybean curd [*doufu*]); Barbieri-Low, *Artisans in Early Imperial China*, 70–71, Fig. 3.3 (sword-making; discussion p. 69). This practice of including several production processes in a single realistic (but technologically not very informative) scene can be found right down to the late imperial period; for a good example from the late Ming, see Wu Pin's painting *Silk Production in a Village*. Cahill, *The Compelling Image: Nature and Style in Seventeenth-century Chinese Painting* (Cambridge, MA/London: Harvard University Press, 1982), 78, Fig. 3.6.

181. Hay, "Along the River," 301.

182. Liu, "*Water Mill*," 567–69, 575, 582, 584, 590n3 and 590–91n23; Whitfield, "Material Culture," 55 and 61. Kuhn identifies it, without comment, as a "copy from the early decades of the thirteenth century." Dieter Kuhn, *The Age of Confucian Rule: The Song Transformation of China* (Cambridge, MA: Harvard University Press, 2009), 223.

183. Liu, "*Water Mill*," 557.

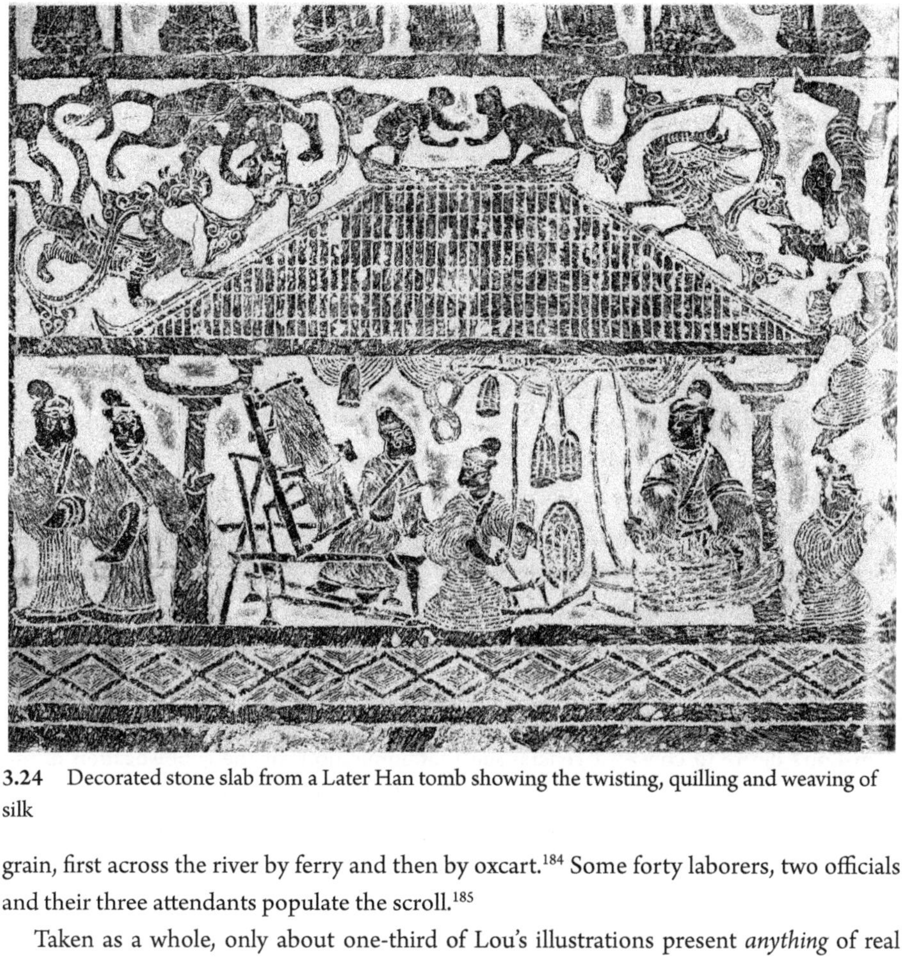

3.24 Decorated stone slab from a Later Han tomb showing the twisting, quilling and weaving of silk

grain, first across the river by ferry and then by oxcart.[184] Some forty laborers, two officials and their three attendants populate the scroll.[185]

Taken as a whole, only about one-third of Lou's illustrations present *anything* of real technological interest, and of these, only a handful picture equipment of some complexity. By and large, one finds only very generalized depictions of farming and silkmaking activities with which the great majority of rural people in south China (though not necessarily emigrants from the north) would be familiar. As an example of unrealized possibilities even in the depiction of quite simple technology, we can compare Lou's illustration of reaping with the version found in an eighteenth-century Korean work that shows the reaping of rice with a specific kind of sickle apparently never pictured in traditional drawings in China

184. Zheng, "Transport Carts at the Mill," 17; Liu, "*Water Mill*," 557. Hulling is not shown; see Dieter Kuhn, *Die Song-Dynastie (960–1279): Eine neue Gesellschaft im Spiegel ihrer Kultur* (Weinheim: Acta Humaniora, VCH, 1987), 177–78. Illustrations combining production and related processes continued to be produced in later centuries. For one example, see the *Ao bo tu* 熬波圖 illustration of the production of sea salt in Yoshida, *Salt Production Techniques*, 92, Illus. A.

185. Kuhn, *The Age of Confucian Rule*, 224. A wall painting from a temple in northern Shanxi dating from 1167 and portraying a similar scene shows the shaft from the waterwheel also powering a hulling apparatus. Kuhn, *Song-Dynastie*, 180–81.

despite that fact that it was in wide use (Figs. 3.25 (a) and (b)). The Korean illustration also allows us to identify three steps in the sequence of reaping that show exactly how the sickle was manipulated, again something that seems never to have been pictured this precisely in any Chinese illustrations in the various versions of the *Pictures of Tilling and Weaving* or in other works on agriculture.[186] Other striking examples of Lou's lack of focus on technology per se are the *three* highly repetitive illustrations of weeding and the *three* moltings of the silkworms, all totally devoid of any technological interest.[187] It is of course not difficult to come up with possible reasons for what might at first seem like overkill in these cases. These were, after all, important stages in the farming and silkmaking processes that extended over a considerable period of time.[188] Perhaps Lou intended to stress the tedium people experienced especially at these times and felt he could do it more effectively with multiple illustrations and multiple lyrics.

That Lou did not draw more deeply on his knowledge and experience to make his illustrations technologically more informative strongly suggests that, insofar as he had a didactic purpose in mind for the *Pictures of Tilling and Weaving*, it had more to do with encouraging a healthy rural order where peasants conscientiously devoted themselves to their work while those who enjoyed a good life because of peasant production appreciated the crucial role of the hardworking masses and concerned themselves about their welfare. In other words, what Bray has described as "an ideal social contract."[189] Even in those rare illustrations where technology of some complexity appears, the incompleteness of the depictions may result from Lou's desire to convey a general sense of admiration for the sophistication of the machines without feeling a need to explain them.[190] In actual historical fact, it would not be until 600 years later, in the 1739 Qianlong edition of the *Pictures of Tilling and Weaving*, that an effort would be made, though with only very qualified success, to provide a more clearly technological portrayal of the processes and tools that appear in Lou's illustrations.[191] Despite the fact that these Qing explanations can be difficult to decipher, for the most part they give us an enormous amount of practical information, often drawn from earlier agricul-

186. Bray et al., *Graphics and Text*, 533 and 555, Fig. 2; Pelliot, "A Propos du *Keng Tche T'ou*," Planche XXIV.

187. For information on the different techniques used in the three weedings, virtually none of which lent themselves to visual depiction, see Franke, *Ackerbau und Seidengewinnung*," 92 and 120–21, nos. 11–13. See also Elvin, "Technology of Farming," 20–21.

188. The extreme importance attached to weeding in Chinese agriculture was pointed out already in 1843 in a pathbreaking article by Edouard Biot. The article is translated from the French into English in Legge, *The She Ching*, 142–71, with the comments on weeding on p. 150.

189. Bray et al., *Graphics and Text*, 534. For a concise but superb summary of the Chinese social contract between the state and the people, see Bray and Métailié, "Who Was the Author of the *Nongzheng quanshu*?" 350.

190. Francesca Bray, "Agricultural Illustrations: Blueprint or Icon?" in Bray et al., *Graphics and Text*, 533.

191. Franke, *Ackerbau und Seidengewinnung*, 90 and 109–38. Interestingly, in Fang Guancheng's 方觀承 *Mianhua tu* 棉花圖 (Illustrations of cotton production) which was presented to the emperor just two-and-one-half decades later in 1765, Fang also supplemented the poems he composed for each of the sixteen paintings on the production of cotton cloth with prose passages that provided detailed technical information and advice. Bray, "Agricultural Illustrations," 548; Arthur W. Hummel, *Eminent Chinese of the Ch'ing Period* (Washington: United States Government Printing Office, 1944), 1, 234–35.

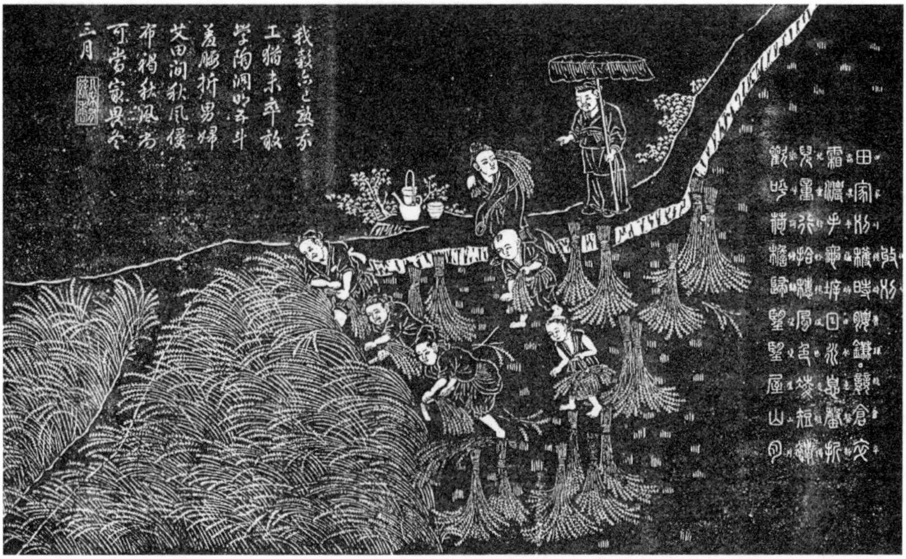

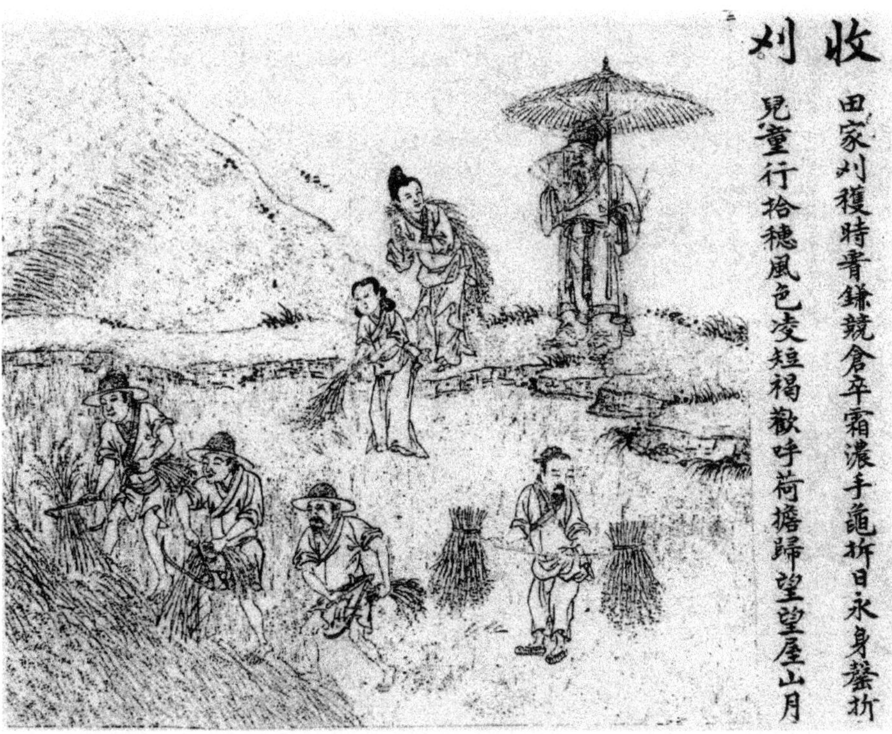

3.25 Reaping as portrayed (a) in a copy of the *Pictures of Tilling and Weaving* that is probably the closest we have to Lou's original, and (b) in an eighteenth century Korean copy.

tural works. We *do* learn when the different "weedings" took place and their very different purposes. We are also reminded that, here as elsewhere, words can often serve much better than illustrations to elucidate technological processes.

Which brings us to Lou's lyrics. Perhaps even less than with the illustrations can one make the case that the lyrics accompanying each painting had any aim to provide information about the technology in use. In the illustrations, as Bray points out, despite the highly detailed portrayal of certain aspects of farming and silkmaking activities, the equipment used is often drawn in a kind of visual shorthand or even partially hidden in the background or buried realistically but uninformatively in mud (Fig. 3.26)![192] As for the lyrics, we have noted above the assessment by Lou's grandson that they were meant to give expression to the emotions of rural folk as they carried on their daily tasks. Here Lou met with considerable success. For whatever the reasons (and, sadly, we know nothing of the music of the songs), at least some of his lyrics seem to have enjoyed great popularity. Despite Lou's inclusion of highly literary elements, he apparently was able to speak with special persuasiveness to the feelings of ordinary people and thus has as much claim to be called a "songwriter" as a poet or a painter. It is for this reason I have avoided the usual word "poem" to describe Lou's comments on the illustrations, substituting instead "lyric."[193] If we see these *shi* 詩 in large measure as songs to be sung, we are less surprised to find that they provide little information of a technological nature, indeed hardly anything that could not be gotten simply from the pictures and the identification of the activity in the title of the lyric.[194] Similarly, the descriptions of the circumstances and environment in which these activities take place are highly poetic with a strong moral tone emphasizing the importance of hard work, but they are not very informative on details as to how that work was carried out.

In 1210, forty-eight years after the death of Lou Shu, his nephew, grandson and others, "fearing that the pictures would fall into oblivion with the passing of years, decided to have them carved on stones together with [Lou's] lyrics so that they could be passed down for all time."[195] Then, in the late 1220s or early 1230s, the first book consisting of the pictures and their lyrics appeared in a printed edition.[196] It was destined to become, along with Wang Zhen's *Agricultural Treatise*, one of the two major illustrated agricultural texts in China down to the twentieth century, giving rise to many editions, especially in the later Ming and in the Qing periods.[197] Its extraordinary popularity, despite what may seem to modern viewers as

192. Bray, "Agricultural Illustrations," 531–32.

193. Francesca Bray's "ballad" serves the same purpose.

194. Even specific references to technology might be couched in metaphor; see Roslyn Hammers, "'How Can It Be That a Crow's Tail Can Hold Water?': The Square-Pallet Pump in Lou Shu's *Pictures of Tilling and Weaving*," *Technology and Culture* 46.1 (January 2005), 132. The poem cited here is of special interest because of its at least general recognition of technological advance.

195. Wang Chaosheng 王潮生, *Zhongguo gudai geng zhi tu* 中國古代耕織圖 [Farming and weaving pictures in ancient China] (Beijing (?): Zhongguo nongye chubanshe, 1995), 36 (translation modified). For a more literal translation, see Hammers, *Pictures of Tilling and Weaving*, 216.

196. Franke, *Ackerbau und Seidengewinnung*, 72; Bray, "Agricultural Illustrations," 527.

197. Dieter Kuhn, "Marginalie zu einigen Illustrationen im Nung-shu 農書 des Wang Chen 王禎," in R. Goepper

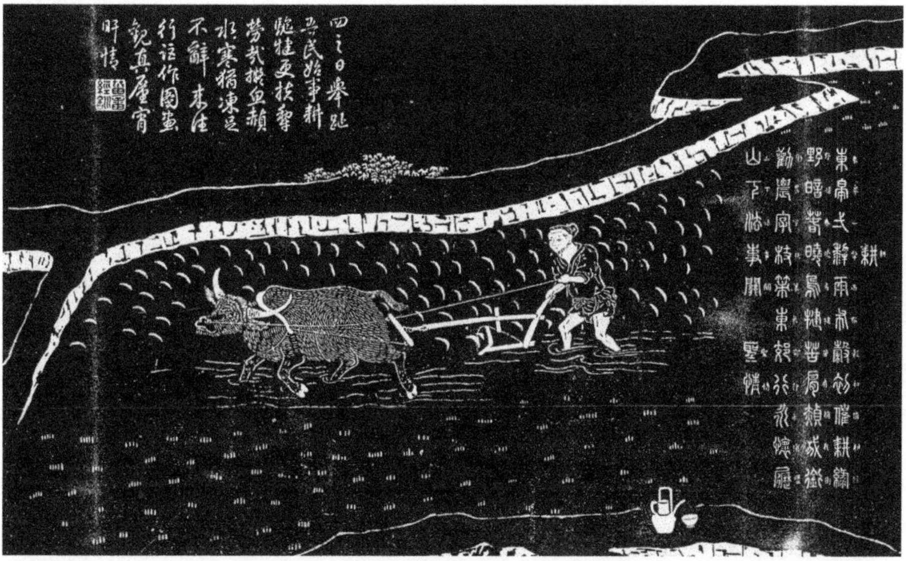

3.26 "Plowing," *Pictures of Tilling and Weaving*. Only an absolute minimum of information is provided on the construction of the plow.

the limitations of its presentation of agricultural and sericultural technology, in fact derived from a number of innovations that left no doubt about the uniqueness of the work in its own time and later. One of those innovations has perhaps not been sufficiently stressed: the sheer number of the illustrations. Lou's rather perfunctory handling of technological details may nevertheless have been an improvement over most other contemporary portrayals of farming and silkmaking, but it was his decision to break down the tasks of agriculture and sericulture into twenty-one and twenty-four individual scenes respectively that enabled him to present a fuller picture of farming and silkmaking than had ever been attempted before. Indeed, in their many editions, the illustrations served as a model for a great variety of similar portrayals of people engaged in extended production tasks involving multiple stages. These included the growing and processing of cotton, hemp and tea; the making of ceramics, paper, inksticks and lacquerware; and mining and the processing of iron and other metals.

Another way in which Lou's work was innovative was its focus, for the first time as far as we know, on farming and sericulture as practiced in southeast China. Francesca Bray points out that, despite the fact that the south had been undergoing great economic and technical advance for at least three centuries, earlier technical works had largely ignored these developments, concentrating instead on rural production in the northern heartland. Now viewers of these illustrations could get a sense for the hard work, the skills and technical

et al. (eds.), *Zur Kunstgeschichte Asiens* (Wiesbaden: Franz Steiner, 1977), 143; Wang Chaosheng, *Farming and Weaving Pictures*, Foreword (1), p. 2; Bray, "Agricultural Illustrations," 531.

sophistication that had taken a potentially rich area and, step-by-step, brought that potential to fruition.

A final innovation, and arguably the most important of all, was the close linkage between Lou's illustrations and their accompanying lyrics. Here was a format that could maximize the advantages of both visual and textual presentations of information, where the limitations of one could be immediately made up by the other. If Lou Shu did not take full advantage of these possibilities to discuss in detail the technologies he was presenting, he can still be seen as a precursor of another approach to linking text and image in the discussion of technology, one that would lead in the *Nongshu* (Agricultural treatise) of Wang Zhen to the first true technical drawings in China.

The *Agricultural Treatise (Nongshu)* of Wang Zhen

The *Nongshu* 農書 or *Agricultural Treatise* of Wang Zhen 王禎, with its preface dated 1313, offers a crowning example of what was probably the main means by which interested officials encouraged the spread of better forms of technology in traditional China.[198] During the Song, a period of impressive advances in productive technologies of all kinds, numerous local officials acting on their individual initiative turned their brushes to the writing of manuals intended to promote better practices in rural production, particularly farming and clothmaking.[199] By now, the spread of woodblock printing made it far easier and less expensive to reproduce and distribute such manuals. Wang Zhen took this opportunity much farther than anyone else, producing a work with an unprecedented range of illustrations "depicting every conceivable agricultural implement or machine."[200] The accompanying text explained their construction, a first in China,[201] as well as how and where they were used. Moreover, the illustrations, some possibly drawn by Wang himself, constituted the first true technical drawings or blueprints in Chinese history. Text and illustrations worked in close harmony to describe in unprecedented detail the technology of farming and silkmaking.[202]

Just like Lou Shu, Wang Zhen lived in a period (late 1200s–early 1300s) when much of China was still suffering from the effects of foreign invasion, in this case by the Mongols. He too had as one of his main concerns to provide knowledge about improved farming methods and implements to local officials who could then pass on this knowledge to

198. For a brief summary of its early editions, see Bray, "Agricultural Illustrations," 536. See also Hammers, *Pictures of Tilling and Weaving*, 127 and Kuhn, *SCC* 5:9, 11–12, fn. c and the references there, especially Wang Yuhu, *Wang Zhen Nong Shu*, "Jiaozhe shuoming," 2–5.

199. Golas, "Rural China," 310; Bray, *SCC* 6:2, 59; Kuhn, *SCC* 5:9, 10, 366. The now indispensable treatment of this topic is Bray, "Chinese Literati and the Transmission of Technological Knowledge."

200. Bray, *SCC* 6:2, 62; Bray, "Agricultural Illustrations," 538.

201. Nathan Sivin, "Review of Bray et al., *Graphics and Text*," *China Review International* 15.4 (2008), 458.

202. But, of course, not always. For a striking example of an illustration (of the cradled sickle for harvesting wheat) that completely fails to picture the implement described in the textual description, see Bray, *Technology and Society*, 32, Fig. 25. It is hard to imagine that Wang Zhen so much as looked at this illustration before it was published.

farmers not only in areas devastated by war and occupation but also to other areas where appropriate.[203] For him that meant not only increasing agricultural production and thereby bringing about improved living conditions generally but also introducing methods that might lighten the tasks and increase the efficiency of rural work. For example, in discussing weeding by hand, Wang not only comments eloquently on how hard this is on farmers but also uses the description to prod officials to spread knowledge of improved techniques.[204] In this case, that meant the use of a hand harrow, which he describes and illustrates (Fig. 3.27).[205] Also, if Otto Franke is correct, the original version of the *Agricultural Treatise*, like Lou Shu's *Pictures of Tilling and Weaving*, had songs for the people to sing while working, another way of making their work a little easier.[206] Finally, Wang the realist also hoped that his book would discourage officials from doing harm by pushing misguided practices.[207]

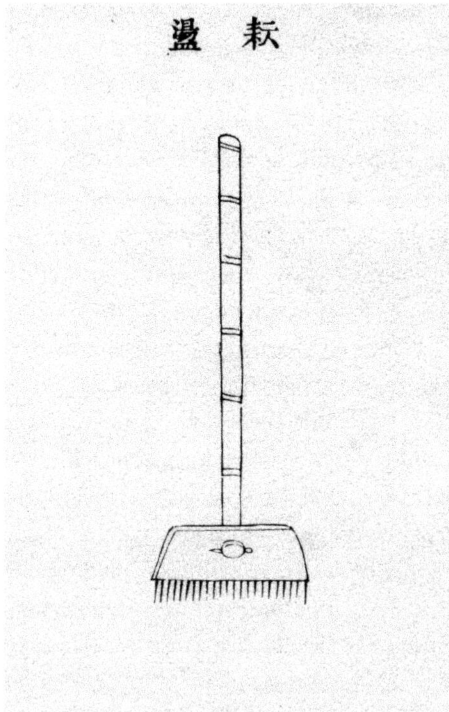

3.27 The hand harrow, a fourteenth-century innovation from the Yangtze region, as depicted in Wang Zhen's *Agricultural Treatise*

Francesca Bray has characterized the *Agricultural Treatise* as "in many respects a conventional, text-based agronomic treatise."[208] It differed however from earlier similar works in dealing with tools and implements of both north and south China, as well as discuss-

203. Bray, *SCC* 6:2, 60. Wang himself, unlike Lou Shu, was a northerner, from Dongping 東平 in Shandong and therefore presumably especially affected by the devastation of his homeland. But it must also be noted that a case can be made that Wang's *main* goal was to promote the spread of labor-saving northern farming equipment in the south; Bray and Métailié, "*Nongzheng quanshu*," 341.

204. Bray, *SCC* 6:2, 61–62; Golas, "Technological Illustration," 51.

205. Bray, *SCC* 6:2, 61–62. It is interesting to see that in the late Ming *Tian gong kai wu*, its author, Song Yingxing, reverts to a scene of hand weeding (while also adding an illustration of foot weeding); compare Bray, *SCC* 6:2, 61–2, 314 and 318 and Sun and Sun, *T'ien-kung k'ai-wu*, 9, Fig. 1–3 and 10, Fig. 1–4.

206. Franke, *Ackerbau und Seidengewinnung*, 46. Poetic verses, either by Wang himself or by other figures who had composed poems on agriculture (including Lou Shu) regularly concluded the written entries on the illustrations; Bray, "Agricultural Illustrations," 537.

207. Bray, *SCC* 6:2, 60.

208. Bray, "Agricultural Illustrations," 521. The following discussion draws extensively on Bray in *SCC* 6:2, 59–64 and her more recent discussion of the work in Bray et al., *Graphics and Text*, especially 66–68, 535–43 and 557–63.

ing polder fields, terraces and irrigation methods, the first work to discuss these topics systematically.[209] It was in the illustrations, however, collected in a separate section of the work entitled "Illustrated Register of Farming Implements" (*Nongqi tupu* 農器圖譜), that Wang showed himself a consummate innovator. Not only is Wang's work the first surviving treatise on agriculture to contain illustrations[210] but it also incorporates them on a lavish scale, with no less than 265 separate drawings and diagrams of implements and machines used in rural production.[211] The illustrations are also unprecedented in their focus on "things." That includes not only tools, machines and various equipment but also types of fields, ceremonial vessels and grain storage facilities.

Wang's approach here is all the more remarkable since it represents about as far as the Chinese ever went in writing about and illustrating the equipment used in farming.[212] As Bray points out, there were no monographs on this subject in traditional China.[213] The Chinese did have a deep fascination with seeds and planting methods (most of their experience here could not be presented visually) but ". . . farm tools and machines were not considered interesting. It is hard to imagine an educated Chinese attempting to benefit mankind by improving the design of the plow, and there is none of the technical, experimental literature in China that we find in the West on such topics."[214] Much of the information on agricultural implements and machines that appears in Chinese works is in a cataloguing mode, often without illustrations.

Wang, by contrast, not only wanted to present accurately the provenance of his implements and machines and tell just how they were used but also, as noted above, unmistakably often intended his illustrations to serve, together with his descriptions, as guides for their actual construction.[215] In the case of the hand harrow, he states explicitly: "It seems to me a great shame that this tool is not more widely known and used, so I have drawn

209. Bray, *SCC* 6:2, 64.

210. This breakthrough may have been inspired in part by the fact that Wang Zhen was familiar with and influenced by Lou Shu's *Pictures of Tilling and Weaving*. Kuhn, "Marginalie," 144; Dieter Kuhn, "Some Notes Concerning Textile Technology Pictured in the *Keng-chih-t'u*," *Zeitschrift der Deutschen Morgenländischen Gesellschaft* 2 (1980), 414n25.

211. Needham and Wang, *SCC* 4:2, 169. Zhang Baichun and Tian Miao ("Visual Presentation," 94) give a figure of "over 280."

212. In order to assure the drawings were accurate, Wang sometimes even had craftsmen disassemble and then reassemble implements so that he could be sure he understood their structure. Bray, "Introduction," 66–67.

213. Even an apparent exception proves to be not much of an exception after all. One seventeenth-century scholar did write a fresh (i.e., not just quoting the *Agricultural Treatise*) account of agricultural tools but it focused on things such as local variants of names *and included no illustrations!* Bray, *SCC* 6:2, 75.

214. Bray, *SCC* 6:2, 74–75.

215. Wang may have borrowed illustrative elements from Lou Shu's *Geng zhi tu* illustrations, but nothing in the *Geng zhi tu* suggests Lou ever intended them as practical construction drawings. To be sure, even Wang's illustrations and explanations together did not unfailingly make possible, on the basis of this information alone, the construction of the indicated machine. See for example his treatment of the seed drill (Bray, *SCC* 6:2, 63, 256 and 257, Fig. 95). But we should not forget the presence here of an important element usually passed over in silence: the knowledge and experience brought by those who would do the constructing. Bray cites a very illuminating statement that comes at the end of Wang's discussion of the square-pallet chain pump: "This pump involves many joints, but if you have a carpenter, it can easily be made." Bray, "Agricultural Illustrations," 540.

and described it on purpose that those of a philanthropic bent (*ai min zhe* 愛民者) may disseminate the practice."[216] This is presumably the major reason why Wang often illustrates individual tools and pieces of equipment by themselves, as in the case of the hand harrow or the cradled scythe (Fig. 3.28). They *are* true technical illustrations or technical diagrams.

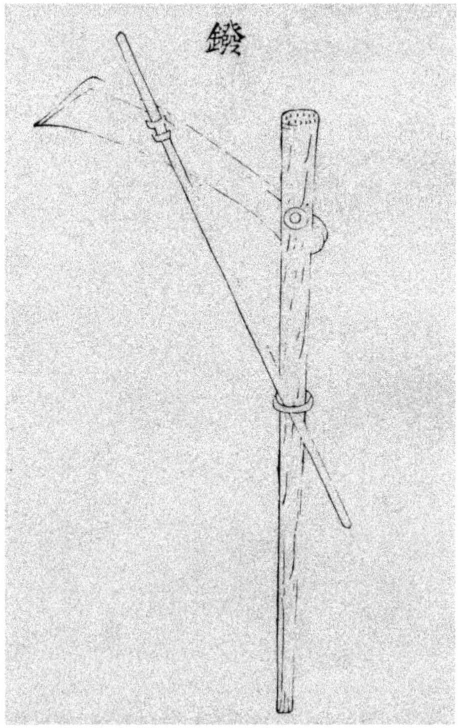

3.28 Cradled scythe (*po* 鏺) from Wang Zhen's *Agricultural Treatise*

The omission of human actors from the illustrations, where it occurred, undercut any inclination of viewers to draw symbolic messages from the illustrations and encouraged instead a focus on the actual technological functions performed by the tools and machines.[217] On the other hand, in the case of larger and more complicated machines, Wang usually adhered to the common practice of including workers operating them, perhaps feeling that this would make their complexities more comprehensible to the reader. Indeed, it does appear that the inclusion of people operating the machines at least sometimes encouraged more precise and complete depictions. When we compare for example Wang's portrayals of the silk-reeling frames used in northern and in southern China (Fig. 3.29), we note that, while the northern frame illustration benefits from a liberal use of informative captions,[218] the southern frame is better drawn to give a clearer picture of how a reeling frame actually worked. In particular, the presence of the woman in the center of the drawing has encouraged the illustrator to include the treadle that turned the reel, something barely hinted at in the drawing of the northern machine.[219]

216. Bray, *SCC* 6:2, 62; Wang Yuhu (ed.), *Agricultural Treatise, juan* 13, 198 (translation slightly modified). This passage does seem to say that Wang himself drew this illustration.

217. More human actors were added in some later editions, possibly reflecting a growing emphasis on aesthetic preferences; Bray, "Agricultural Illustrations," 536.

218. For a translation of these captions, see Kuhn, *SCC* 5:9, 368–69.

219. Francesca Bray, *Technology and Gender: Fabrics of Power in Later Imperial China* (Berkeley: University of California Press, 1997), 227. Even in the case of simpler devices, the inclusion of their operators could make the their functioning clearer; indeed, without such figures, the illustrations even with their accompanying text could be unclear to incomprehensible. A dramatic example from another technology is provided by the ultimately inconclusive efforts of Needham and his collaborators to determine just how the "whip arrows" (*bian jian* 鞭箭) portrayed in the *Military Essentials* were actually launched. Needham, Ho, Lu and Wang, *SCC* 5:7, 149–52.

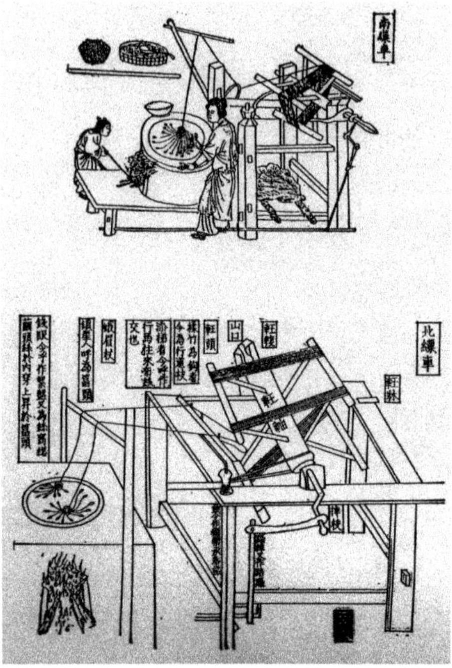

3.29 Southern (top) and northern (bottom) silk-reeling frames from Wang Zhen's *Agricultural Treatise*

♦ ♦ ♦

The wide circulation of books made possible by the spread of woodblock printing led to certain works becoming enormously popular and influential, sometimes for centuries. In the case of works on technology, this could lead to the emergence of what were effectively "standard accounts," relied on, often slavishly, by later authors. This is precisely what happened with both the illustrations of the *Pictures of Tilling and Weaving* and with both the illustrations and the text of the *Agricultural Treatise.* Bray points out how this has sometimes skewed our knowledge of later Chinese technology. While agreeing with the widely accepted view that there were few important advances in agricultural technology in China "between 1300 and 1950," she also notes that the great popularity of Wang's book especially could well have "obscured many minor developments which would otherwise have been separately documented." Moreover, in the case of textiles, where there were some quite important improvements in the Ming and Qing, authors continued to quote supposedly relevant passages from the *Agricultural Treatise* and to reproduce outdated illustrations, thus making it difficult or impossible to reconstruct those improvements.[220]

220. Bray, SCC 6:2, 64.

Four

The New Confucian Paradigm

The masterworks on which we focused in the previous chapter were the product of a remarkably vital period of Chinese cultural history, a time perhaps unique in its ability to meld culmination and change. Central to the changes taking place at this time was a revival of Confucian philosophy that had begun already in the Tang. By the end of the Song, it had so completely won the day that the primacy of this interpretation of Confucian ideas and values would remain dominant in China until the nineteenth century when threats from beyond her borders called into question the entire Chinese order, institutional and intellectual. Our purpose in this chapter is to suggest that the revival and recasting of Confucianism during these centuries can contribute part of the answer to one of the central questions in this study, namely, why it was that the highest achievements in technological illustration in the Song and Yuan centuries were so seldom improved upon–indeed usually not even matched–in later centuries.

Much of our focus will be on painting. As we have pointed out earlier, the techniques and values adopted by Chinese painters powerfully influenced the development of drawing techniques. In the centuries after about 1200, this influence became generally negative for the effective, informative portrayal of technological subjects.

Realism in Retreat

Historians of Chinese art are largely in agreement that the Southern Song (1127–1279) marks a watershed in the history of painting in China. From this time forward, representational values and the inclusion of accurate and detailed realistic elements tended to be downplayed, at first by scholar painters and later in Chinese painting generally.[1] As James Cahill summarizes this development, Chinese painting "moved from painting as pictorial

1. The "shorthand" techniques increasingly seen in landscape paintings from the Song on are reminiscent of conventionalized painting of landscape elements in the earliest Chinese paintings which, if Cahill is correct, expected the viewer to be able to metamorphose in his imagination a few primitively painted trees into a thick forest or low hillocks into lofty peaks. Cahill, *Chinese Painting*, 26–27. Wen Fong has pointed out ("Chinese Bronze Age Arts," 32) that a recurring characteristic not only of Chinese art but also of literature and thought was the appearance, after a "primary evolutionary phase," of *fugu* 復古 or a "return to the archaic." Of course, this is not at all to say that the earlier primitive painting provided actual models for what came after the Song.

description of appearances to painting as an expressive art concerned with its own conventions and its own past . . ."[2] Why so much of Chinese painting after the Song forsook realism, specificity and similitude in favor of an emphasis on suggestion, impressionism and expressivity has been quite properly characterized by Michael Sullivan as "one of the most fascinating questions in the study of Chinese art."[3] Cahill, in a comment that points to the importance of certain developments within painting itself, suggests that by the early Southern Song Chinese painting "had progressed as far in the direction of descriptive realism as it was ever to go and, on the whole, afterward could only move away from realism . . ."[4] The painters themselves may have sensed that the vocabulary of realism had been largely exhausted. Certainly, when we look at some of the realistic masterpieces of the Five Dynasties and Northern Song such as Zhang Zeduan's *Along the River at the Qingming Festival* (see Figs. 3.1 and 3.2 in Chapter 3), or the famous *Flour Mill* painting from the very early Song (see Plate 9),[5] or the anonymous late tenth or early eleventh century *Bamboo and Old Tree Growing by Rocks* (see Fig. 2.2 in Chapter 2) or many of the paintings emanating from Huizong's painting academy and indeed from the hand of the emperor himself, we are compelled to conclude that, to a remarkable degree, what the Chinese sought for in realism had been achieved in these masterpieces. This was true even though Chinese painters had hardly even begun to explore certain techniques such as the use of shadows to convey a sense of mass.

2. James Cahill, "Chinese Painting: Innovation after Progress Ends," in Sherman Lee (ed.), *China, 5000 Years: Innovation and Transformation in the Arts* (New York: Solomon R. Guggenheim Museum, 1998), 181. Roslyn Hammers discusses this development in connection with paintings of farming and the making of silk. Hammers, *Pictures of Tilling and Weaving*, "Epilogue."
3. Sullivan, *Symbols of Eternity*, 71. See also Wen Fong who describes a "return to symbolic representation, with an emphasis on calligraphic and surface abstraction" (Fong, *Beyond Representation*, 6) and Max Loehr who speaks of "a subjective, introspective, expressionistic, or intellectualized art . . . no longer concerned with the image of nature or external reality" (cited in Mark Elvin, *The Pattern of the Chinese Past* (Stanford: Stanford University Press, 1973), 225). Western painting went through its own period of rejecting realism with the rise of Christian art after the Roman Empire. March, "Linear Perspective," 122.
4. James Cahill, *The Distant Mountains; Chinese Painting of the Late Ming Dynasty, 1570–1644* (New York and Tokyo: Weatherhill, 1982), 4–5. It should be recalled that this was by no means an entirely new perspective. Excessive display of skill and overemphasis on detailed accuracy had been warned against by critics such as Zhang Yanyuan as early as the Tang: "Now in painting (any) subject the things that one should especially avoid are a methodical completeness in delineation and coloring, (as well as) extreme carefulness, extreme detail, and the display of skill and finish. . . ." Acker, *Some T'ang and Pre-T'ang Texts*, 1, 185; see also Yee, "Cartography in China," 135.
5. Liu Heping, "*Water Mill*," 568–69. On the basis of meticulous research though without conclusive evidence, Liu Heping suggests (584) that this is a composition by a "painter-engineer-architect" in the early Northern Song court. Zheng Wei, on the basis of equally serious research, attributes it to the tenth-century painter Wei Xian 衛賢, an attribution accepted by Anita Chung. Zheng, "Transport Carts at the Mill"; Chung, *Drawing Boundaries*, 21 and caption to Plate 1. Cahill says tenth to eleventh century (*Painter's Practice*, 117 and 121, Fig. 4.7). In any case, this seems to be the first appearance of a waterwheel in a surviving text or illustration, some three centuries earlier than the date given by Needham who was unaware of this painting and based himself on a rather inferior copy from around 1300; Needham and Wang, *SCC* 4:2, Pl. CCXLI b, facing p. 405; Chung, *Drawing Boundaries*, 24 and 25, Fig. 1.8.

Elite Dominance of Painting

Many other influences, most of them related in one way or another to the Confucian revival, also impacted the direction of Chinese painting. The reaffirmation of the centrality of Confucian values stemmed in part from the mid-Tang revival of imperial political institutions whose philosophical foundations had always been the teachings of Confucius and his followers. The emergence of an elaborate official recruitment system, whose centerpiece was examinations based on these teachings, set the stage for a strong reaction by most of China's educated elite against the prominence Buddhist and Daoist teachings had enjoyed in China in the period between the third and eighth centuries. The new Confucianism offered an alternate and comprehensive worldview that provided at one and the same time an almost religious calling for the new elite of scholar officials as well as a strong ideological cement for the new bureaucratic and meritocratic political order.[6]

That the political, social and, to a degree, economic dominance of this elite also brought with it cultural preeminence was only to be expected in the Chinese context. For example, the idea that the finest calligraphy and painting was inevitably produced by members of an intellectual and moral (and of course, political and social) elite had a long history in China. In the ninth century, it was given one of its most uncompromising expressions by Zhang Yanyuan 張彥遠 (fl. 847–74), author of China's first comprehensive history of art and the foundation work for so much of later Chinese painting criticism: "From ancient times those who have excelled in painting have all been men robed and capped and of noble descent, rare scholars and lofty-minded men . . ."[7] Just as literature came to be emphasized as a means to the moral transformation of readers, painting came often to be seen as a way by which the artist of high moral attainment could express his loftiest self in painting and thereby exercise a beneficent influence on his viewers. There was thus something of a shift in emphasis from the painting to the painter, and therefore from objectivity to subjectivity. As the new elite increasingly came to view themselves both as the moral and also the aesthetic arbiters of the society, those who concerned themselves with defining the values of fine painting formulated an aesthetic that disparaged objective pictorial representations, decorative richness and technical proficiency as vulgar. In so doing, they not only influenced the direction of scholar-painting but also the efforts of court painters and even professional artists well outside of literati ranks.[8]

And beyond the ranks of painters, too. Among all the surviving Ming and Qing depictions of technology, we see that relatively few were paintings.[9] Most were in the form of

6. Clunas' point that Buddhist "philosophical disdain for the sensory world, perceived as an illusion" (*Art in China*, 118) also contributed to the literati devaluation of representation reminds us that the new painting values by no means resulted exclusively from the Confucian revival.

7. Acker, *Some T'ang and Pre-T'ang Texts*, 1, 153. The citation comes from Zhang's *Lidai minghua ji* 歷代名畫集 [Record of famous painters through the ages] of 847. For the whole question of the image(s) of the artist in China, a good starting point is Cahill, *The Painter's Practice*, 123–26.

8. Cahill, *Three Alternative Histories*, 33.

9. Indeed, it is remarkable how even scenes of agricultural work are largely absent from the hundreds upon

outline drawings meant to serve as book illustrations. Nevertheless, developments in painting such as those discussed above remained central to the story. The artist-craftsmen most often responsible for these illustrations regularly underwent an apprenticeship that included training in painting, and many may even have made most of their living from painting.[10] Though they were generally disparaged by the more prestigious literati painters and critics who determined what made for painting as a fine art, the way book illustrators went about their work of designing images and the visual vocabulary they drew upon overlapped considerably with that of the literati painters.[11] As a result, changes in painting tastes and standards inevitably found reflection in illustration practices.

"Historical" Painting

The new Confucianism of the Song (commonly referred to in English as "Neo-Confucianism") also reinforced in more than one way the historical cast of mind that has so dominated Chinese culture. In painting, we see beginning in the eleventh and twelfth centuries a new value assigned to paintings that supposedly display an antique flavor. Unfortunately, given the very few "antique" paintings that have survived in the original, it is difficult to be certain just how this feeling of antiquity was expressed in paintings.[12]

More important and easier to define, however, was a radically new definition of the relationship between history and painting. History since very early times had been mined as a repository of people and actions that set standards of good and bad conduct, especially in the political sphere. In painting, there had grown up since the Tang a significant historical-critical literature having as its main focus assessment of the strengths and weaknesses of earlier painters. Now, however, the act of painting by scholar painters took an increasingly historical turn in which, as summarized by Max Loehr, "styles of the past began to function as subject-matter and to be explored as if they were the primary reality." Instead of close observation of the natural world, painters came increasingly to rely for inspiration on past paintings by recognized masters. By Ming times, this would lead to most painting becoming "a kind of humanistic discipline, an art blended with learning; it was painting about painting, allusive, commentarial, and increasingly abstract because its content was

---

hundreds of landscape paintings surviving from the Song and later; Bray, *SCC* 6:2, 77.

10. Robert E. Hegel, *Reading Illustrated Fiction in Late Imperial China* (Stanford: Stanford University Press, 1998), 196–97.
11. Jerome Silbergeld (*Chinese Painting Style*, 45–47) characterizes this way of doing paintings as "performance art". Robert Hegel cites Silbergeld's remarks as providing "the general basis for [a] convergence of formal painting with . . . book illustrations." The convergence is seen in a common "vocabulary of forms for trees, rocks, and landforms in landscape pictures, as well as a grammar including the presentation of figures in three-quarters view, . . . depth represented by the elevation of the subject in the picture, and a perspective afforded the viewer from a position somewhat above the subjects." Hegel, *Reading Illustrated Fiction*, 257.
12. Even the original paintings of the tenth-century Guo Zhongshu, noted for this antique quality (as well as their "aloofness, simplicity, profundity and a spirit free from worldly contamination"), have failed to survive. Bush and Shih, *Early Chinese Texts on Painting*, 202; Chung, *Drawing Boundaries*, 31–32.

thought."[13] This was a little as if European painters in the nineteenth century had opted to direct their painting to further explorations of the styles exemplified in the masterpieces of the Renaissance, relying greatly on learning by copying. The obsession with immersing oneself in the styles and spirit of the past did little to stimulate advances in technique that might have been useful in the production of better depictions of technology. Though certain drawing and painting skills could be honed by copying earlier paintings, the kind of craftsmanship that emphasized the accurate and convincing representation of three dimensional objects on a flat surface was not one of them.

Philosophic Contributions to the New Painting Aesthetic

One of the most original contributions of those who rethought and reformulated Confucianism in this period was to provide a metaphysical underpinning to how educated Chinese viewed the world around them, including but not limited to the natural world. Confucians had from earliest times been convinced that this world of which humans are an integral part manifested an orderliness and harmony that was real, meaningful, good *and* comprehensible by humans, and that should even serve as a model for human institutions. In this conception of an ever-changing assemblage of patterns such as the interaction of yin and yang and the progressions of the five elements/phases (*wu xing* 五行), the Chinese found what was for them a satisfying explanation of a reality ever in flux. Now Confucian thinkers supplemented this long-established outlook with a metaphysical/ontological dimension that held each and every thing in the world to be the manifestation of a principle or "pattern of construction and operation" (*li* 理) that constituted its essence.[14] All objects of a given class shared in the same principle. Extending our knowledge of the world, a highly moral activity,[15] meant engaging in the "investigation of things" (*ge wu* 格物). This referred to the empirical study of not only the material universe but also and even more importantly of the moral universe of human conduct.[16] All of these "things," tangible and intangible, were to be intensively studied in themselves in the hope that such study would be followed by an intuitive moment when one would grasp what we might call the heart of things, that is, the principles or very general patterns that lay behind all the multiplicity and variability of reality.[17] The result was an immensely complicated "system of symbolic correlations whose

13. Loehr, "Some Fundamental Issues in the History of Chinese Painting."

14. Elvin, *Pattern*, 225.

15. John B. Henderson, *The Development and Decline of Chinese Cosmology* (New York: Columbia University Press, 1984), 151.

16. Bodde, *Chinese Thought, Society and Science*, 265–66 and 268.

17. Bodde, *Chinese Thought, Society and Science*, 266–67. This contrasts profoundly with the Western search for fixed laws laid down by an all-powerful creator that governed the functioning of physical reality. See Needham and Wang, *SCC* 2, 518–83; Bodde, *Chinese Thought, Society and Science*, 332–68. Bodde's discussion of *ko wu* (*ge wu*) is indispensable; see pp. 264–68.

divinat[ory] and quasi-philosophical applications were more sought after than anything like 'scientific' understandings of nature."[18]

The concept of *li* was intrinsic to how most educated Chinese from the Song onward comprehended the universe. It also informed their efforts to portray the objects of that universe in their painting, especially in the landscapes that were the quintessential subject matter of scholar painting. The scholar-painter in his quest to comprehend the natural world aimed to capture the essences that united objects rather than portray objects in all their detailed individuality.[19] As Su Shi 蘇軾 (1037–1101) expressed it, "the artisans of this world may be able to capture the forms perfectly, but when it comes to the principles, only a superior man of outstanding talent can discern them."[20] A painting depicting bamboos with great realism, for example, could never be a great painting unless it somehow conveyed the principle of "bambooness." For most painters, realism therefore was limited, contingent and never complete. In the course of time, realistic details increasingly came to be seen by many as an impediment to painting that would capture the underlying essences of things.[21] This emphasizing of "the essential and the typical" led in turn to the emergence of a highly formalized, conventionalized pictorial language.[22] It was an approach that relied heavily on intuitive understanding to create a general, synthetic view and was not likely to encourage careful observation of details or mechanical analysis.[23]

But at least tools and machines, like everything else that exists, had their *li*. What they did not have was *qi* 氣, a concept that had a long and important history in Chinese thought but which now gained further importance in painting criticism because of its prominence in the new metaphysics. *Qi* embodies a whole range of ideas including "the vital force and substance of which man and the universe are made."[24] In connection with painting, it is perhaps best translated as "life-force." Already in the early fifth century, the scholar-painter Zong Bing 宗炳 (375–443) expressed in his "Introduction to the Painting of Landscape" (*Hua shanshui xu* 畫山水序) a sophisticated Daoist approach to landscape painting that helped prepare the way for later Confucian insistence on the need that a painting embody *qi*. As Michael Sullivan describes Zong's theory:

18. Kenneth R. Stunkel, "Technology and Values in Traditional China and the West," *Comparative Civilization Review* 23(1990), 88n9.
19. On this point, see especially Hegel, *Reading Illustrated Fiction*, 319–20. On the overall question, I have relied especially on: Sullivan, *Symbols of Eternity*, 71–74; Cahill, *The Compelling Image*, 45; Cahill, *The Painter's Practice*, 5–7; Bush, *Chinese Literati on Painting*, esp. Chap. 1; Maeda, "Spatial Enclosures," 385–7.
20. Wen C. Fong, *Images of the Mind* (Princeton: Princeton University Press, 1984), 5.
21. Wen Fong sees this as happening mainly from the Yuan onwards; *Images of the Mind*, 5.
22. Hegel, *Reading Illustrated Fiction*, 319–20, drawing on Sullivan, *Symbols of Eternity*, 71–72. See Hegel's further, elegant discussion there.
23. Sullivan, *Symbols of Eternity*, 71–72; Hegel, *Reading Illustrated Fiction*, 319–20; Needham and Wang, *SCC* 3, 163.
24. Conrad Schirokauer, *A Brief History of Chinese Civilization* (San Diego: Harcourt Brace Jovanovich, 2006), 149.

> [T]he forms of nature visible to the artist's eye are indeed the outward manifestations of the working of *[D]ao* [and] the artist is uniquely endowed with the faculty of apprehending these forms and delineating them in such a manner that the cosmic vitality which they embody is not lost but transferred to the surface of the painting.[25]

Or, as later Confucian critics or theorists would insist (essentially substituting *qi* for *Dao*), any painting to be regarded as a success must capture and reflect *qi*. Already in the ninth-century Zhang Yanyuan contended:

> With regard to terraces and pavilions, trees and rocks, carriages and palanquins, utensils and objects in general, they have no liveliness that can be imitated or *ch'i-yun* [*qiyun* 氣韻, "the breath of life and its reverberation," that is, sense of life] that can be matched. They only require placing and alignment and that is all.[26]

For those who held this view, and they were the majority of elite painters and critics in the post-Song period, there was obviously little payoff in satisfaction or reputation from the effort to develop the skills needed to picture these things realistically.[27]

Painting and Calligraphy

The growing emphasis on the personal qualities that literati painters were expected to reflect in their works rather than on their technical proficiency dovetailed nicely with the thrust of an overall education which worked to produce generalists, not specialists.[28] For example, we have spoken in Chapter 2 ("The Dominance of Brush and Line") of the transfer of the values of calligraphy to painting, which increasingly made it possible for painting to share in the unchallenged prestige calligraphy had previously achieved in China as a visual art. Painting, originally considered a craft, acquired a similar prestige only very slowly. It did so in large measure by adopting many of the aesthetic standards of calligraphy, especially the non-mimetic emphasis on line and brushwork.

The literati painters found these values most congenial since all of them out of necessity developed at least some proficiency in the techniques of calligraphy, especially the manipulation of the brush. They were naturally attracted by an approach to painting that allowed them to draw on that proficiency, all the more since painting was almost exclusively done

25. Sullivan, *The Birth of Landscape Painting in China*, 103–4, as cited in Acker, *Some T'ang and Pre-T'ang Texts*, 2, 122.
26. Bush, *Chinese Literati on Painting*, 16 (Bush's modification of Acker's translation: *T'ang and Pre-T'ang Texts*, 1, 150). Though Zhang does not say so, his prejudice against "utensils and objects in general" as subjects in painting likely derived in part from the fact that they often had to be portrayed with straight lines which were seen as necessarily lacking in *qi*/vitality. Yee ("Cartography in China," 154) notes that the same prejudice against straight lines led to avoidance of planimetric projections in Chinese cartography.
27. This was but one aspect of a general narrowing of range of subjects "respectable" literati painters from Song on were expected to deal with. It went hand-in-hand with declining interest among scholar painters in using their painting to explore the physical world. Cahill, *The Painter's Practice*, 115–23, esp. 117–18.
28. Cahill, "Approaches to Chinese Painting," in Yang, Barnhard et al., *Three Thousand Years*, 8.

with the same kind of brush as used in calligraphy. Added to this, many of the scholar painters simply did not have the time or the inclination to work hard for years to become highly proficient in painting techniques. Nothing could be more natural for them than to belittle the importance of technique and craftsmanship in favor of lively brushwork.[29] Moreover, their experience with calligraphy helped them to achieve the spontaneity that was central to the kind of expressivity they aimed for in their paintings.

Many literati painters now also took a technical cue from calligraphers and increasingly used paper instead of silk for their painting. Painting on silk required a relatively slow and cautious technique since a dry or fast-moving brush often did not leave a clear track on a silk surface. It therefore invited attention to fine details. By contrast, paper made possible more rapid and lively brushwork. In painting, therefore, it encouraged spontaneity over precision, the impressionistic over the objective.[30]

By the Yuan, these influences had so changed the definition of what constituted fine painting that Wen Fong can speak of "a new and different kind of pictorial art" and "transforming painting into the equivalent of calligraphy."[31] But the new ideals of literati painting, though largely triumphant by the late thirteenth and early fourteenth centuries,[32] were never entirely beyond challenge, then or later.[33] Throughout the Mongol Yuan dynasty, the demise of the imperial painting academy gave the scholarly elite a largely free rein to define the standards of fine painting.[34] The academy's re-establishment in the early Ming, however, brought a revived challenge from the professional painters it tended to favor.[35] By the end of the Ming, with declining court patronage of professional painters, scholarly values were again on the rise.[36] But as specifically painterly skills in China broadly fell into a decline,

29. Chou Ju-hsi, "In Defense of Qing Orthodoxy," in Barnhard et al., *The Jade Studio* (New Haven, CT: Yale University Art Gallery, 1994), 39. This attitude was uncompromisingly expressed by Dong You 董逌 the famous early twelfth-century connoisseur and author of the *Guangchuan huaba* 廣川畫跋 (Dong You's colophons on painting): "The artisan-painter makes his work salable by his skillful craftsmanship; by giving pleasure to the vulgar ones of his time, he hopes to make his picture easier to take. He is afraid only that the world will not want his pictures because they are different." Cahill, "Confucian Elements," 138.

30. Silbergeld, *Chinese Painting Style*, 8–9. Paper was also less expensive than silk, certainly a consideration in an environment where very large numbers of paintings were hardly more than casual sketches tossed off very rapidly. For different kinds of paper for different kinds of painting strokes, see Julia Hutt, *Understanding Far Eastern Art* (New York: E. P. Dutton, 1987), 23–24.

31. Fong, *Beyond Representation*, 440.

32. Fong, *Beyond Representation*, 7, 431.

33. Cahill, *The Painter's Practice*, 14–15. Kathlyn Maurean Liscomb, *Learning from Mount Hua* (Cambridge: Cambridge University Press, 1993), *passim*; Clunas, *Art in China*, 153–54. Liscomb's Chapter 5 is especially good for a subtle analysis of the Yuan and Ming debates between those favoring expression and those who favored representation. She shows that it was frequently not an either/or question but rather a dispute over the relative importance to be attached to one or the other quality. Thus the scholar-painter Zhu Tong 朱同 (1338–85) could recognize the importance of brushwork in painting but then ask "if one devotes all of one's efforts to calligraphic methods and slights representation, how does this pertain to painting?" (p. 100).

34. Fong, *Beyond Representation*, 431.

35. Bush, *Chinese Literati on Painting*, 154.

36. Bush, *Chinese Literati on Painting*, 181.

sometimes to a very low level, trenchant criticisms of contemporary literati painting arose from various directions.[37] For example, the amateur painter Li Rihua 李日華 (1565–1635) and the collector Xie Zhaozhi 謝肇淛 (1567–1624) mounted attacks against "sloppiness" and lack of skill in paintings.[38] Still, even at this time, outstanding literati artists such as Mo Shilong 莫是龍 (1552–87) and Dong Qichang 董其昌 (1555–1636) felt the need to launch yet another critique of the inclusion of too much detail in paintings, in this case, landscape paintings.[39] In the Qing, at least through the seventeenth century, professional painting continued to be widely disparaged by those whose voices most influenced what was accepted as quality painting.[40] So it was that the values of literati painting, with their resolute denigration of similitude and realism, came to form a remarkably powerful and enduring orthodoxy highly successful in defining for the five centuries of the Ming and Qing what kind of painting was most worthy of admiration. It was not painting likely to produce distinguished depictions of technology.

37. Barnhard, *Painters of the Great Ming*, 300.

38. Cahill, *Three Alternative Histories*, 97–98 for these and other comments. "Sloppiness" is Cahill's word.

39. Nelson Wu, "Tung Ch'i-ch'ang [Dong Qichang]," in Arthur F. Wright and Denis Twitchett (eds.), *Confucian Personalities* (Stanford: Stanford University Press, 1962), 272–74.

40. Hegel, *Reading Illustrated Fiction*, 250–53; 270; Wu, "Tung Ch'i-ch'ang," 275.

Five

Late Ming and *The Exploitation of the Works of Nature*[1]

A Chinese living in the mid-or late 1600s, on entering a bookstore in some southeastern Chinese city, perhaps Suzhou, could well have come across a volume bearing the somewhat cryptic title *The Exploitation of the Works of Nature* (*Tian gong kai wu* 天工開物). On examining it, he would have found on the cover page an advertisement (Plate 10) claiming that the book contained "all kinds of money-making and ever-useful professional secrets and essential instructions on agriculture, weaving, manufacture, mining and working of metals and treasures."[2] If this piqued his interest, he might look further into the book and find out from the preface by the author, a man named Song Yingxing 宋應星, that the book aimed to provide a straightforward account of those everyday working activities that provided people with food, clothing, housing, vehicles, tools. Skimming the pages, his eyes would be caught repeatedly by attractive illustrations, over a hundred of them. Perhaps he would decide to buy the book. If so, he would then possess the most complete and competent account of traditional Chinese technology that had yet been written, one indeed that would never be surpassed.

1. This chapter is a revised version of Golas, "Like Obtaining a Great Treasure: the Illustrations of Song Yingxing's *The Exploitation of the Works of Nature*," in Bray et al., *Graphics and Text*, 569–614.

2. Pan Jixing 潘吉星. *Tian gong kai wu jiao zhu ji yanjiu* 天工開物校注及研究 [Studies on an edited and annotated edition of the *Tian gong kai wu*] (Chengdu: Ba Shu shushe, 1989), 143, Fig. 1–9A; Yang Lien-sheng, "Review of Yabu'uchi Kiyoshi, *Tenkô kaibutsu no kenkyû*," *Harvard Journal of Asiatic Studies* 17 (1954), 307–16, 311. The translation is L. S. Yang's. Insofar as we know, it was in a revised version of the second edition of Song's work that this advertisement first appeared. For the editions of the *Tian gong kai wu*, see Golas, "'Like Obtaining a Great Treasure': The Illustrations of Song Yingxing's *The Exploitation of the Works of Nature*," in Bray et al., 570, fn. 5, 594–97 and 614. For an excellent sample of the depth and range of topics in just one area of technology (weaving and dying textiles), see Cheng Weiji, *History of Textile Technology of Ancient China* (Rego Park, NY: Science Press New York, 1992).

The Exploitation of the Works of Nature: A Culmination of Sorts

The *Tian gong kai wu,* which we opt here to translate as "The Exploitation of the Works of Nature,"[3] appeared in print in 1637, only seven years before the fall of the Ming dynasty.[4] Treating not only the activities mentioned in the advertisement above but also many other productive activities such as ceramics, papermaking, weapons, deep drilling, shipbuilding, and the making of vehicles, it represented the author's attempt to explain most of the practical productive processes in China as he understood them, sometimes from personal experience, sometimes from the accounts of informants, sometimes from books.[5] Its often detailed textual explanations, supplemented by its profuse illustrations–122 line drawings in the original edition–provide much of the information on which our understanding of traditional Chinese technology rests.[6] The illustrations are an even more useful resource because, for every one of the illustrations of the original edition, we have at least one later illustration, typically a copy, of the same subject. Often the differences between the earlier and later illustrations are relatively minor. Sometimes, however, they are considerable, both in content and in quality of execution. Identifying and trying to account for these differences can tell us a great deal about the capacities and the limitations of Chinese illustration of technology from the seventeenth to the nineteenth centuries.

3. Craig Clunas' translation, "Heaven's Craft in the Creation of Things," crucially misses Song's point that it is the joint activity of Heaven and humans that creates all useful things; see Clunas, *Superfluous Things: Material Culture and Social Status in Early Modern China* (Cambridge: Polity Press, 1991), 166. The same can be said for Dagmar Schäfer's "The Works of Heaven and the Inception of Things" which, as a literal rendering, is unassailable; Schäfer, *Crafting,* 2–3; Bray, "Introduction," 58–59. Clunas later modified his translation to "The Creations of Nature and Craft" ("Luxury Knowledge," 35), a much more satisfying rendering. In contrast to most Chinese works with difficult titles where the author explains, usually in his preface, why he chose that title, Song gives no such explanation; Yabu'uchi Kiyoshi, *Tenkô kaibutsu* [(Translation of the) *Tian gong kai wu*] (Tokyo: Heibonsha, 1969), 362. I have chosen to follow Joseph Needham and use "The Exploitation of the Works of Nature," a translation with which Yabu'uchi agrees. The sense is essentially human beings utilizing resources provided by nature to devise and produce useful things; Pan, *Studies,* 209. See also the discussion in Saigusa Hiroto (ed.), *Tenkô kaibutsu* 天工開物 [The *Tian gong kaiwu* of Song Yingxing] (Tokyo: Juichigumi shuppanbu, 1943), 18–21 and Yang Lien-sheng, *Studies in Chinese Institutional History* (Cambridge, MA: Harvard University Press, 1963), 62.
4. For references to the original edition of the *Tian gong kai wu* (i.e., TGKW), I have used the excellent 1959 Beijing facsimile printing by Zhonghua shuju.
5. By one count, the book deals with over thirty different technologies. Pan Jixing, *Song Yingxing pingzhuan* [A critical biography of Song Yingxing] (Nanjing: Nanjing daxue chubanshe, 1990), 457. Endymion Wilkinson accurately characterizes it as China's "most comprehensive work on industrial and agrarian arts." Wilkinson, *Chinese History: A Manual* (Cambridge, MA: Harvard University Press, 2000), 667. The terminology is well-chosen: as Françoise Saban emphasizes, the book includes much more information on productive techniques than on "technology" with a narrow focus on implements, machines etc. Saban, "L'industrie sucrière, le moulin a sucre et les relations sino-portugaises aux XVIe–XVIIIe siècles," *Annales, Histoire, Science Sociales* 49.4 (July–August 1994), 834.
6. This is especially true both because traditional Chinese technology had to a large extent reached its maturity by the early seventeenth century and because no other author before the twentieth century ever again attempted a technological compendium of such scope.

Song Yingxing and His World

Before turning to the illustrations themselves, a look at conditions in early seventeenth-century China will help us to understand Song Yingxing and how he was led to produce a work as unique as *The Exploitation of the Works of Nature*. The world in which Song grew up and spent most of his adult life coincided with the last half-century of the Ming dynasty. In hindsight, it was a period marked by a series of crises that culminated ultimately in the collapse of the dynasty in 1644. Since at least the 1590s, the government in Beijing, nominally headed by the feckless Wanli emperor (r. 1572–1620),[7] had been fiscally crippled by a free-spending court and imperial nobility, as well as by bloated military expenses. Its effectiveness was further undercut by pervasive corruption and bitter factional infighting among its officials and between officials and eunuchs, many of whom used their powers to manipulate government policies to their own financial advantage. Much of the countryside was sunk in poverty with farmers suffering the dual depredations of exploitative landlords and high taxes as well as unusually capricious weather conditions.[8] Threats from waves of internal rebellion[9] were exacerbated by the ever-deepening shadow of rising Manchu military power on the northern borders.

Yet all of this was happening in a period of considerable economic prosperity and luxurious living for many in the cities who profited handsomely from unprecedented levels of economic productivity as well as from a great influx of new world silver brought to China to pay for the silks, porcelains and other luxury goods that had captured the fancy of wealthy Europeans.[10] It was this flourishing urban sector that made possible the vibrant and growing book publishing industry that, among other things, produced the "new urban culture that sought its inspiration outside scholarly traditions."[11]

Song himself was born into a family with a tradition of government service but which now found itself in reduced circumstances.[12] He spent long years of study preparing for

7. A striking example was his allowing some half of local government posts to remain unfilled; Konrad Herrmann (trans.), *Erschliessung der himmlischen Schätze* (Bremerhaven: Wirtschaftsverlag NW, Verlag für neue Wissenschaft, 2004), 306.
8. These conditions seem to have been partly the result of a worldwide "little ice age" at this time; Joanna Waley-Cohen, *The Sextants of Beijing: Global Currents in Chinese History* (New York: Norton, 1999), 59.
9. Even as Song was finishing his *Tian gong kai wu*, rebel leaders were forging more unified command structures that would enable them in a mere three or four years to take over control of whole provinces. Gernet, *History of Chinese Civilization*, 434. See also the very useful maps in Frederick W. Mote and Denis Twitchett (eds.), *The Cambridge History of China (CHC)*, vol. 7, part 1, "Ming Dynasty, 1368–1644" (Cambridge and New York: Cambridge University Press, 1988), 624–25.
10. See William S. Atwell, "The T'ai-ch'ang, T'ien-ch'i, and Ch'ung-chen Reigns," *CHC*, 7: 1, 587–88; Evelyn S. Rawski, "Economic and Social Foundations of Later Imperial Culture," in David Johnson, Andrew J. Nathan and Evelyn S. Rawski (eds.), *Popular Culture in Late Imperial China* (Berkeley and Los Angeles: University of California Press, 1985)," 3–4.
11. Catherine Jami, "'European Science in China' or 'Western Learning,'" *Science in Context* 12.3 (1999), 423.
12. Herrmann, *Erschliessung*, 308–12; Sun and Sun, *T'ien-kung k'ai-wu*, vii; Hummel, *Eminent Chinese*, Vol. 2, 690; Pan, *Critical Biography*, 66 ff.; Christopher Cullen, "The Science/Technology Interface in Seventeenth-Century China," *Bulletin of the School of Oriental and African Studies* 53.2 (1990): 299; Schäfer, *The Crafting of*

the civil service examinations that could open for him the door to an official career. This preparation, as we have noted above, focused almost entirely on Confucian teachings, among them the necessity of good, moral government for the maintenance of a decent society. His studies may have been facilitated by what appears to have been a photographic memory.[13] Nonetheless, he failed in five attempts to pass the highest level examinations that alone would have qualified him for high office.[14] Condemned at best to a mediocre career in low level posts, he decided to follow the precedent of centuries of officials before him who, despite undistinguished political careers, had gained renown and respect as thinkers and writers.[15] His choice may have been made easier by the thriving market for books that offered at least the possibility of significant earnings to supplement his official salary.[16]

Book Illustration in the Late Ming

In the sixteenth century, some eight or nine centuries after the invention of woodblock printing, printed books for the first time displaced manuscripts as the dominant form of reading matter throughout most of China.[17] The exploding demand for books was fueled by a rapidly growing and increasingly literate population.[18] High among the books in demand were textbooks for use in proliferating educational institutions (private academies, community schools, charity schools) and works designed to help increasing numbers of candidates prepare for the official examinations. For the first time, many students were able to own their own printed versions of the crucial classics as well as other study aids. The reading

---

the 10,000 Things, 24–29.

13. Pan, *Critical Biography*, 232.
14. Pan Jixing, *Tian gong kai wu daodu* [A guide to reading the *Tian gong kai wu*] (Chengdu: Ba Shu shushe, 1988), 9–10.
15. Indeed, most of his more than ten works (dealing with scientific and technological topics, politics and society, history and literary creativity, as well as his own literary essays and poems) were completed in the years from 1634 to 1638 when he served in a low-ranking and clearly not very demanding position as an education official in Fenyi, Jiangxi. Even more striking, the five of his works that survive were all written in 1636–1637. Schäfer, *The Crafting of the 10,000 Things*, Chap. 2; Pan, *Guide*, 10; Pan, *Critical Biography*, 234–35, 274 and all of Chap. 6.
16. Golas, "Like Obtaining a Great Treasure," 573.
17. Joseph McDermott's article on "The Ascendance of the Imprint in China" consolidates a great deal of research (by scholars such as Ming-sun Poon, Susan Cherniak, Jean-Pierre Drège, Ôki Yasushi and, especially, Inoue Susumu) that has as a major thrust the revision downward of the extent to which printed books were available to Chinese in earlier periods such as the Song. Overall, this article superbly locates late Ming developments in the long history of book culture in China. See also Tobie Meyer-Fong, "The Printed Word: Books, Publishing, Culture and Society in Late Imperial China," *Journal of Asian Studies* 66.3 (August 2007), especially 792–93; and Lucille Chia, *Printing for Profit: the Commercial Publishers of Jianyang, Fujian (11th to 17th Centuries)* (Cambridge, MA: Harvard University Press, 2002), 12–13.
18. For a discussion of how population growth need not have resulted in a greater demand for books but in fact did, see Brokaw, "History of the Book," 10–11. Indeed, it was much earlier, in the twelfth and thirteenth centuries, another period of rapid population growth, that a vigorous, highly competitive book market first made its appearance in China. Chia, "Text and *Tu*," 242.

population at large could also access a wide range of books, many of them now written in the vernacular language. They included medical manuals, travel books and accounts of foreign countries, household encyclopedias, manuals for writing letters, popular law books, handbooks on gardening, almanacs, and texts on religion, morality, music and good manners. Also popular were books produced solely for entertainment: collections of stories and anecdotes, texts of plays, poetry anthologies and historical novels.[19]

As a true mass market emerged, the cost of printing books fell to one-tenth of what it had been two or three centuries earlier.[20] The great Jesuit missionary, Matteo Ricci, attributed the "ridiculously low prices" of books in China in the late Ming to the simplicity of the Chinese woodblock printing process.[21] He was at least partly right. Other reasons included the low cost of materials, especially paper, and of labor; the designing of new fonts that made block carving quicker and easier;[22] and the fact that a single carved block might be used for as many as 30,000 impressions and one laborer might pull from 1000 to 6000 copies in a day.[23]

But if seventeenth-century China was a world of print, it was also a "world of images"[24] in which the public increasingly had access to books with illustrations. In a highly competitive market, illustrations could increase a book's market appeal.[25] In some cases, illustrations were probably indispensable simply for a book to hold its own against competitors. Even the growing absence in book titles of indications that they contained illustrations may well imply that illustrations were increasingly taken for granted.[26]

In quality editions for elite readers, it was often specialized artists and carvers who collaborated to produce increasingly refined and complex full-page illustrations that at their best were never surpassed in the following centuries.[27] Given a significant population of highly trained and even renowned professional painters who applied their skills now to

---

19. Chow, "Writing for Success: Printing, Examinations, and Intellectual Change in Late Ming China," *Late Imperial China*, 17.1 (June 1996): 123.
20. See Timothy Brook, "Communications and Commerce," in *CHC* 8:2, 636–37, 649; Tadao Sakai, "Confucianism and Popular Educational Works," in Wm. Theodore de Bary, *Self and Society in Ming Thought* (New York: Columbia University Press, 1970), 331–39; Chow, "Writing for Success," 124.
21. Rawski, "Economic and Social Foundations," 17, citing Gallagher, *China in the Sixteenth Century*, 21.
22. Francesca Bray, *Technology and Society in Ming China (1368–1644)* (Baltimore: Society for the History of Technology, 2000), 11.
23. Evelyn Sakakida Rawski, *Education and Popular Literacy in Ch'ing China*,(Ann Arbor: University of Michigan Press, 1979), 120; Hegel, *Reading Illustrated Fiction*, 123. See also Joseph McDermott, "The Ascendance of the Imprint in China," in Brokaw and Chow, *Printing and Book Culture in Imperial China*, 59, 78–80. By never mechanizing woodblock printing, which would have increased the wear and tear on the woodblocks, the Chinese maximized the number of copies they could draw from a single woodblock.
24. Clunas, *Art in China*, 177.
25. Clunas, *Pictures and Visuality*, 32–33.
26. Clunas, *Pictures and Visuality*, 34; Chia, "*Mashaben*," 317; Bray, "Introduction," 45.
27. Cahill ("Chinese Painting: Innovation after Progress Ends," 186) sees the late Ming as China's "peak period of pictorial woodblock printing." See also Cahill, *The Painter's Practice*, 107ff.; Wang Bomin, *Zhongguo banhua shi* [A history of Chinese woodblock illustrations] (Shanghai: Renmin chubanshe, 1961), 83; Hegel, *Reading Illustrated Fiction*, 192–93 and 407n23. Some carvers also designed illustrations; Wang, *Chinese Woodblock Illustrations*, 73.

painting, now to designing book illustrations, we should probably not be surprised that this was precisely the period when professional painting drew fierce attacks from many of the more prestigious scholar-amateur painters whose emphasis on the calligraphic aspects of painting along with their disdain for visualistic illusion had brought their representational techniques to something of a nadir.[28] On the other hand, such disdain was far from universal and we have many examples of major painting masters who were directly involved in the making of book illustrations.[29] Again, the absence in China of a clear distinction between drawing and painting facilitated a crossing of borders. Thus, a painter like Du Jin 杜堇 (fl. c. 1465–1500) could be primarily renowned for his *baimiao* or ink paintings using strong outlines essentially the same as the technique used almost universally for book illustrations (Plate 11).[30] Books aimed at the lower end of the market presented quite a different aspect (Fig. 5.1).[31] A large proportion of the cheaper books were of very poor quality, both in appearance and in content.[32] Here considerations of costs, sales and profits predominated, leading to cost-cutting measures such as reliance on less able artists for designs, the use of poorer quality materials, especially paper, and settling for perfunctory or no proofreading. The growing numbers and mobility[33] of woodblock carvers offered an option for reducing costs on what was usually the major production expense in publishing books.[34] Carvers varied considerably in their skill levels but, once they had gone through two or three years of training,[35] all were basically capable of transferring both text and images to woodblocks for printing. Publishers could save on labor costs by selecting less skilled but adequate carvers and providing them with designs that were clear and simple; the illustrations that resulted might be quite weak on detail and realism but still be an attraction for readers.[36]

28. Barnhart, *Painters of the Great Ming*, 293, 295, 300. The criticisms were undoubtedly intensified not only by the fact that commercial painting, along with other commercial arts such as ceramics, often shared stylistic parallels with works done by the literati (Hegel, *Reading Illustrated Fiction*, 250) but also by the prestige enjoyed by the best of professional painters especially if they were classically educated (Hegel, *Reading Illustrated Fiction*, 109–10, 192, 281) or, as was sometimes the case, they were so adept at imitating great literati paintings of the past that their copies were often accepted as originals even down to very recent times (Escande, "Perspectives et limites," 233).
29. Fong, *Beyond Representation*, 496; Wang Fang-yu, "Book Illustration in Late Ming and Early Qing China," in Edgren, *Chinese Rare Books*, 31; Thorp and Vinograd, *Chinese Art and Culture*, 321.
30. Hegel, *Reading Illustrated Fiction*, 277.
31. An indispensable reference for lower-market books is Chia, *Printing for Profit*.
32. For a scathing criticism by the mid-sixteenth century scholar and connoisseur, Lang Ying 郎瑛, of the general quality of books being produced in his time, see Clunas, *Pictures and Visuality*, 32–33.
33. Clunas, *Superfluous Things*, 14.
34. Cynthia Brokaw, "Commercial Publishing in Late Imperial China: The Zou and Ma Family Businesses of Sibao, Fujian," *Late Imperial China* 17.1 (June 1996): 80; Chia, *Printing for Profit*, 39. As many as a hundred carvers might be employed to produce a single book, each carving perhaps 100 characters per day. Chia, *Printing for Profit*, 37 and 333n52. Roger Cartier notes that the situation in China differed significantly from that of Europe where paper represented the major cost–rising as high as almost fifty percent–of printing an edition. Roger Cartier, "Gutenberg Revisited from the East," *Late Imperial China* 17.1 (June 1996), 3.
35. Chia, *Printing for Profit*, 34 and 332n47.
36. Flawed illustrations could result not only from lack of skill on the part of illustrators but also from

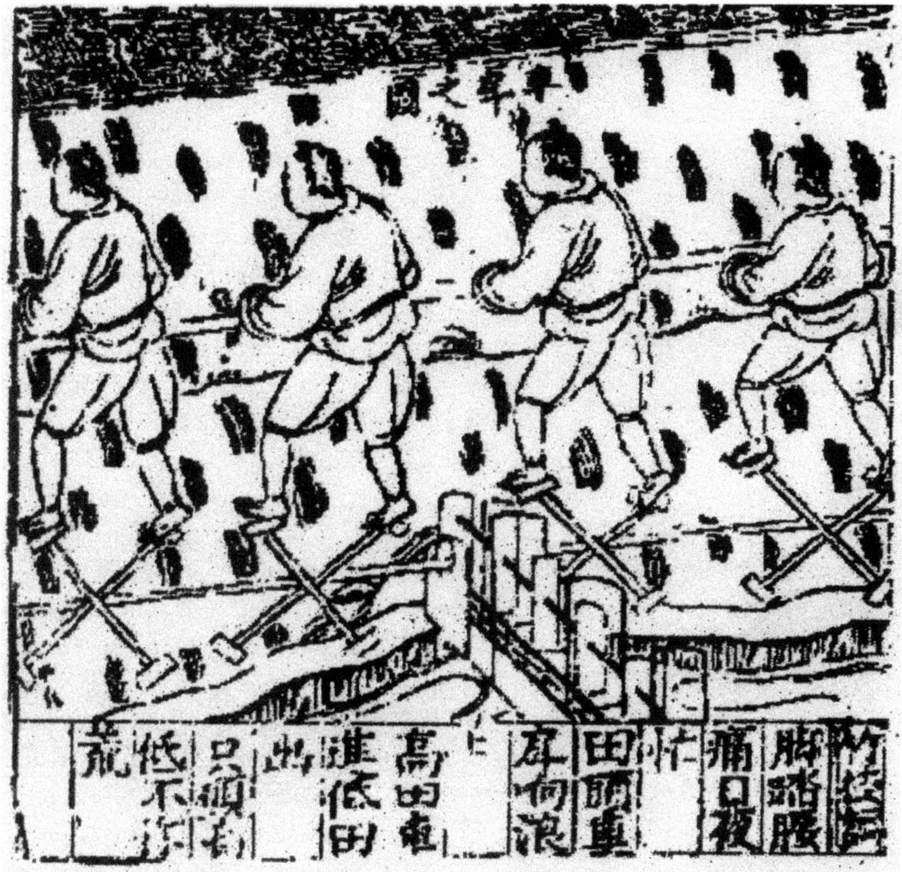

5.1 An example of the poor quality of technical illustrations in so many of the popular encyclopedias of the Ming and Qing, in this case a portrayal of a square-pallet chain pump from the 1607 *Seas of Knowledge and Mines of Jade: Encyclopedia for Convenient Use* (*Bianyong xuehai qun yu* 便用學海群玉).

We hardly need be surprised then that most of the technological subjects appearing in book illustrations were also treated cursorily and conventionally. Robert Hegel has noted that, in works of fiction, ". . . the quality of book illustrations was adjudged by the degree of perfection of the variations on the conventional rather than by uniqueness; repetition of familiar stock elements allowed quick recognition and ease of comprehension and appreciation."[37] The designers of book illustrations usually found no reason to treat technological subjects any differently or with any special care. The effort to do so would have been of no particular interest to them, nor would *technologically* more successful illustrations have impressed most prospective readers. Song Yingxing's work, even though it was the most

misunderstanding on the part of carvers. Ruitenbeek provides an example of a poor illustration that resulted because the carver who did a recutting had "no understanding of perspective." Ruitenbeek, *Carpentry and Building*, 126n8.

37. Hegel, *Reading Illustrated Fiction*, 312. See also Cahill, "Approaches to Chinese Painting," 8.

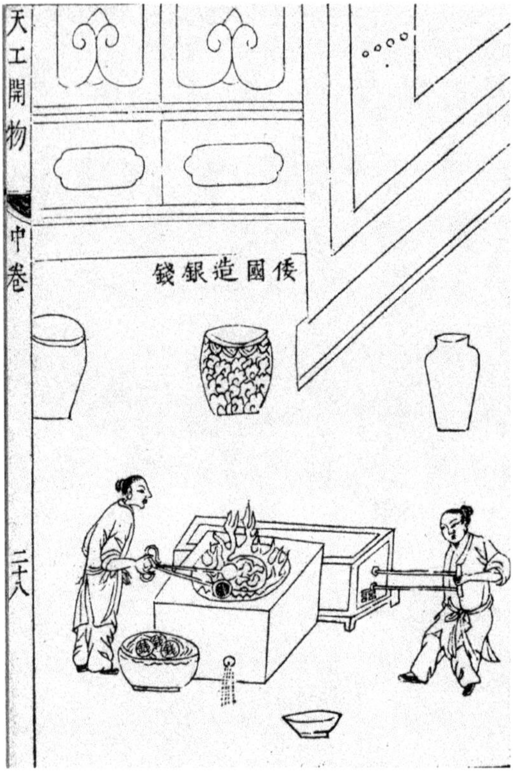

5.2 Illustration from Song Yingxing's work purporting to show use of the box bellows *in Japan(!)* in the casting of individual silver coins; Sun and Sun, *T'ien-kung k'ai-wu*, 168, Fig. 8–7. For its use in other processes, see ibid., 161, Fig. 8.2; 164, Fig. 8–4; 166, Fig. 8–5; 239, Fig. 14–2; 240, Fig. 14–4; 244, Fig. 14–6.

successful compendium of Chinese technology written in traditional times, was not immune to similar limitations in its illustrations. We see just how pronounced these limitations could be by comparing a traditional style Japanese illustration of the working of a double-acting box bellows with any of the many illustrations of the same process found in the *Exploitation of the Works of Nature* (Plate 12 and Fig. 5.2).[38]

Why This Work?

Determining an author's motives for writing can sometimes be relatively straightforward. For *The Exploitation of the Works of Nature,* it is not.[39] Song's preface states explicitly his intent to give readers of wealth and high position a keener awareness of what went into providing them with their daily necessities as well as their luxuries.[40] He also clearly wanted to offer his thoughts

38. It is worth noting that, in a period when Jesuit missionaries introduced Renaissance artistic techniques on a significant scale in both Japan and China, the Japanese for whatever reasons showed themselves extraordinarily more adept at imitating Western visual arts. As one of the missionaries, Pedro Gómez, commented in 1594, some of the young Japanese at an art academy set up by the Jesuits "draw most naturally paintings of the finest quality which the Japanese [ambassadors] brought from Rome, with such perfection both in color and form, that afterwards, among our own fathers and brothers, many could not tell which were the ones they made and which had been done in Rome. And some declared that those made by the Japanese were the ones which had come from Rome." Gauvin Alexander Bailey, *Art on the Jesuit Missions in Asia and Latin America, 1542–1773* (Toronto: University of Toronto Press, 1999), 68. Though, in Bailey's words, "the historical sources fall over each other praising the skills of the Japanese painters in copying" (ibid., 73), I know of no example of such praise by Westerners for any Chinese artist or artists, at this time or later.

39. Dagmar Schäfer's *The Crafting of the 10,000 Things* (Chicago: University of Chicago Press, 2011) goes into this question in much greater depth than we can here.

40. It is useful to note the distinction, newly prominent in the late Ming, between works that sought to convey how something was done and those that explained how to do it. Craig Clunas, "Luxury Knowledge: the *Xiushilu* (Records of lacquering) of 1625," *Techniques et cultures* 29 (1997), 34. Song's aim was mainly to tell his readers how things were done.

on what we would today call the philosophy of technology. These are usually ignored by contemporary scholars but ideas such as the view of technology as a tripartite composite of methods 法, skill 巧 and implements 器, with human skill seen as the most crucial/active/energetic (*jiji* 積極) component,[41] the role of nature's creations as models for technology (clouds and flowers inspiring the invention of dyes),[42] and the conviction that technological creation is basically a superhuman achievement that derives from the power of Heaven but can then be drawn on by humans[43] were clearly of great importance to Song.[44] Add to this the archaic and abstruse vocabulary that Song often employs and it seems arguable that he was writing *in the first instance* for people much like himself insofar as they were members of the highly educated elite and had, for whatever reasons, some interest in technological matters.

One of those reasons could well have been the changing intellectual climate at this time. Spurred on by the host of problems China was facing, many in the elite were turning their attention to what they could do, either in an official role or acting privately, to improve social and economic conditions. The new attitudes ranged from a greater emphasis on philanthropic activities,[45] to a new willingness to accept the legitimacy of profit-taking by merchants,[46] to a growing interest in scientific and technological knowledge that had practical applications.[47] It is this last trend that no one exemplified better than Song Yingxing in his desire to promote better technology that could raise overall prosperity as well as perhaps ease the labor demands on those who used it.[48] Even absent any such indication in the preface, Song's book was surely directed also and in no small measure at other scholar-officials who shared his concerns for the economic welfare of the people and the country.[49]

41. Pan, *Critical Biography*, 455–56.
42. Sun and Sun, *T'ien-kung k'ai-wu*, 73.
43. Yabu'uchi, *Tenkô kaibutsu* , 364.
44. I would therefore qualify the assessment of Mote and Twitchett (*CHC*, 8:2, 4) that the *Exploitation of the Works of Nature* is "altogether practical in its focus." Overwhelmingly practical to be sure, but Song was also quite concerned to place technology in a larger philosophical context, as is apparent in the brief essays that introduce each chapter (*juan*) and in many comments throughout the text.
45. Jacques Gernet, "La société chinoise a la fin des Ming," *Recherches de science religieuse* 72.1 (1984), 27–36, 33–34.
46. Timothy Brook, "Weber, Mencius, and the History of Chinese Capitalism," *Asian Perspective* 19.1 (1995), 79–97. For the limits of such change, however, see Brook's *Confusions of Pleasure*.
47. Gernet, *History of Chinese Civilization*, 441–46; Pan, *Studies*, 92, 102–3; Gernet, "La société chinoise a la fin des Ming," 32–33. Even publishers sometimes shared with their authors a desire to produce books that would "improve the world." Kenneth Pomeranz, *The Great Divergence: Europe, China and the Making of the Modern World Economy* (Princeton: Princeton University Press, 2000), 44, drawing on the work of Ellen Widmer.
48. Yang, Lien-sheng, *Studies*, 63. Song's very arrangement of the topics of the book is chosen, as he makes clear in the preface, to rank technologies according to their importance for the well-being of the society at large. The book discusses them in that order. Agriculture, of broadest importance for the lives of all the people, comes first. It is followed by clothmaking and then other technologies of lesser importance, though the order at times seems to be rather arbitrary. The book ends with a chapter on jade and pearls, luxury adornments of the elite. Pan, *Studies*, 52.
49. Scholars like Wang Zheng 王徵 who, ten years earlier, had made clear in the preface to his *Qiqi tushuo* that

On the other hand, there is no lack of evidence suggesting that Song also had in mind a broader audience beyond the elite, however defined. This was an audience much less erudite and so it is probably of some relevance that Song rarely quotes from or even refers to earlier writings on the topics he discusses.[50] This broader audience would also have been much closer to the actual production of goods than most of the elite. Although almost surely not written by Song, the advertisement referred to at the beginning of this chapter certainly seems aimed at an audience with practical reasons for reading the book.[51] That might include merchants, artisans and craftspeople, and even farmers, many more of whom were now literate and buying books. Concern for this audience is also suggested by the resolutely practical, down-to-earth thrust of most of the book. For example, the topics Song regularly discussed about any given technology included raw materials, areas of production, the technical processes involved, essential operations, equipment used, consumption of raw and processed materials, special characteristics of the products and their uses.[52] Readers keenly interested in this information would presumably care less about the aesthetic qualities of the illustrations. But they would no doubt have very much appreciated the punctuation that was added to the second edition of the work, published between 1650 and 1680.[53] This makes one wonder whether the book, once published, might not have had a greater attraction for this broader audience than Song had anticipated.

Hints as to how Song viewed his intended audience may also be found in what he chose to de-emphasize or omit. He seems to have had limited interest in technology itself as an object of intellectual investigation and thus tends not to devote much attention to its narrowly mechanical aspects. The construction and functioning of the implements and machines is typically treated cursorily or omitted altogether. He also makes it clear that he will not discuss technologies that are on the way out such as "Japanese" satin, which is losing favor because it is not durable and soils easily.[54] Nor does he necessarily give a good or complete picture of the scale of a given technology. For example, his portrayal of papermaking (which we shall discuss in some detail later in this chapter) focuses on a small-scale craft operation, which is fine as far as it goes. This was probably how most paper was produced in

he wished to introduce only those Western machines that could be of use in people's daily lives (*minsheng er yong* 民生而用) and only those innovations for which there was a real need in Chinese production processes (*guojia gongzuo zhi jixu* 國家工作之急需). *Qiqi tushuo*, 奇器圖說 [Illustrations and explanations of wonderful machines] (Siku quanshu ed.) 10 (3b), 11 (4a). We discuss Wang further in the next chapter.

50. Christopher Cullen comments that this would have been regarded by highly educated readers an "an odd and unscholarly way to write." Cullen, "Science/Technology Interface," 298, fn. 11.

51. Note that this advertisement need not have been aimed only at potential readers. It could also have been meant to appeal to book merchants who visited printing shops to purchase books that they would take to other areas to sell. Cynthia J. Brokaw, "Commercial Publishing," 73–74.

52. By contrast, Song does not ordinarily indicate just where the processes he describes were used or provide economic information on costs and the like. Most frustrating for modern scholars is his general silence, apart from telling us he relied both on books and observation, about how and where he got his information. Wagner, "Iron Production in Three Ming Texts"; Sun and Sun, *T'ien-kung k'ai-wu*, xiv.

53. Golas, "Like Obtaining a Great Treasure," 594–95.

54. Sun and Sun, *T'ien-kung k'ai-wu*, 60.

his day. However, we know that at the end of the sixteenth century, in addition to the many small production operations, there were also some thirty paper factories in Jiangxi employing perhaps around 50,000 workers.[55] One learns from Song's work nothing about the practices in these factories. Finally and for whatever reasons, the *Exploitation of the Works of Nature,* despite the breadth of its coverage, still omits, without explanation, some very important technologies: printing, irrigation engineering, construction of buildings, and the growing and processing of tea.[56] It is hard to avoid the conclusion that, to some extent, Song simply included what he felt like including or, to put it more gently, he never developed clear standards for inclusion or exclusion of technologies just as he never thought as hard as he might have about the technological content of his illustrations.[57]

Assessing the Illustrations

The overall perspective on technology that Song brought to his writing is well reflected in the book's illustrations. The key idea that technology involved human beings employing their skills to make use of the resources provided by nature gives us a clue for understanding something of a contrast between his illustrations and those, for example, of Wang Zhen in his *Agricultural Treatise* three centuries earlier. A clear majority of illustrations in the *Agricultural Treatise,* as we have noted earlier, are stand-alone pictures of individual tools and machines. This was about as close as the Chinese came in the traditional period to what we tend to expect of true "technical illustrations." Landscape elements are rarely included, and then typically have some direct connection with the technology being portrayed (farming, water control, etc.). Portrayals of people actually using tools or operating machines constitute a clear minority of all the illustrations in contrast to the *Exploitation* illustrations, nearly every one of which contains one or more figures doing something productive. In an extreme case, thirteen workers cooperate in making an anchor (Fig. 5.3). Though often involving the use of tools and machines, the primary emphasis tends always to be on the process, and not on the equipment used to carry it out. Even the illustrations of the *Exploitation* that show tools and machines (there are many that do not) should be viewed *mainly* as attempts to portray the employment of technology by humans. To use Bray's terminology, they are

55. Gernet, *History,* 426.

56. By contrast, this work includes the only substantial treatment of Chinese mining technology that has survived from before the mid-nineteenth century.

57. It is also a little difficult to know what to make of the fact the Song originally prepared two further chapters on astronomy and music, and then decided to omit them, saying that their *dao* (principles) were too refined and he was not sure of his competence to deal with them. Sun and Sun, *T'ien-kung k'ai-wu,* xiv. Pan Jixing suggests that Song also realized they did not fit in well with the subject matter of the rest of the book. Pan, *Critical Biography,* 246. Unfortunately, these chapters were later lost and so we do not know how Song intended to treat these subjects. Conceivably, they could have focused on astronomical and musical instruments, in which case they might have fit in quite well. If Song, however, had in mind more theoretical treatments, as that refined *dao* may suggest, the chapters would indeed have been something of an anomaly since these were activities that did not result in the direct production of material objects.

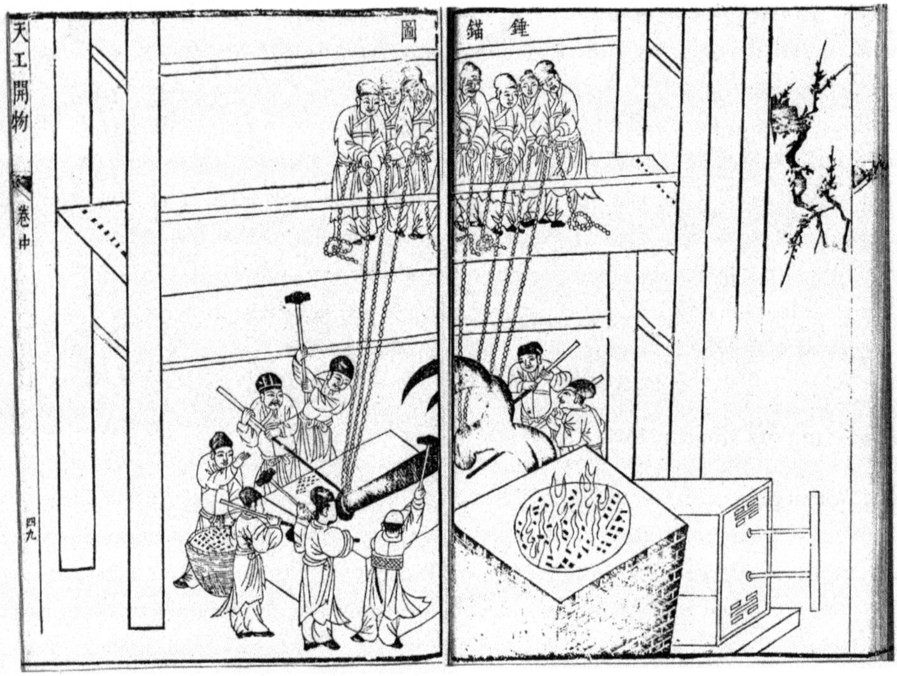

5.3 Forging an anchor; Sun and Sun, *T'ien-kung k'ai-wu*, 191, Fig. 10–12. (Use of a crane could almost have halved the number of workers needed!)

"social documents" more than "technological representations."[58] This view is hardly likely to encourage an emphasis on precise construction details or discussions of mechanical functioning although the inclusion of the human figures does have the unintended advantage of providing at least an approximate scale to which the other objects in the illustration tend to conform rather well.[59]

Song not only stressed the crucial role of human participation in technology but also believed that technology's primary purpose was to make people's lives better. He thus felt no compulsion to describe or illustrate all the technology with which he was familiar. For the most part he gives information only on the best standard practices found in current production technologies, omitting obsolescent or fading technologies such as the "Japanese" satin mentioned above.[60] He might also omit an illustration of a machine with which he felt people were generally familiar, such as a loom for weaving cotton; as he says, "It is not

58. Bray, *Technology and Gender*, 95.

59. Just as Chinese mapmakers found no pressing need to draw maps to scale since geographical details, including distances, could be included in the text accompanying the maps, so Song's work also reserves information on dimensions for the text, thus precluding the need for any precise or consistent scale in the illustrations. See the comments of Sivin and Ledyard in Harley and Woodward, *The History of Cartography*. vol. 2, bk. 2, 29 and Richard J. Smith, "Mapping China's World: Cultural Cartography in Late Imperial Times," in Wen-hsin Yeh, *Landscape, Culture and Power in Chinese Society* (Berkeley: Institute of East Asian Studies, 1998), 95.

60. This too is a reflection of Song's relatively limited interest in technology as an object of intellectual investigation for its own sake.

necessary to illustrate the loom for weaving cotton fabrics since, in any ten houses, one such is to be found."[61]

The quality as well as the quantity of the illustrations may also have been influenced by the expectations of Song's potential readers. Song, like most authors in this flourishing publishing culture surely wanted his book to sell well, not least because of his family's far from comfortable financial situation.[62] Illustrations, even not very good ones, could be a great help here.[63] Just the fact that the book had illustrations in abundance might for many readers be more important than their intrinsic quality.[64]

That would of course be less true for the highly educated elite whom we have identified as Song's main audience. Regardless of the intrinsic value they might assign to printed illustrations, they were much more accustomed to illustrations that were well done according to their tastes.[65] Conceivably it was in part a decision to appeal to elite tastes that led to some of the "anomalies" we find in the *Exploitation* illustrations like the landscape screens in working areas such as this tin-smelting workshop where one would not expect to find such an elaborate piece of decorative art (Fig. 5.4); or the clothing of the craftsmen forging an anchor which seems inappropriate for their work and perhaps above their station (see Fig. 5.3 above). At the same time, allowance must be made for the considerable growth of conventions, not to say stereotypes, which printing had promoted in the making of images of all kinds. This reliance on conventions was reinforced by the emergence of more or less well-defined schools of book illustration, each with its own stylistic markers and manuals to teach aspiring painters how to incorporate them.[66] By late Ming, the method for learning how to draw and paint in China was already well on its way to the situation that prevailed

---

61. That is, these looms are in common use. *Tian gong kai wu, shang*, 42a, drawing on translations in Sun and Sun, *T'ien-kung k'ai-wu*, 63 and Bray, *Technology and Gender*, 214.
62. Pan, *Guide*, 10–11.
63. Correspondingly, a lack of illustrations might well hinder sales of the book especially to that broader audience of merchants and craftsmen. Hegel, *Reading Illustrated Fiction*, 289; Clunas, *Pictures and Visuality*, 35.
64. Hegel, *Reading Illustrated Fiction*, 289.
65. The attitude of the elite toward illustrations in books (if indeed one can speak of a single elite with clearly definable tastes; Clunas, *Pictures and Visuality*, 40) is a matter that has only begun to attract serious study. For the difficulty of teasing out firm conclusions on the basis of the evidence that has so far been assembled, see Clunas, *Pictures and Visuality*, 33–37. See also Hegel, *Reading Illustrated Fiction*, 133, 164, 201, 204, 253, 257–58 and 270. Hegel tends to attribute to late Ming elite readers a higher appreciation of book illustrations than Clunas, arguing that "there was no necessary distinction in the perceived aesthetic value–or of the essential artistic features–of literati painting and of the more refined of the commercial arts, including fine woodblock printing" (253).
66. Clunas, *Art in China*, 177, 191; Clunas, *Pictures and Visuality*, 51, 136–38; Hegel, *Reading Illustrated Fiction*, 196. For a fascinating look at what has often been considered the first of these painting manuals, see Maggie Bickford, "Stirring the Pot of State: The Sung Picture-Book *Mei-Hua Hsi-Shen P'u* and Its Implications for Yüan Scholar-Painting," *Asia Major* (Third Series) 6.2 (1993), 169–226 and Figs. 1–5. Blunden and Elvin make the important point that these painting manuals inhibited the ability of Chinese painters to observe nature directly in its own right. *Cultural Atlas*, 146.

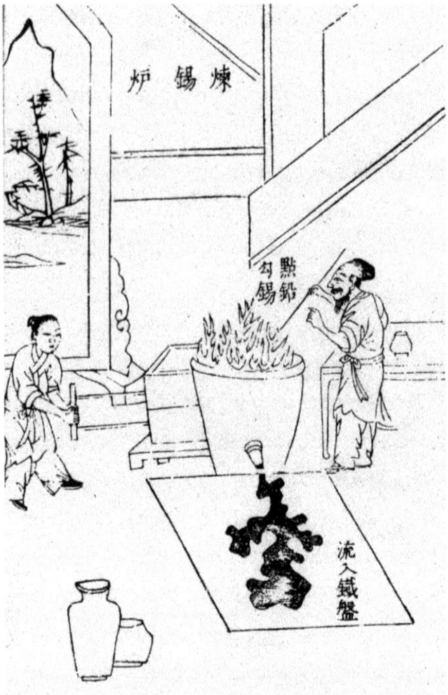

5.4 Smelting tin with the addition of lead; Sun and Sun, *T'ien-kung k'ai-wu*, 255, Fig. 14–13.

in later centuries when handbooks of style and technique came to be universally used for learning to paint.[67]

With all of their limitations, the illustrations of the *Exploitation* have nevertheless provided a rich lode of information for students of the history of technology in China, especially in the later imperial period. Although artistically they display a certain primitive quality and usually fail to measure up, for example, to some of the fine drawing of technological subjects being done in Japan around this time, there is no doubt that, even from the perspective of artistic technique, the illustrations often have much to recommend them. For example, the common if not universal practice of including a minimum of decorative elements serves to place the focus very strongly on the technological process being portrayed.[68] We can see this by comparing a typical illustration, here a weaver of silk (Fig. 5.5), with a rather untypical one (Fig. 5.6) of a man ginning cotton where the banana tree certainly adds to the aesthetic effect. At the same time, the presence of the tree somewhat subordinates the cotton ginning by moving it deeper into the picture, thus making it less immediate to the viewer. This tendency for the technical process to get lost is even more striking in the version of the same scene given in the 1927 Tao edition (Fig. 5.7).[69]

The original *Exploitation* illustrations of 1637 have sometimes elicited excessive praise. E-tu Zen Sun and Shiou-chuan Sun, in the preface to their English translation of the work, extol the original illustrations for their "simplicity and clarity" and dismiss the clear

67. Hegel, *Reading Illustrated Fiction*, 438n48.

68. Brian Scott Baigrie, *Picturing Knowledge: Historical and Philosophical Problems Concerning the Use of Art in Science* (Toronto: University of Toronto Press, 1996).

69. Dong, *Jiaozheng Tian gong kai wu*, 231. For the convenience of readers of English, I have provided references for the original illustrations to the easily available translation by the Sun and Sun's, which includes all the original illustrations. The Dong volume is a widely available edition in Chinese that reproduces all the illustrations of the 1927 Tao edition which substituted later illustrations for all the illustrations of the original edition while also adding 33 illustrations (most of which are included in the Sun and Sun's volume) with no counterparts in the original edition. See Golas, "Like Obtaining a Great Treasure," Appendix 1. It should be noted that the *Exploitation* illustrations in the *SCC*, 4:2 volume, though dated 1637, are actually illustrations from the Tao edition.

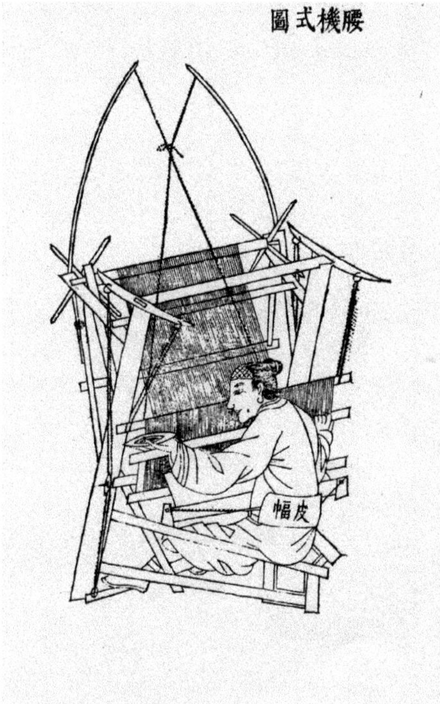

5.5 A relatively small waist loom for weaving silk; Sun and Sun, *T'ien-kung k'ai-wu*, 57, Fig. 2–14.

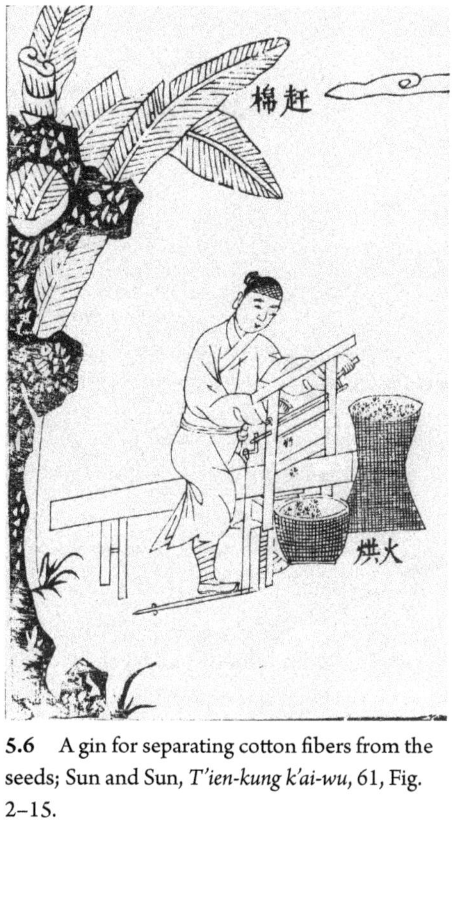

5.6 A gin for separating cotton fibers from the seeds; Sun and Sun, *T'ien-kung k'ai-wu*, 61, Fig. 2–15.

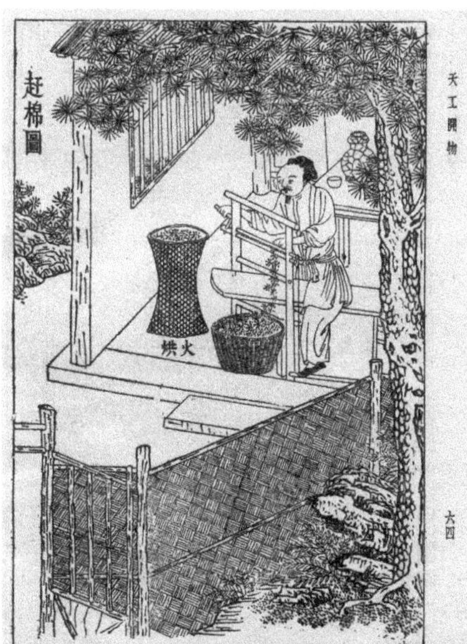

5.7 The process of ginning cotton as redrawn for the 1927 Tao edition of the *Exploitation*.

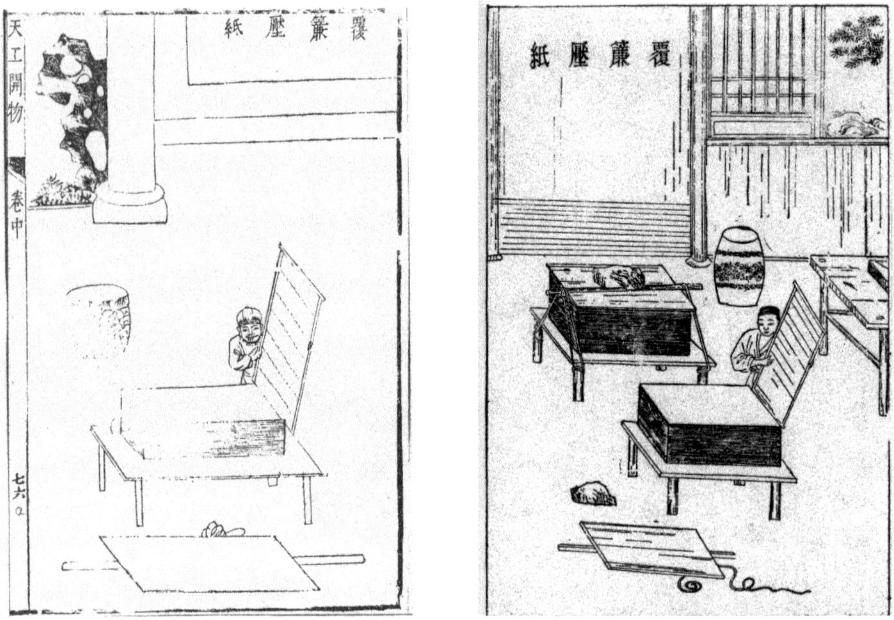

5.8 Removing and pressing paper sheets, from (a) Sun and Sun, *T'ien-kung k'ai-wu*, 228, Fig. 13–4; and (b) Dong, *Jiaozheng Tian gong kai wu*, 231.

superiority of many later versions as "perhaps to some eyes more pleasing." More explicitly, they also incorrectly claim that the later illustrations "add nothing to the technical subjects being illustrated."[70] While it is true that the later illustrations do tend to add aesthetic rather than technological detail, this is by no means always the case. Figs. 5.8 (a) and (b) represent the same papermaking scene from, respectively, the original edition and the 1927 Tao edition.[71] The scene shows the papermaker flipping the screen on which the pulp has been collected in order to add a sheet of paper to the stack of sheets waiting for pressing to remove the excess liquid. The original deals with the pressing only by including in the foreground barely identifiable implements used in the pressing process. By contrast the Tao illustration makes very clear what those objects were and how they were used by including in the middle ground a stack of paper undergoing pressing. We also realize that the implements in the foreground (including the rock omitted in the original) are waiting to be used on the stack that the papermaker is in the process of assembling.[72]

70. Sun and Sun, *T'ien-kung k'ai-wu*, x.
71. Sun and Sun, *T'ien-kung k'ai-wu*, 226, Fig. 13–3; Dong, *Jiaozheng Tian gong kai wu*, 231.
72. Other striking examples of superior portrayals of technology in the Tao edition are the illustration of a grain freighter with more precise rigging and with the rudder included, as it was not in the original (Sun and Sun, *T'ien-kung k'ai-wu*, 173, Fig. 9–1; Dong, *Jiaozheng Tian gong kai wu*, 182–83) and the several illustrations dealing with salt production which, with also the inclusion of several scenes not found in the original edition, provide a much better sense of the variety and scale of salt-making technology in China than one gets from the illustrations in the original edition (Sun and Sun, *T'ien-kung k'ai-wu*, 108–21; Dong, *Jiaozheng Tian gong kai wu*, 103–25).

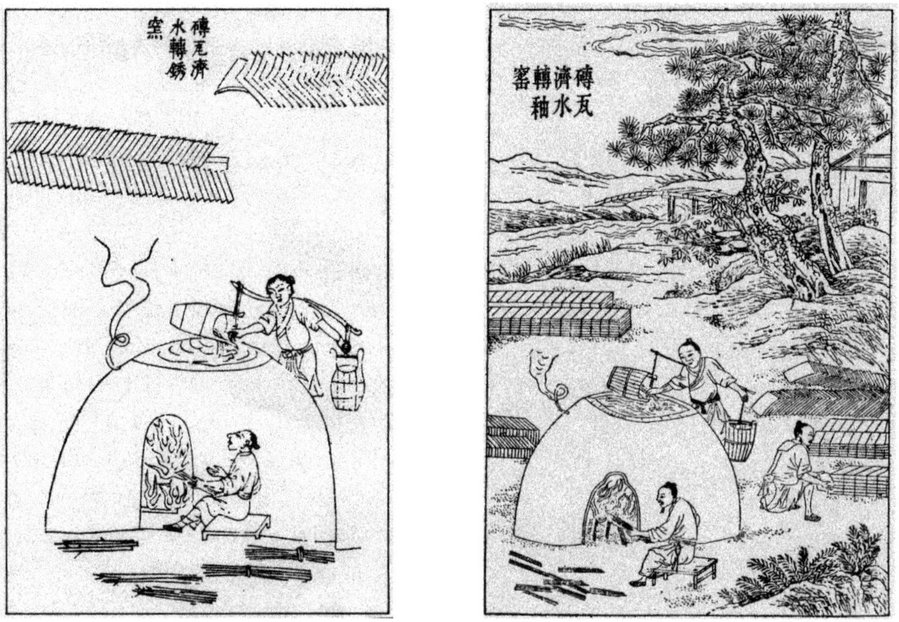

5.9 Glazing of bricks and tiles by water-quenching, from (a) Sun and Sun, *T'ien-kung k'ai-wu*, 140, Fig. 7–3 (left); and (b) Dong, *Jiaozheng Tian gong kai wu*, 145 (right).

Pan Jixing also thinks highly of the original illustrations, even pointing to what he sees as a feeling of three-dimensionality.[73] In fact, a better case can often be made for a kind of flatness they often display. Figs. 5.9 (a) and (b) illustrate the glazing of bricks and tiles. In the latter illustration, both the shape of the kiln as well as the placement of the bricks and tiles, not to mention all of the scenery, convey a much more effective sense of three-dimensionality than one sees in the original drawing.[74] It is typical of Chinese drawings in general that, like the works of painters, they did not use diminution of size or foreshortening to portray smaller objects or shorter distances in perspective (in contrast, for example, to larger landscape views).[75] For the most part, the lack of three-dimensionality or of a coherent perspective did not of themselves make an illustration incomprehensible.[76] But they did make it harder to read and discouraged careful reading. Moreover, the lack of a clear rendering of perspective became an increasingly serious problem the more complex machines

73. Pan, *Guide*, 98. Pan also criticizes the later illustrations that appear in the Tao edition, seeing them as aesthetically pleasing but technologically inferior. Pan, *Studies*, 150. This assessment is accurate for relatively few of the Tao illustrations.

74. For another excellent example, compare the illustrations of an ox-powered square-pallet chain pump in Sun and Sun, *T'ien-kung k'ai-wu*, 19, Fig. 1.10 and Dong, *Jiaozheng Tian gong kai wu*, 21.

75. March, "Linear Perspective," 117.

76. A nice example is the illustration in the fourteenth-century *Ao bo tu* 熬波圖 (Illustrations of salt production techniques) which is drawn in linear perspective but where the iron pans are portrayed in plan. The important point is that, in this case, a consistent scale is maintained for both views. Vogel, "Diagrams and Illustrations," 8; Yoshida, *Salt Production Techniques*, 233. For another example of effective mixed use of perspective and plan, see Golas, "Technical Representation in China," 32.

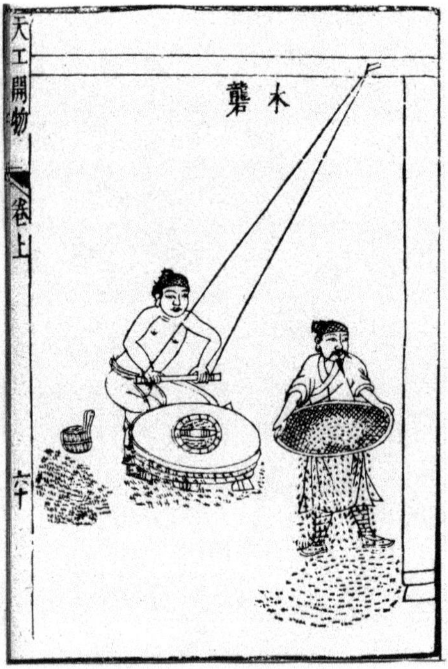

5.10 A strikingly defective portrayal of a hand-powered wooden hulling mill; Sun and Sun, *T'ien-kung k'ai-wu*, 87, Fig. 4–6. For a somewhat superior portrayal as early as the Han (but without the suspending rope), see Chen, *Illustrations*, 200, Fig. 14–98.

became.[77] This was not a serious problem in the *Exploitation* because, apart from a few illustrations of machines (especially looms) involved in cloth production, illustrations of complex machinery are largely absent.

Saigusa Hiroto, who judges the *Exploitation* illustrations by the accuracy with which they show how the machines portrayed were built and how they worked, is perhaps only a bit too harsh when he rates them a "failure" (*rakudai* 落第, the Japanese word commonly used to indicate a student's failure to pass an examination).[78] It was a similar negative assessment (as well as a desire for stylistic consistency) that influenced Tao Xiang to redraw all of the "original" illustrations in his 1927 edition.[79]

In fact, the illustrations of the *Exploitation* provide abundant examples of virtually all the kinds of limitations that are typical of earlier Chinese illustrations of technology and that would

77. Ultimately, in the West, the challenge of portraying the increasingly complex machines of the Industrial Revolution was partly met by the development of standardized projections that provided three separate views: plan, side or profile, and front. Fred Dubery and John Willats, *Perspective and Other Drawing Systems* (New York: Van Nostrand and Reinhold, 1972), 17. For an excellent example of how disappointing the results could be when Chinese illustrators sought to portray machines of real complexity, compare the heroic efforts of Mark Elvin and Dieter Kuhn to reconstruct a hemp-spinning machine pictured in Wang Zhen's *Agricultural Treatise* (Elvin, "The High-Level Equilibrium Trap," 26–30 and 60–63; Kuhn, *SCC* 5:9, 226–35). Elvin's "Postscript" underlines how different results may be arrived at (and then sometimes only tentatively) depending on what particular depiction of a machine one relies on.

78. Saigusa, *Tenkô kaibutsu*, 64. One might perhaps say the same for Clunas' unqualified generalization that the illustrations are of "limited accuracy." Clunas, "Luxury Knowledge," 35–36.

79. Strictly speaking, Tao did not have access to the original illustrations but only those of the second edition which, however, are for the most part identical to those of the first edition. He presumably agreed with the characterization of the illustrations by Ding Wenjiang as "crude and sketchy" (*culie jianlüe* 粗劣簡略). Dong, *Jiaozheng Tian gong kai wu*, postface, 1. It is worth recalling that, in the history of Chinese illustrations of technical subjects, later versions may be clearer or more accurate, as is generally true for the Tao illustrations of the *Exploitation* and sometimes true for those included in Chen Menglei's 陳夢雷 1726 *Gujin tushu jicheng* 古今圖書集成 [Compendium of books and illustrations past and present]; Bray, "Introduction," 68. More often, however, they will be inferior, as for example in the case of the 1530 and 1774 editions of Wang Zhen's *Agricultural Treatise*. Kuhn, "Marginalie," 144.

continue to be seen throughout the Qing period. We see a whole range of mistakes deriving from carelessness, indifference[80] or incomprehension, including particularly the omission or the misdrawing of mechanical details. Sometimes, within the *Exploitation* itself, we see a clearly different level of skill even in two quite similar illustrations; this is true for example in the drawing of the powering linkage for two hulling machines (Figs 5.10 and 5.11). To some extent, this can be seen as an effect of the long-term trend toward devaluing detail in most Chinese painting. Fewer details also meant an illustration easier to draw, as mentioned above, as well as easier to carve. Indeed, in some cases, it would have been impossible to carve the appropriate details in the space available in the illustration. We can see this clearly by comparing the rather good 1742 portrayal of a silk-reeling machine by Yang Shen with the superb reconstruction by Dieter Kuhn (Figs. 5.12 (a) and (b)). Although one can speculate that the designer of this illustration was either careless or simply lacked the skill to draw a right-angle gearing ((a), far left), it is easy to see that, given the scale of the drawing, he could never have portrayed an accurate and comprehensible version of the gearing as appears in the reconstruction ((b), right). Nor could the carver have gotten that level of detail onto the block.

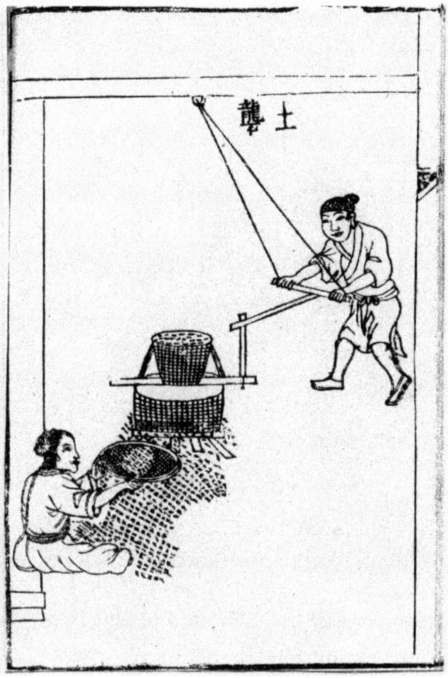

5.11 A hand-powered earthen hulling machine; Sun and Sun, *T'ien-kung k'ai-wu*, 88, Fig. 4–7.

This reminds us that none of Song's illustrations are devoted specifically to mechanical details. There are for example no assembly or component parts drawings of the kind to be seen in the *System Essentials* or the *Building Standards*. Except perhaps for the people who actually used the machines, Chinese did not share the same interest in the workings of mechanisms that we see in medieval Europe where there was sometimes even a preference for illustrating details rather than complete machines.[81]

As in the other works we have examined, Song's work is also not short of sometimes minor, sometimes egregious inconsistencies between what the text says and what the

80. Needham, Ho, Lu and Wang, *SCC* 5:7, 284 on the attitudes of woodblock artists, suggesting they may even have disdained working on technological illustrations.

81. Hall, "Production et diffusion," 153.

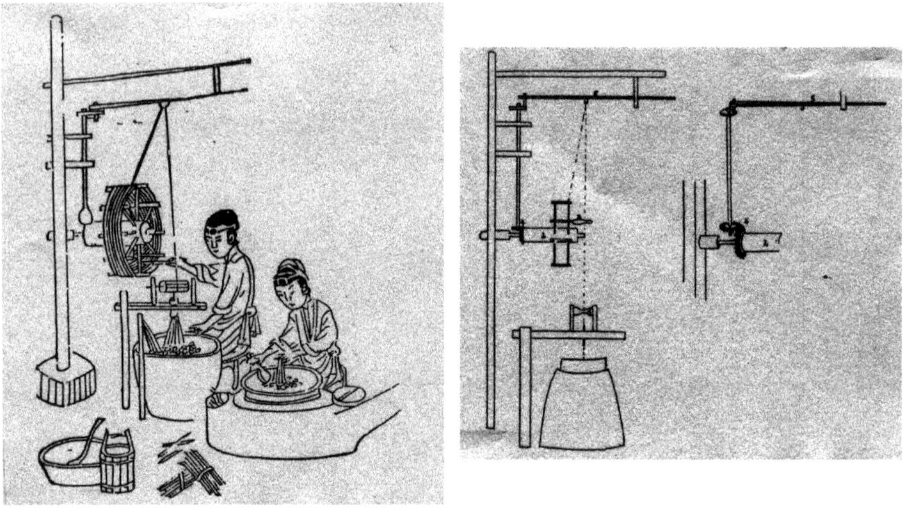

5.12 The "soundless roller" silk-reeling machine: (a) original illustration of Yang Shen 楊屾 from 1742, and (b) reconstruction of the mechanism by Dieter Kuhn.

illustrations show.[82] Related to this is the fact that the closest the text comes to linking an illustration to a specific explanation is an indication after the title of a section that a (relevant) illustration is included. It can therefore be unclear just what part of the text an illustration refers to. To be sure, the frequent use within the illustrations of labels (often consisting of technical terms that would otherwise be incomprehensible) instead of the reference letters or numbers common in European illustrations did provide some relatively effective linkage between text and illustration.[83] In many cases, however, these labels seem to have encouraged less than numbers and letters the identification of *all* important elements of an illustration. The *Exploitation,* in comparison with some other works, is actually quite sparing in the use of labels. We can compare, for example, the silk-reeling machine (Fig. 3.29 in Chapter 3) with its fourteen labels and captions[84] with the similar machine from the *Exploitation* with its mere three labels (Fig. 5.13).[85] It is often impossible to detect

82. Golas, "Like Obtaining a Great Treasure," 590n99. For another example, compare also Sun and Sun, *T'ien-kung k'ai-wu,* 260, Fig. 15–1 with the textual description of how to measure the pull of a bow on p. 263; the illustrator seems either not to have understood just what the "pull" of a bow was or not to have cared. In his excellent examination of the illustrations of iron production, Donald Wagner could draw on his expertise to provide evidence for yet another reason why text and illustrations might not correspond, i.e., Song's misinterpretation of the written sources (very possibly unillustrated) on which he was totally dependent when he had not personally seen the technology about which he was writing (assuming that he did not have access to a reliable informant). Donald B. Wagner, "Song Yingxing's Illustrations of Iron Production," in Bray et al., *Graphics and Text,* 615–32.

83. The writing of labels and even captions on illustrations may have been encouraged by the practice of writing poems on paintings.

84. See Kuhn, *SCC* 5:9, 368–69, Fig. 227, which also gives a translation for each of the labels.

85. Only rarely, probably beginning about this time, do the Chinese take the caption idea a step further and include in the illustrations themselves very extended captions that make possible a highly effective marriage of text

a principle of selection to explain why some parts are labeled while others, seemingly just as deserving of labels, are not. We do not know what role may have been played by the desire to make the block(s) easier to carve or by aesthetic considerations. In any case, it is a plausible hypothesis that, in general, more use of captions was found in those illustrations where there was a greater emphasis on clarifying the technology itself, as seems to be the case in Wang Zhen's silk-reeling machine. Unfortunately, captions may also have undercut the verisimilitude of technological drawings insofar as they encouraged an attitude of: "If the label makes clear what it is, the picture doesn't have to."

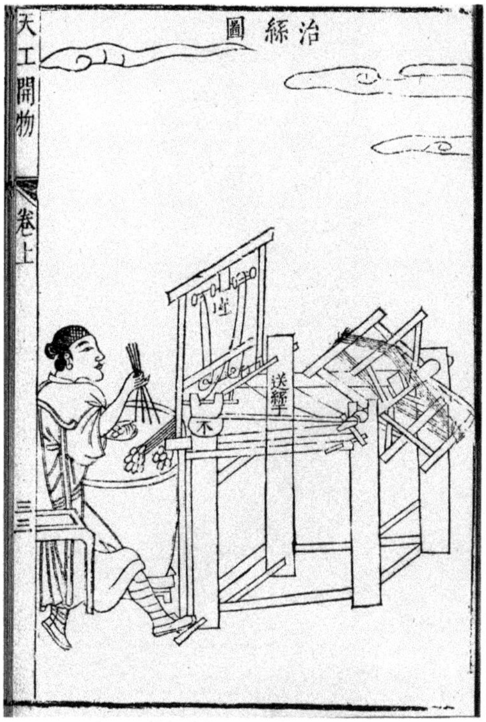

5.13 Reeling silk fibers; Sun and Sun, *T'ien-kung k'ai-wu*, 46, Fig. 2–6.

In common with most other Chinese illustrations of technology before the nineteenth century, no dimensions are provided in the *Exploitation* illustrations;[86] moreover, there is no attempt to introduce any explicit scale into the drawings, even though precedents for this procedure were available (especially in architectural painting) had the artist(s) chosen to use them. Likewise, we find no plan, elevation or section views (as, for example, in the *Building Standards*).

Finally, many of the limitations of the illustrations also derive from a general inability or unwillingness to draw on useful, available artistic techniques such as hatching, modeling

---

and illustration. See, for example, the illustrations of the "Turkish" musket from the 1598 manual "Treatise on Extraordinary Weapons [i.e. Firearms]" (*Shen qi pu* 神器譜) by Zhao Shizhen 趙士楨. Joseph Needham, Ho Ping-yü (Ho Peng Yoke), Lu Gwei-Djen and Wang Ling, "Chemistry and Chemical Technology: Military Technology; The Gunpowder Epic," in *Science and Civilisation in China (SCC)*, ed. Joseph Needham, vol. 5 part 7 (Cambridge: Cambridge University Press, 1965), 445–52, Figs. 174–79. Maps, though not technological illustrations as we have been using the term, provide some of the most thorough examples of the integration of text and illustrations; see for example, Yee, "Taking the World's Measure: Chinese Maps between Observation and Text," in J. B. Harley and David Woodward, *A History of Cartography*, vol. 2, book 2 (Chicago: Chicago University Press, 1994),103, Fig. 5.6.

86. We cannot be sure that the idea of doing so never occurred to Song Yingxing. At one point (TGKW *shang*, 30b; Sun and Sun, *T'ien-kung k'ai-wu*, 48; Herrmann, *Erschliessung*, 55), he seems to suggest that the dimensions of his silk-reeling machine will be included in the illustration, but they are not.

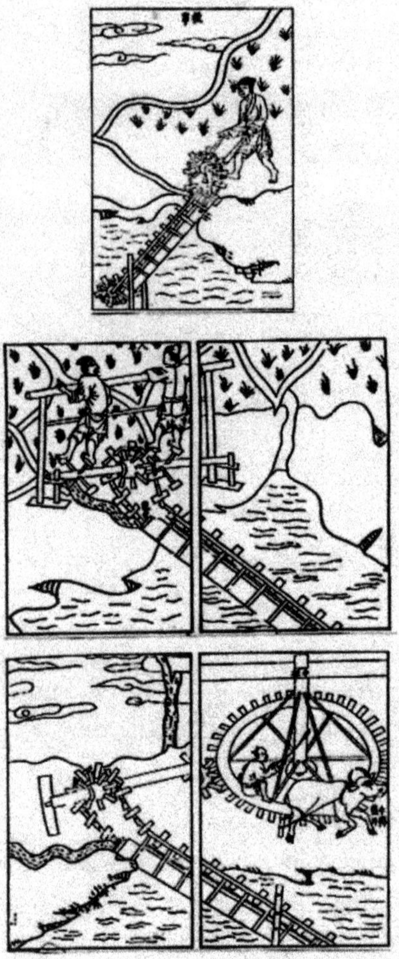

5.14 Hand-powered (top), treadle-powered (middle) and ox-powered (bottom) square-pallet chain pumps; Sun and Sun, *T'ien-kung k'ai-wu*, 19–21, Figs. 1–10, 1–11 and 1–12.

strokes or shading/chiaroscuro to give a sense of mass.[87] Song's illustrations did however make use of one technique, the departure from realistic depiction in favor of schematizing. An excellent example is the three portrayals of square-pallet chain pumps (Fig. 5.14).[88] If one looks at the upper assembly where the chain of paddles passes over the sprocket wheel, one sees that the drawing is very similar in all three illustrations.[89] At the same time, this is *not* a schematic rendering that eliminates extraneous information in order to focus on the technological essentials. It would hardly make the actual working of the pump or even the relative size of its components clear to anyone who had never seen one. I doubt that this concept of schematizing ever crossed the mind of a Chinese artist-designer. Insofar as he thought along these lines, he was more likely to ask himself how he could simplify the drawing in order to make it easier to accomplish; hence the almost universal failure of Chinese artists to portray gearing or similar linkages of any complexity clearly and convincingly.

87. Hegel (*Reading Illustrated Fiction*, 314) notes that this reliance exclusively on outlines to portray subjects contrasted with paintings that usually gave at least some suggestion of surface and texture, and therefore a certain sense of depth not present in a pure line drawing. He suggests that Europeans, when they adopted woodblock printing, incorporated more grain into the illustrations and that this in turn led them to substitute relatively quickly engravings for woodblock prints because engravings made possible "greater illusion of depth and texture."

88. Needham assesses the second of these illustrations as "probably" the best traditional Chinese illustration of the square-paddle chain pump (Needham and Wang, *SCC* 4:2, 339). Actually, it is not as we see by comparing it with the version in Hammers, *Pictures of Tilling and Weaving*, Fig. 5.21. Note also that Needham here as elsewhere reproduces not the original TGKW illustration, as indicated in the caption, but rather the improved illustration from the 1927 Tao edition.

89. This suggests that all three of these illustrations may have been done by the same artist, either working for Song's book or the creator of an earlier series of illustrations copied for Song's book.

Production Challenges

Keeping in mind the various audiences to whom Song might have directed his book and the attitudes and expectations these readers might have brought to the reading of the book provides additional clues to help us account for some of the limitations of the illustrations. For example, elite readers in Ming China were not accustomed to illustrations packed with information. Aesthetic preference for less rather than more detail and precision colored their expectations. Nor did they insist on up-to-date information. The editors of the eighteenth century imperial book collection *Siku quanshu* 四庫全書 (Complete collection of the four treasuries) catalog, in their assessment of a major compendium on water control from around 1590, give a very forthright statement of this attitude: "[A]lthough changes in methods were afterwards necessary to fit changing circumstances, yet experts in river control always take this book as a standard guide."[90] Most readers outside the elite, on the other hand, had considerable experience with illustrations that were not of very high quality by any standards.[91] They would probably have found very appealing the aesthetic vigor of the *Exploitation* illustrations.

We have also mentioned above Song Yingxing's relatively limited interest in technology as an object of intellectual investigation for its own sake. This was an attitude virtually ubiquitous among the elite and also may help to account for Song's acceptance of illustrations that lacked accuracy, completeness and verisimilitude. It may also have made it easier for Song and his illustrators to harbor without questioning mistaken assumptions such as that the washing of gold and of ironsand was carried out by the same method.[92]

We could further account for the shortcomings of the illustrations if we could answer a number of questions about the actual production of Song's book. For example, we do not know the source or sources of the illustrations from which the carvers worked.[93] Significant stylistic variations are to be found in the original illustrations and they sometimes strongly invite the conclusion that a certain group of illustrations, and maybe only those illustrations,

90. The translation is Joseph Needham's; see Needham, Wang and Lu, *SCC* 4:3, 325. This making-do with less than up-to-date information was undoubtedly encouraged by all the time the literati spent reading old–not to say ancient–books; the rarity, even in a society where books were relatively plentiful, of good presentations of contemporary technology; and the fact that most traditional technology changed very slowly if at all.

91. Lucille Chia writes of a "routinized workshop approach" to the illustrations in many Ming editions aimed at a popular audience. Chia, "Text and *Tu*," 249.

92. Wagner, "Song Yingxing's Illustrations," 616–17.

93. For convenience, we regularly refer to illustrators *and* carvers. In fact, it was in just this period that the tasks of designing the illustrations and carving the blocks were increasingly being divided between two persons–an artist and an artisan, as it were–whereas previously it had been more common for both tasks to be performed by the same person. Wang, "Arts of Ming Woodblock-printed Images," 57. It was also not uncommon at this time for illustrations to be designed by artists of considerable reputation. See, for example, Julia K. Murray, "Didactic Illustrations in Printed Books," in Brokaw and Chow (eds.), *Printing and Book Culture in Late Imperial China* (Berkeley: University of California Press, 2005), 429. Unfortunately, we have no clear information on just who performed these tasks in the production of Song's illustrations.

were done by a single artist.[94] Considering also that the book contained over 120 illustrations and that its printing was something of a rush job (more on this below), it seems highly likely that at least several people were engaged in the designing and drawing of the illustrations.[95] Labor costs may well have been an important consideration here. This was a period when good professional painters enjoyed considerable prestige[96] and presumably were able to command a commensurate compensation when producing book illustrations. If so, money could be saved by hiring less competent artists and the designs would suffer accordingly.[97]

Song's personal role in the production of the book is also a crucial question. Did he himself do at least some of the drawings?[98] Or did he perhaps play some role in supervising their production? Or was he rather completely "out of the loop" as it were when it came to all or some of the illustrations? Pan Jixing opines (without providing any evidence) that the "true-to-life" illustrations of the *Exploitation* might have been produced by relying on sketches Song made during his extensive travels.[99] We have seen that Donald Wagner also raises the possibility that some mistakes in the illustrations may have resulted from Song's misinterpretation of a text without illustrations that he relied on for a given technology.[100] In any case, if Song did have some kind of direct involvement in producing the illustrations, his own abilities as a draftsman as well as his ideas of what constituted a good illustration could have been crucial for the nature of the illustrations.

---

94. Even if all of the illustrations were stylistically more consistent, this would not rule out participation by a number of artists since there were at this time well defined schools or regional styles of illustration with their characteristic techniques, motifs etc. that made illustrations by various members of the school very similar. Hegel, *Reading Illustrated Fiction*, 196. Indeed, Hegel's description of the Jinling school of book illustration, centered on Nanjing, with its "naive, archaic" style, its elongated faces and its animated living figures, both human and animal (Hegel, *Reading Illustrated Fiction*, 233, 237), reads as quite a good listing of characteristics found in many of the *Exploitation* illustrations. (One need only glance at a number of the farming illustrations to see that the artists regularly have given the animals as much or more vivacity as the human figures; e.g., Sun and Sun, *T'ien-kung k'ai-wu*, 5, 7,19, 26, 27, etc.)
95. In one striking example of this kind of collaboration, a set of 200 illustrations for an edition of the novel *Jin Ping Mei* included twenty-seven illustrations signed by five different illustrators/carvers; Hegel, *Reading Illustrated Fiction*, 193, 196. One can only presume that many artists and artisans worked on the remaining 173 illustrations, perhaps sometimes with even two or more artists collaborating on the same illustration; Hegel, *Reading Illustrated Fiction*, 255; also 109–10, 192.
96. Hegel, *Reading Illustrated Fiction*, 281.
97. Similarly, highly skilled carvers of illustrations enjoyed especially high prestige in the late Ming (Hegel, *Reading Illustrated Fiction*, 109–10, 192); there might have been great temptation to save money here too.
98. If Song did have a personal hand in the production of the illustrations, it is a bit surprising that he makes no mention of that fact in his preface; Sun and Sun, *T'ien-kung k'ai-wu*, xiii–xiv.
99. Pan, *Guide*, 53.
100. Wagner, "Song Yingxing's Illustrations," 621. Wagner is probably right in suggesting that Song could hardly have seen personally all the technologies he describes; ibid., 615. Vogel also raises doubts about whether Song had ever seen a deep-drilled well along the lines of those used for collecting salt brine in Sichuan. Vogel, "Mechanical Knowledge in the Context of Pre-modern Chinese Salt Industry," in Zhang and Renn, *Transformation and Transmission*, 103.

In addition to the costs of designing and carving the illustrations, many other questions concerning the actual printing of the book suggest themselves: other costs that added to the overall cost of the printing;[101] number of copies printed,[102] selling price of the book. At present we lack the evidence to answer these questions with any specificity. What we can note is the considerable carelessness with which the book was produced, whether out of a desire to get it done as quickly as possible, to cut costs, or for some other reason or combination of reasons. One sees the results in the more than four hundred mistaken characters in the text;[103] in the misplacing of some of the illustrations;[104] and perhaps also in the ultimate arrangement of the chapters that may not have been what Song originally had in mind.[105]

There is no question but that Song was extraordinarily busy at this time. In addition to his responsibilities at the county school, he completed *six* writing projects in 1636–37.[106] Moreover, not only the *Exploitation* but also a book the previous year were printed with the help of an old friend, Tu Shaokui 涂紹煃, who had gone on to a successful official career and was back in his home area only for about two years during the mourning period following the death of his mother.[107] Song almost certainly wanted to have the printing completed before Tu left for his next post. Then there were also the family financial considerations

101. Chia (*Printing for Profit*, Chap. 2) is very good on this question.

102. Editions could run as few as 100 copies or as many as twenty-five thousand or more copies. Rawski, *Education and Popular Literacy*," 120; Hegel, *Reading Illustrated Fiction*, 126; Chia, *Printing for Profit*, 30–31. All we can say for sure about the first edition of the *Exploitation* is that it was large enough so that significant deterioration of the blocks could occur during its one or more print runs. Three copies of the original version are known to have survived; they are held by the Beijing Library, the Seikadô Bunko in Tokyo and the Bibliothèque Nationale in Paris. The Beijing Library copy was used for the excellent 1959 Zhonghua shuju (Shanghai) photo offset reprint. Unfortunately, however, the Beijing Library copy is the least good of the three surviving versions, having apparently been pulled relatively late in the print run when some of the blocks were beginning to show wear from repeated use. The Seikadô Bunko copy, which I have not personally examined, seems from reproduced illustrations to be the best, presumably coming therefore from early in the print run. The quality of the Bibliothèque Nationale copy seems to be intermediate between the Beijing and Seikadô versions.

103. That works out to an average of between one and two mistakes every page. These were presumably mainly made by the scribe(s) who wrote out the text that was used for the carving. But it is not inconceivable that many were slips made by Song himself. Insofar as the latter is true, one wonders how far they may have resulted out of a carelessness that came from having too good a memory. That may be too harsh, however, for Song complains in his preface that he lacked funds to purchase "rare artifacts" that he would have liked to study carefully in connection with writing the book; Sun and Sun, *T'ien-kung k'ai-wu*, xiv. He may also have lacked sufficient funds to have in his own library, and therefore easily accessible for checking, many of the books on which he was drawing. We do not know anything of the holdings of the library at Fenyi where he was serving at this time. Cullen, "Science/Technology Interface," 298. In any case, the key point is that the mistakes were able to pass through though they would have been easy to correct even after the blocks had been carved. See Golas "Technical Representation in China," 38–40. The *Exploitation* thus stands as a good example of many late Ming printed books in which "textual errors were particularly common." Tsien, *SCC* 5:1, 188.

104. Pan, *Biography*, 247. This was a problem that some publishers sought to avoid or at least minimize by grouping illustrations at the end of a chapter or at the end of the book; Bray, "Introduction," 63.

105. Pan, *Studies*, 63–66.

106. Pan, *Guide*, 65–66; Pan, *Critical Biography*, 247; Schäfer, *Crafting of the 10,000 Things*, 2 and passim.

107. Pan, *Critical Biography*, 194–95.

we have already mentioned and which may also have encouraged getting the book out as quickly as possible. Thus a very busy Song Yingxing quite possibly exercised relatively little supervision and control over the details of preparing the book for printing.[108]

That his illustrations often failed to measure up to the completeness of the textual discussion may also indicate a confidence on Song's part that, for practical purposes, his text would convey most of the relevant information with sufficient clarity.[109] I have shown elsewhere how, in Song's presentation of a sugarcane-crushing machine, it was not only the illustrations that suffered from a lack of clarity and completeness. The clarity and completeness of the textual descriptions could also sometimes be more apparent than real.[110] Nevertheless, the combination of text and illustration might still work reasonably well, depending on how it was used. For example, one can imagine an official posted to a sugarcane producing area that did not yet know of this kind of mill. Having his own copy of the *Exploitation*, he could show the illustration to some sugarcane producers and explain on the basis of the text how it was built.[111] Actual construction of an experimental mill would be done by experienced artisans able to combine the information in the *Exploitation* text and illustration with their own experience to produce a version that would stand a good chance of operating effectively.[112]

108. That even easily noticeable discrepancies between text and illustrations were not corrected seems to testify to inadequate oversight by Song. An example is the illustration of the first soaking of bamboo in the papermaking process (Sun and Sun, *T'ien-kung k'ai-wu*, 222, Fig. 13–1). The scene is set in a courtyard while the text explicitly says that this process is carried out in the mountains where the bamboo is cut. Also, though the text emphasizes that the soaking is done in running water, the illustration has only a vat with no indication of an inflow or outflow of water. Both of these mistakes are corrected in the 1927 version of this illustration. See Dong, *Jiaozheng Tian gong kai wu*, 228.

109. Indeed, rather than being disconcerted by mistaken or incomplete portrayals of the technology in the illustrations, Song was probably satisfied when the illustrations generally succeeded in providing essential information that could not be conveyed by the text alone. Or perhaps "satisfied" understates it: his characterization of the illustrations as things of "great value" (Sun and Sun, *T'ien-kung k'ai-wu*, xiv), suggests considerable pride in them. Of course, we do not know whether this was his genuine assessment or whether it was mainly an effort to promote sales of the book. Perhaps a little of both?

110. Moreover, in Ming China no less than in medieval Europe, much of the knowledge of artisans in any case was simply incapable of being expressed in *either writing or illustrations.* Hall, "Production et diffusion," 155. See also Golas, "Like Obtaining a Great Treasure," 576–82.

111. For examples of this kind of use of illustrated books, see Wang Chaosheng, *Farming and Weaving Pictures*, 170, 179–81.

112. It cannot be assumed, of course, that the appropriate artisans were always available. An impediment to the spread of more advanced machines was sometimes local scarcities of skilled artisans, especially in relatively backward areas. Timothy Brook has shown, for example, that even as late as the eighteenth century, the further spread of square-pallet chain pumps would have required the importation of carpenters with relevant skills both to construct the pumps and to maintain them after their introduction. Brook, "The Spread of Rice Cultivation and Rice Technology into the Hebei Region in the Ming and Qing," in Li Guohao, *Explorations*, 685–86.

The Later History of the Illustrations

Much of the later history of Song's illustrations is intimately associated with the history of the text in which they appeared, but those associations are by no means uncomplicated. I have tried in a "Table of the Editions and Reprints of the *Tiangong kaiwu*" to present an overview of how various sets of illustrations made their way into various versions of the text.[113] But, over time, many of the illustrations also enjoyed independent reincarnations that constitute another history, much more complicated to trace but also of considerably broader if more diffuse influence. We begin with the history of the text.

Perhaps in part because of his anti-Manchu sentiments, perhaps too because he had never become a high official,[114] perhaps also because of the unsettled conditions in the years immediately after the book appeared, two-thirds of Song Yingxing's total writings had been lost by the first half of the eighteenth century. The lack of new editions or printings of the *Exploitation* during the Qing makes it clear that few copies of the book were in circulation. One author has speculated that the information the book contained on government monopolies such as coinage, salt making, and weapons manufacture, information that the government wanted kept secret, impeded its circulation.[115] Pan Jixing refers back to Song's statement in his preface that those people studying to pass the civil service examinations would find nothing of use in this book and suggests that a general disinterest of the literati in what Song was writing about also contributed to the virtual disappearance of the book in China.[116]

This situation did not change until the twentieth century. In 1927, the eminent bibliophile, Tao Xiang 陶湘 (1870–1940), produced a new edition with 162 illustrations, some of which we have referred to above. In addition to the original illustrations, Tao decided to add some forty new illustrations from various Qing works[117] to present aspects of the technologies that had not been illustrated in the original edition or later editions in China and Japan. He had these illustrations redrawn for his new edition. Then, since the newly drawn illustrations were stylistically so different from and aesthetically so superior to the original illustrations, he decided also to have the original illustrations redrawn in a compatible style.[118] Tao's edition stimulated a number of new editions leading to increasing recognition of the great value of Song's work.

The influence of the *Exploitation* during the Qing period, however, was considerably greater than that capsule history of the book itself might suggest. When one examines various compendia such as the early eighteenth-century imperial encyclopedia *Gujin tushu jicheng*

113. Golas, "Like Obtaining a Great Treasure," 614.

114. He failed even to get a biography in the *Official History of the Ming*. Pan, "Critical Biography," 235–36.

115. Tu Lien-che, "Sung Ying-hsing," in *ECCP*, 691.

116. Pan, "Critical Biography," 90, 236.

117. Pan, "Studies," 150.

118. Pan, "Studies," 150, 153, 169; Sun and Sun, *T'ien-kung k'ai-wu*, x.

古今圖書集成 (Compendium of books and illustrations past and present) and the mid-eighteenth-century agricultural compendium *Shoushi tongkao* 授時通考 (Comprehensive study of the farming year), many of them widely read, we find they included parts of the *Exploitation* and many of the illustrations (sometimes redrawn and even improved). The same is true also for other more specific works. Song's discussion on making steel by cementation was incorporated by his younger contemporary Fang Yizhi 方以智 (1611–71) into the seventh chapter of his *Wuli xiaoshi* 物理小識 (Small encyclopedia of the principles of things).[119] The section on silver mining and smelting was copied into the *Yunnan tongzhi* 雲南通志 (Comprehensive gazetteer of Yunnan) of 1736 and into the mid-nineteenth century work on copper and silver mining in Yunnan, the *Diannan kuangchang tulue* 滇南礦廠圖略 (An illustrated account of the mines and smelters of Yunnan) of Wu Qijun 吳其濬. All of this suggests that much of what Song Yingxing wrote in the *Exploitation* was relatively well known in Qing China, even by people who had never seen his book. The illustrations on the other hand, in copies that were sometimes quite similar to, at other times far removed from the originals, became simply anonymous components of a much larger body of illustrations with little to distinguish them in viewers' eyes from other similar illustrations.

119. Pan, "Critical Biography," 90, 478.

Six

Qing Developments

Roads Not Taken

The Seventeenth-Century Transition

Perhaps the overriding theme in most of cultural life in the early and middle Qing (roughly from the middle of the seventeenth to the end of the eighteenth centuries), and one that contrasts sharply with the freewheeling cultural experimentation that prevailed in the late Ming, was a renewed emphasis on classical teachings which resulted in what might be seen as a systematization or standardization of cultural knowledge, whether moral, textual or worldly.[1] To a great extent, it was the Manchu court, intent on maintaining a stable and durable rule, that took the lead. But the efforts of the court were echoed in various ways throughout the society.

One example, especially relevant for our discussion, is provided by painting. The seventeenth and eighteenth centuries witnessed a proliferation of manuals focused on cataloguing accepted genres, motifs and techniques. To at least some degree, the result was a choking off of experimentation, as many artists found it easier to conform to a growing consensus, for example by assembling their paintings from stock graphic elements.[2]

Furthermore, through most of the Ming dynasty, technical skill and craftsmanship had enjoyed considerable esteem in painting circles.[3] Highly regarded painters, for example, were often willing to put their drawing skills at the service of book illustration without feeling they demeaned themselves in so doing. The seventeenth century, however, witnessed a growing social gap between elite literati artists on the one hand who took pride in painting that supposedly expressed above all the mind and heart of the cultivated gentleman, and book illustrators or decorative painters on the other whose efforts "had been denigrated to the category of mere craft work."[4] Whereas a renowned painter like Dong Qichang 董其昌

1. Benjamin A. Elman, *On Their Own Terms: Science in China, 1550–1900* (Cambridge, MA: Harvard University Press, 2005), 60.
2. Thorp and Vinograd (*Chinese Art and Culture*, Chapter 9) are especially good on this. See also Cahill, *Three Alternative Histories*, 100–102, for his argument that it was in large measure the growing prevalence of the *xieyi* 寫意 (sketching the idea) style of painting with its "simplified pictures done in relatively loose, fast brushwork" that led, at its worst, to a general decline in the quality of Chinese painting after the early Qing.
3. Richard M. Barnhart (*Painters of the Great Ming*, 300) discerns a change in this attitude already toward the end of the Ming.
4. Hegel, *Reading Illustrated Fiction*, 271.

could disparage his ability to render certain subjects convincingly (and perhaps even take a "perverse" pride in that failing),[5] book illustrators came to be seen as merely transcribers of appearances, the technique for which painters like the academic bird-and-flower painters were regularly and roundly criticized by literati painting critics.[6] This attitude was captured very succinctly in statements such as that by Gong Xian (龔賢, 1618–89): "To insist on a specific subject or the representation of some event is very low class."[7] The unsurprising result was a decline during the Qing in the quality of book illustration in general and of illustrations of technology in particular, as the literati increasingly came to see book illustration as a "culturally marginal craft."[8] Even those rare figures such as Wang Zheng 王徵 who strove to do accurate technological drawings found, as we shall see below, that their training and tradition often did not provide them with the necessary resources in the way of visual approach and painting or drawing techniques.[9]

Here as in many cultural developments that influenced technical illustration, it is often illuminating to compare what was happening in China with conditions at the other end of Eurasia. Even well before the Renaissance, we encounter, beginning in the late fourteenth century, a vibrant and rapidly changing Europe with not only growing numbers of literate and numerate artisans but also a number of highly educated figures who manifest one or another combination of notable intellectual, artistic and technological accomplishments. One of the first and greatest was the famous clockmaker, Giovanni de'Dondi (c. 1330–89), who was also unsurpassed in his knowledge of philosophy, medicine and astronomy.[10] Others such as Leonardo da Vinci (1452–1519) combined engineering ability with artistic talents in painting or sculpture or architecture.

These broad-ranging technologists, answering to demands–and indeed working to create demand–for their talents, played a crucial role in a surge of new kinds of technological writings and illustrations during the Renaissance.[11] Topics discussed included architecture and building, courtly entertainment and displays, weapons and ballistics, machines for hauling and lifting, hydraulics, surgery and anatomy, mining and metallurgy. Nearly all of these works contained illustrations, many of which were produced using newly devised techniques such as cutaway and exploded drawings.[12] Some even explained how to draw in linear

5. Peter J. Golas, "Technological Illustration in China," in Alain Arroult and Catherine Jami (eds.), *Science and Technology in East Asia: The Legacy of Joseph Needham* (Turnhout: Brepols, 2001), 48.
6. Cahill, *The Painter's Practice*, 96.
7. Hegel, *Reading Illustrated Fiction*, 271.
8. Hegel, *Reading Illustrated Fiction*, 270–71.
9. See below, especially Fig. 6.9.
10. Carlo M. Cipolla, *Clocks and Culture 1300–1700* (New York: W. W. Norton, 1977), 45.
11. Previously most writings on technology had been in the nature of an aide-memoire for artisans (to a great extent taking the place of apprenticeship) or had served as technical notes. Hall, "Production et diffusion," 157.
12. Competition to sell books must have played some role in the introduction of new and improved illustration techniques (Thorp and Vinograd, *Chinese Art and Culture*, 321) as did also a growing professionalization among book illustrators (Bray et al., *Graphics and Text*, 72).

perspective.[13] As these increasingly advanced drawing techniques came into wider use, they may well have stimulated a higher quality of non-verbal thinking, the ability to manipulate visual images, among artisans and engineers, leading to greater engineering creativity.[14] In any case, whatever their readership, illustrated manuscript books especially on the mechanical arts were proliferating rapidly even before book printing became widespread.[15]

In contrast to China, there was in Europe nothing comparable to the literati dominance of what they themselves defined as fine painting and in which there was little or no room for technological subjects. Indeed, the finer European illustrations, including many devoted to technological subjects, attracted both growing numbers of literate and numerate artisans as well as upper-class readers, among whom they encouraged a growing interest in science and the mechanical arts.[16] Hence we find a number of striking contrasts in how the learned in the two cultures thought about technology. For example, nothing in China at this time compares with the European gift for what Mark Elvin has called "precisely visualized mechanical fantasy"–the ability to imagine accurately and in detail machines that do not yet exist.[17] Nor do we find in China any evidence for that kind of x-ray vision that enabled Europeans to see machines as "geometry in motion," a perspective that effectively encouraged thinking about ways of improving them.

Nor was this the only way that the thinking of the authors who produced the new technological literature in Europe was apt to foster technological creativity. Whether in Europe or in China, the craftsmen actually engaged in their daily tasks probably did not differ greatly in how they thought about or approached their work. Theirs was a practically oriented kind of thinking where a particular action could be expected to produce a predictable result. The same similarity of thinking, however, did not hold for the highly educated members of both societies. While in Europe cause-and-effect patterns of thought tended to predominate among the intellectually esteemed and, if anything, were growing stronger in the intellectual liberation represented by the Renaissance, the Chinese literati artists were much more at home with numerological, associative (correlative, analogical) and even symbolic ways of thinking. Add to that a more highly developed admiration for ingenuity as well as a propensity for experimentation among educated Europeans and it becomes easier to understand that when they thought about technology, they were more likely to consider it in terms of how it actually worked than the Chinese literati did. That, and the fact that Europeans applied so much of their imaginative powers to technology, helps explain at least in part why highly complex mechanical technology played so much greater a role in Renaissance Europe than in Ming and Qing China.

13. Edgerton, "Renaissance Development," 184.
14. Pacey, *Maze of Ingenuity*, 79.
15. Pamela O. Long, "Power, Patronage and the Authorship of *Ars*: From Mechanical Know-how to Mechanical Knowledge in the last Scribal Age," *Isis* 88.1 (March 1997), 1–2.
16. Edgerton, "Renaissance Development of Scientific Illustration," 194.
17. Elvin, "China as a Counterfactual," 112.

When it came to portraying technology, European artists also benefited from the fact that the reigning artistic values stressed objective representation. This encouraged the development of new techniques such as modeling with light and shadow (chiaroscuro), maintenance of convincing proportions, and creating the illusion of three-dimensionality by means of single-point or convergent perspective. Largely absent in Europe was the disinterest if not disdain exhibited by Chinese literati both for highly representational art as well as for technology.[18] Indeed, it was at this time that *drawings* were increasingly accepted by Europeans as art in their own right, bringing respect and renown to those artists who excelled in this sphere of artistic creation.[19]

Mathematics and Measurement

Mathematics in late medieval and Renaissance Europe assumed an importance unprecedented anywhere in the world up to that time.[20] We see the spread of mathematical learning reflected in the publication of at least 214 books on mathematics in the quarter-century between 1472 and 1500, mostly in Italy.[21] Equally unprecedented was the extent to which *both* theoretical and applied mathematics flourished.

During the earlier Middle Ages in the West, mathematics had been regularly employed in astrology, architecture, calendrical calculations, and even the practice of medicine, which postulated a link between physical health and sickness and the positions of the planets among the constellations.[22] The Church also promoted the study of mathematics because, just as the ancient Greeks saw mathematics as a good preparation for philosophy, it regarded a proper dose (but not too much!) of mathematical training as a useful preparation for the study of theology.[23] The Church's views of nature and the cosmos, resting on the conviction that nature, i.e. God's ways, became intelligible to human beings through the teachings of scripture, also contributed to a rational intellectual environment that proved congenial to a growing emphasis on mathematics.

By the late Middle Ages, even though such scripture-based understandings of the world were losing their persuasiveness for many, the conviction that nature could be understood

---

18. Even in Europe there might be occasional developments reminiscent of the prevailing situation in China. The history of botanic and anatomic illustration provides an interesting example. The rapid maturing of this field in the 1530s and 1540s was followed by stagnation, in part because "the great artists withdrew from the field, leaving it to poorly paid craftsmen who not only were incapable of further innovations but who often were unable even simply to make accurate copies." Ackerman, "Involvement of Artists," 121–22.
19. Hall "Production et diffusion," 169.
20. For mathematics in medieval and Renaissance Europe, besides Crosby, *Measure of Reality*, I have found Morris Kline, *Mathematics in Western Culture* (New York: Oxford University Press, 1953), Chapters 7, 8, and 10 especially helpful.
21. Colin A Ronan, *Science: Its History and Development among the World's Cultures* (New York: Facts on File, 1982), 322.
22. Lynn White, Jr., "The Flavor of Early Renaissance Technology," in B. E. Levy (ed.), *Developments in the Early Renaissance* (Albany, NY: State University of New York Press, 1972), 44–46.
23. Kline, *Mathematics in Western Culture*, 95–96.

by the human mind grew, if anything, stronger, relying mainly on rational/logical methods in which mathematics and a geometric perspective played an integral part. [24] As European thinkers became enamored with what they saw as the certitude of mathematical truths, many of them adopted a new faith that mathematical laws should replace scriptural teachings for understanding the cosmos, with God seen now as the Great Mathematician.[25] He had created a universe that was, as the Greeks had long ago believed, "a harmonious medley of mathematical laws."[26]

China of course also saw harmony as central to the order of the universe, but it was a harmony of *forces* such as yin and yang that constituted "an all-pervading spiritual power that moved the world"[27] and were not susceptible of measurement. In contrast to the more static European view, the Chinese outlook had a dynamic quality that encompassed the complementary ideas that the universe could get out of balance and that humans played a role in maintaining the universal harmony. While the Chinese did not totally reject rationality and certainly not mathematics, they placed more faith in an intuitive and aesthetic nature-knowledge.[28] This faith was bolstered to be sure by considerable reliance on empirical observations, but lacked the notion that a demonstrable mathematical reality underlay the appearances provided by our sense organs.[29] In China, understanding of nature consisted essentially of more-or-less shared intuitions.

Europeans applied mathematically-based notions of harmony and proportion to a wide range of intellectual and artistic pursuits such as astronomy/astrology, architecture, music, painting and sculpture.[30] Of special importance for our concerns, their efforts to work out harmonious proportions, relying on careful measurement, led to two major innovations: scale drawings and perspective views that were geometrically accurate.[31] In these

24. Brant Weller, "The European Discovery of External Cultures and Their Effect on European Expansion," *World History Bulletin* 24.1 (2008 Spring), 39. Needham has described this as Europe's arrival at a "fusion point" of mathematics and "nature-knowledge." Needham and Wang, *SCC* 3, 168.

25. Crosby, *The Measure of Reality*, 123. Alternatively, as Descartes would later picture it, the world was "fabricated by God as a solution to an engineering problem." Baigrie, *Picturing Knowledge*, xxi–xxii.

26. Kline, *Mathematics in Western Culture*, 105, 108 and 110.

27. Blunden and Elvin, *Cultural Atlas*, 146, describing the thought of Fang Yizhi 方以智.

28. Sivin, "Science and Medicine in Chinese History," in Paul S. Ropp (ed.), *Heritage of China: Contemporary Perspectives on Chinese Civilization* (Berkeley: University of California Press, 1990), 169–70; Sivin, "Chinese Conceptions of Time," *The Earlham Review* 1966.1, 82–83.

29. Numerological interpretations of nature, often resorted to especially in earlier times, were of course not "demonstrable" in a mathematical sense. Joseph Needham and Wang Ling, "History of Scientific Thought," in *Science and Civilisation in China (SCC)*, ed. Joseph Needham, vol. 2 (Cambridge: Cambridge University Press, 1956) 271, 272, 287–88 [drawing on Granet] and 484.

30. Because sculpture never became a fine art in China, literati artists were deprived of what might have been a potent stimulus to sketching based on careful observation of the kind that was so conspicuously absent in most Chinese portrayals of technology.

31. Pacey, *Maze of Ingenuity*, 62–63. What can only be called an obsession with measuring things was shared by many of the great figures of the Renaissance including Nicholas of Cusa, Filippo Brunelleschi, and Leon Battista Alberti. Crosby, *Measure of Reality*, 181–83. Most of the great painters in this period worked to incorporate mathematical principles and harmonies into their paintings (Kline, *Mathematics in Western Culture*,

circumstances, many European engineers and technicians developed, as noted above, an unprecedented ability to view machines as "geometry in motion"; they could use that ability to analyze machines, often in drawings, with a view to making them work better.[32]

Nothing like this occurred in China. There can be little doubt that mathematics was more widely used in the Ming and Qing than ever before in China.[33] Moreover, mathematics also drew broader interest even among some literati.[34] But Chinese mathematics retained its overwhelming emphasis on calculation and problem-solving. It played no role in the formation of theories about the cosmos and the natural world as in Renaissance Europe. For most literati, especially those much concerned with maintaining their elite status in a time of considerable social flux, practical knowledge of all kinds, including mathematics, was best avoided.[35] This disdain contributed to a milieu in which, among other things, few painters were attracted to learning the methods of geometrically based linear perspective even when these became available.[36]

Mechanical Ingenuity

Another area of contrast between attitudes and practices in China and those of late medieval and Renaissance Europe is the very different role of mechanical thinking and applications in the two societies. This too was reflected in the way complicated mechanisms were dealt with in Chinese portrayals of technological subjects.

---

132), while geometry became essential for artisans, mapmakers, architects and surveyors. Benjamin A. Elman, *A Cultural History of Modern Science in China* (Cambridge, MA: Harvard University Press, 2006), 18.

32. Elvin, "China as a Counterfactual," 108, 112. It should be noted that the Chinese *were* more successful in adopting "northern European (Dutch and Flemish) modes of rendering space and solid forms" as opposed to the Italian linear techniques generally taught by the Jesuits. James Cahill, *Pictures for Use and Pleasure* (Berkeley, CA: University of California Press, 2010), especially Chap. 3. These were not likely, however, to influence portrayals of technology.
33. Craig Clunas, "Text, Representation and Technique in Early Modern China," in Karine Chemla (ed.), *History of Science, History of Text* (Dordrecht, The Netherlands: Springer, 2004), 110–14; Sivin, "Science and Medicine in Chinese History," 171–77. Even as early as the thirteenth century, if Elvin is correct, China may have led the world in the numeracy of its population. Elvin, *Pattern*, 181. With the continuing economic development in the Ming and Qing, numeracy undoubtedly became even more widespread.
34. Catherine Jami, "Learning Mathematical Sciences during the Early and Mid-Ch'ing," in Benjamin A. Elman and Alexander Woodside (eds.), *Education and Society in Later Imperial China, 1600–1900* (Berkeley: University of California Press, 1994), 223, 224.
35. Clunas, "Text, Representation, and Technique," 114. If we are to believe the assessment of the Kangxi emperor (r. 1662–1722), China seems generally to have suffered in his time from a paucity of good, highly trained mathematicians. Jami, "European Science in China," 427. Even in the best of times, however, most of the officials in the astronomy bureau were not what we think of as accomplished mathematicians but rather technicians who mainly followed step-by-step procedures, not entirely unlike painting by the numbers.
36. See March, "Linear Perspective," esp. 137 and 139, for further discussion of this point. Yee in his study of Chinese cartography makes the further point that the Chinese experience of space as dynamic and unbounded, ever-changing over time depending on one's vantage point, precluded its being fixed or measured. "As a result, no abstract geometrical system governed space, and points within it were not definable or delimitable in any absolute terms." Yee, "Cartography in China," 144.

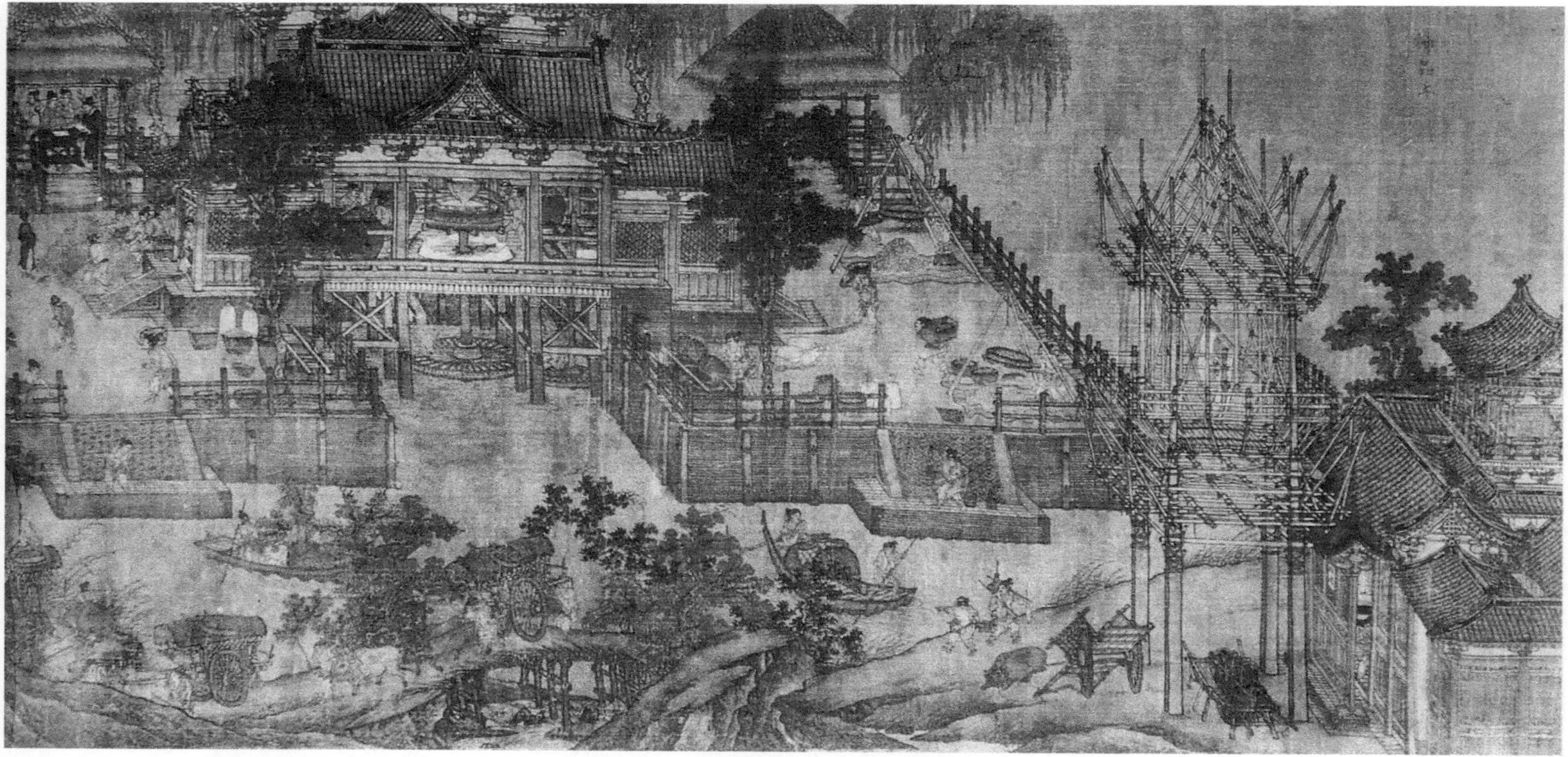

Plate 9 The "water mill" handscroll portraying the processing of harvested grain to flour and its readying for shipment. Ink and color on silk, attributed to Wei Xian (10th century).

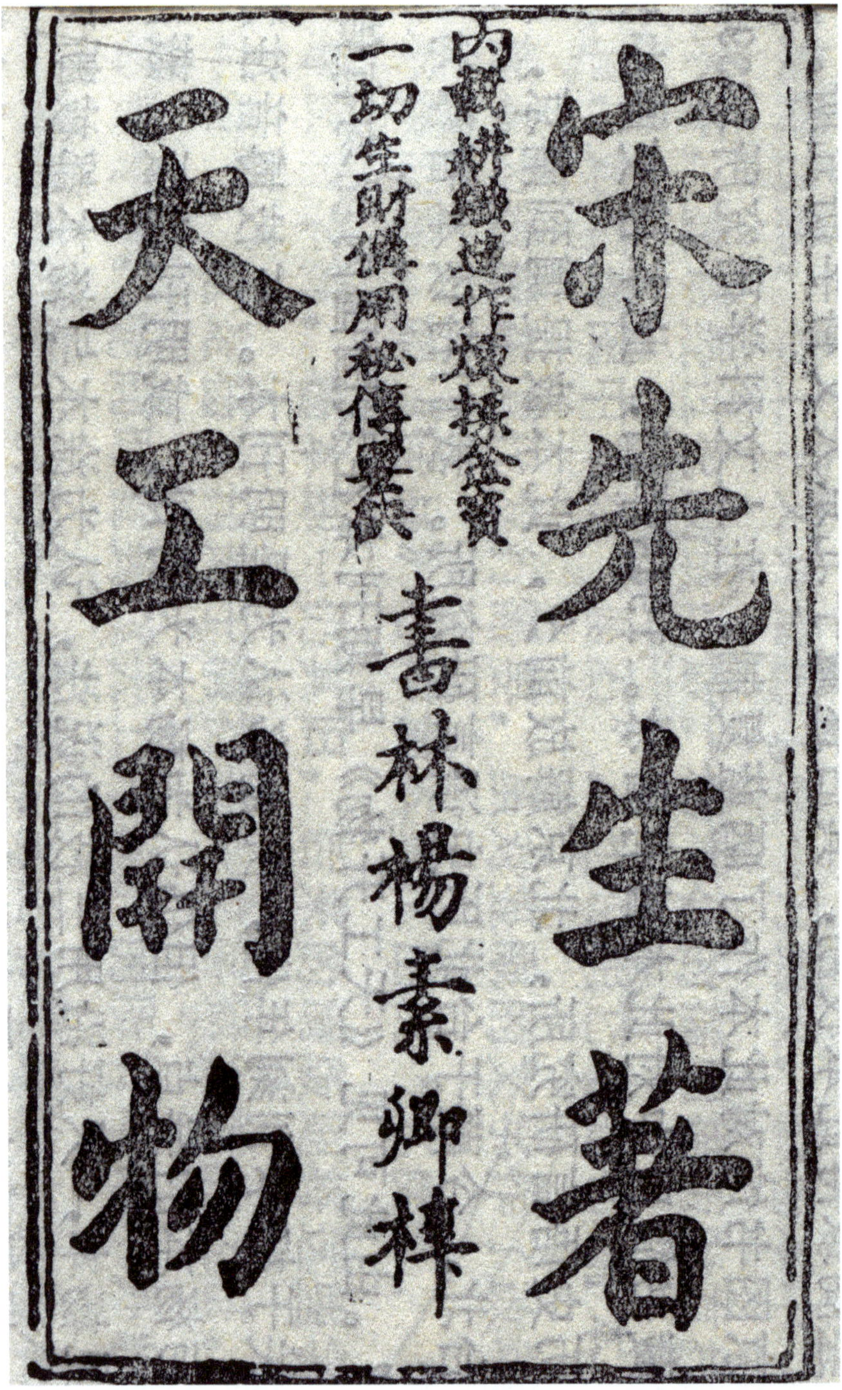

Plate 10 Advertisement (two lines in the upper center) for the second edition of the *Tian gong kai wu*

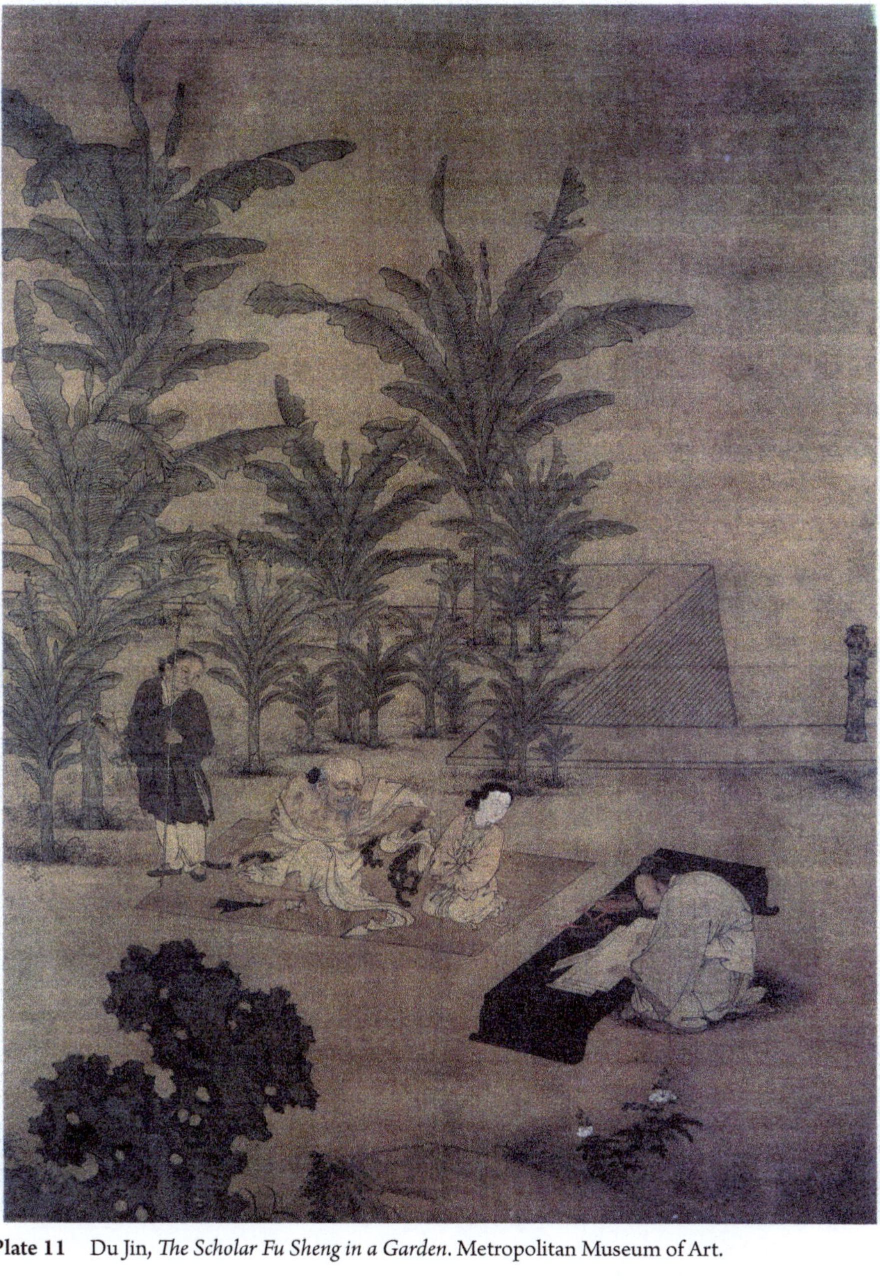

Plate 11 Du Jin, *The Scholar Fu Sheng in a Garden*. Metropolitan Museum of Art.

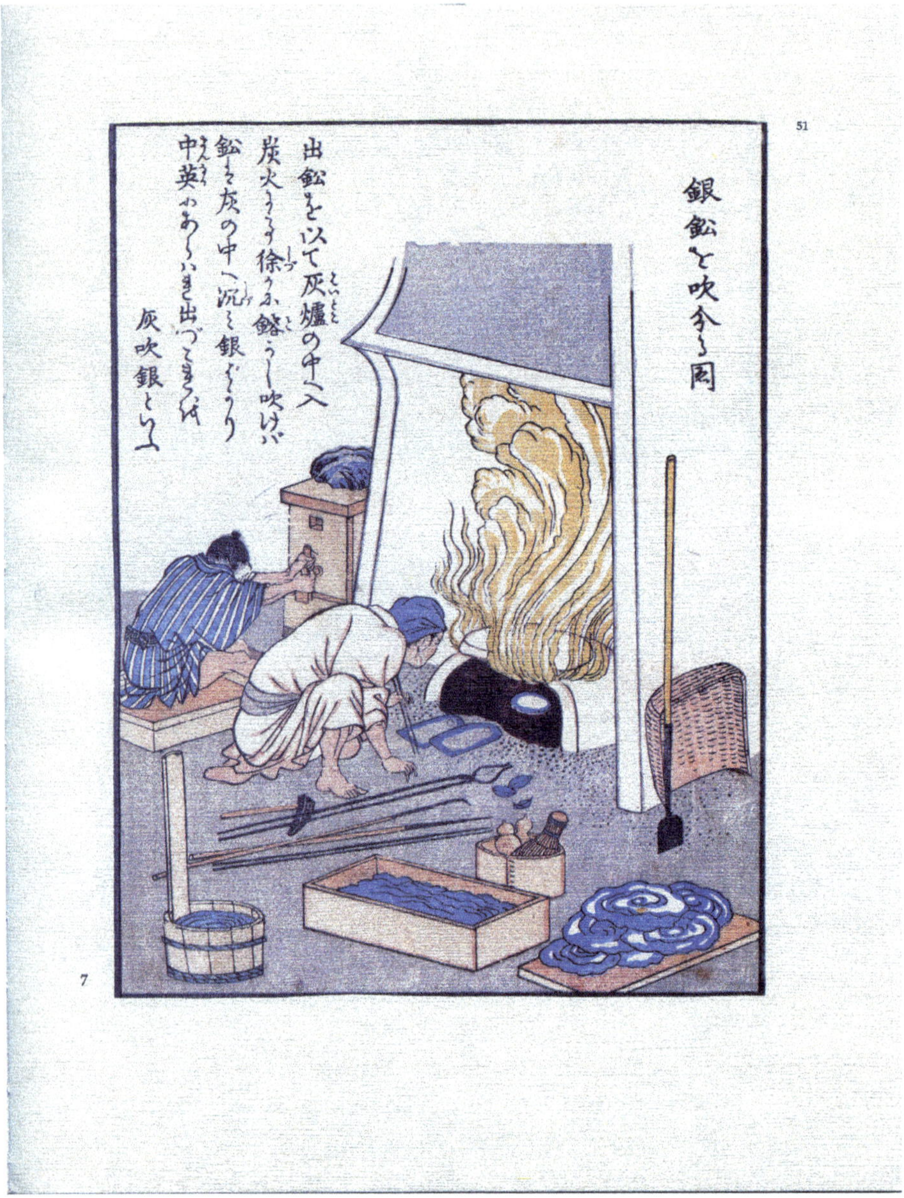

Plate 12 Working a double-acting box bellows furnace to separate silver and lead. Note the careful portrayal of the push-pull handle (as well as the hands manipulating it), round above to facilitate pulling and flat below to facilitate pushing; no handle like this is seen in any Chinese illustrations.

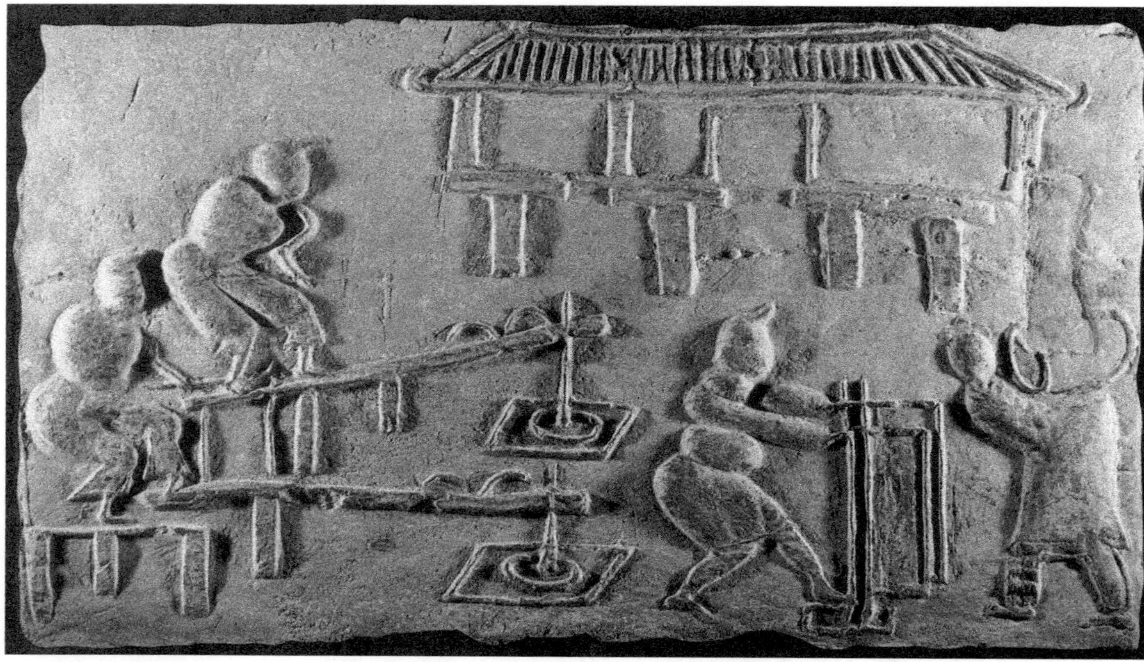

Plate 13 Hulling and winnowing scene from a Later Han (25–220 CE) pottery tomb relief

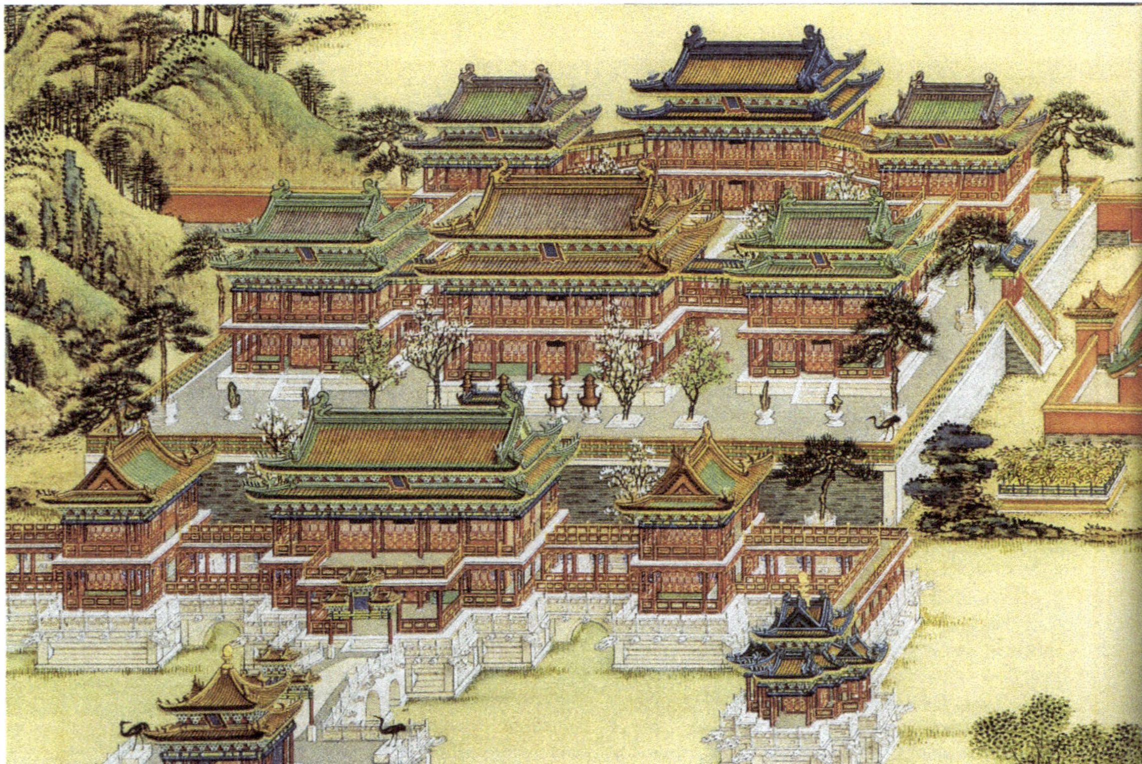

Plate 14 Tang Dai (1673–after 1751) and Shen Yuan (active c. 1736–46), a view of the Yuanming Garden. Tang Dai painted the surrounding landscape, Shen Zhou the buildings.

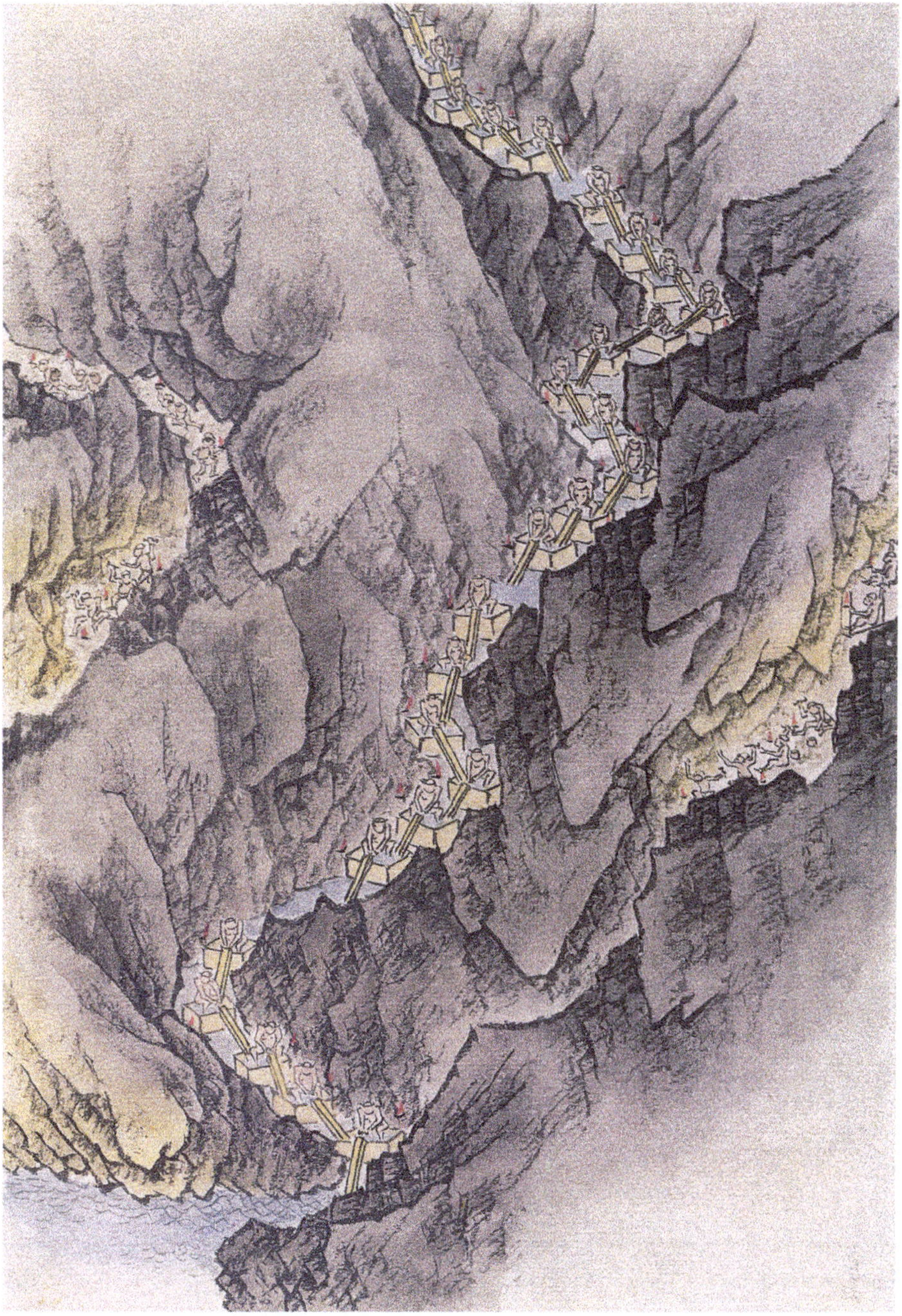

Plate 15 Pumping water out of the Besshi mine

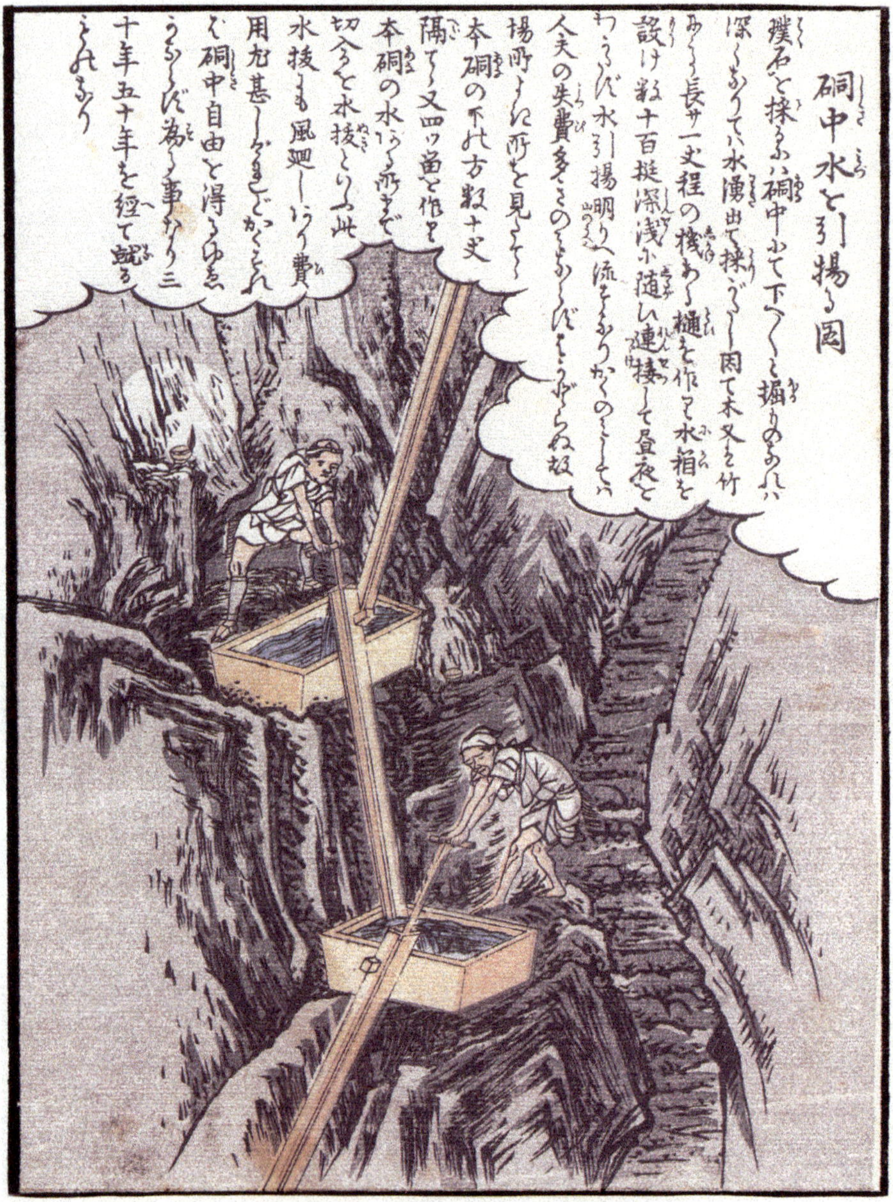

Plate 16 A close-up of water raising from the Besshi mine

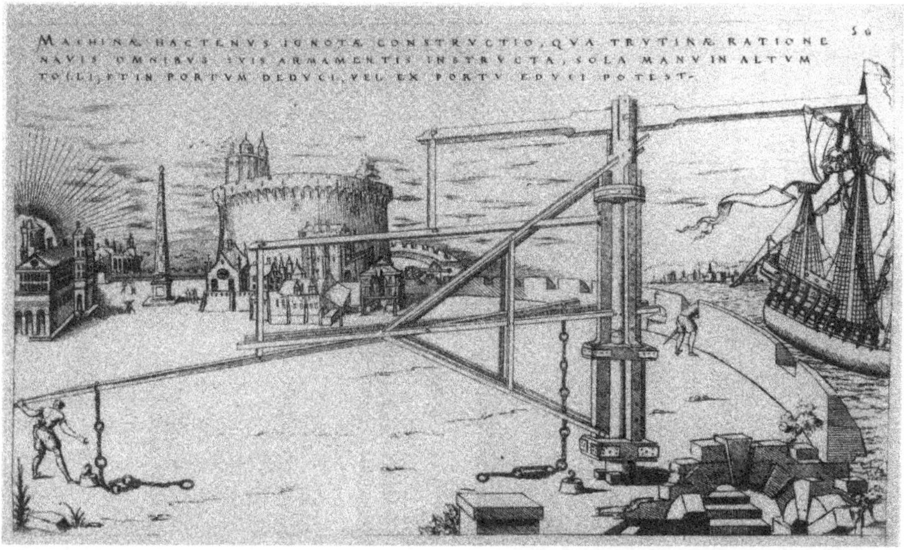

6.1 A fantastical design for a cargo crane by means of which a single hand pulling on a lever could raise a fully equipped ship out of the water; Besson, *Theatrum Instrumentorum et machinarum* (1582).

A keen interest in things mechanical had not always characterized that part of the world that would eventually come to be known as "Europe." In particular, the elites of ancient Greece and Rome shared a bookishness and a widespread lack of interest in mechanical invention much like the attitudes of most Chinese literati throughout the imperial period.[37] But beginning around the thirteenth century, a remarkable gift for mechanical imagination increasingly came to mark European culture, an "inventive spirit [that] achieved self consciousness" in the Renaissance.[38] In the following century, Leonardo da Vinci (1452–1519), partly influenced by dissection work being done in anatomical studies at that time,[39] pioneered a view of mechanisms that gave unprecedented emphasis to the mechanical elements of which they were composed.[40] At its extreme, this perspective opened the way for Europeans to give full rein to a delight in mechanical ingenuity that sometimes even took precedence over mechanical efficiency so that machines became more complex and elaborate than really needed to perform their tasks.[41] At the extreme, wildly imaginative

37. Ivins, *Prints and Visual Communication*, 11.
38. Pacey, *Maze of Ingenuity*, 58–59. One sees this fascination even in the devising of tools for producing art. Albrecht Dürer (1471–1528) for example was a great inventor of graphic and mechanical drawing procedures, the latter making use of new instruments such as his "extraordinary compasses for drawing ellipses, spirals, and new curves not yet defined on the geometric level." Camerota, "Renaissance Descriptive Geometry," 208. Nothing of this sort is to be found in China. The Chinese instead confined their searches to improved brushes, paper and ink for use mainly in calligraphy and painting; see the many examples in Tsien *SCC* 5:1.
39. Galluzzi, "Leonardo da Vinci; Engineer and Architect," 397.
40. Bert S. Hall, "The New Leonardo," *Isis* 67 (1976), 467, reviewing the work of L. Reti. Leonardo's approach assigned unprecedented importance to component parts drawings, as seen throughout his notebooks.
41. Donald Cardwell, *Wheels, Clocks and Rockets; A History of Technology* (New York: W. W. Norton, 1995), 42.

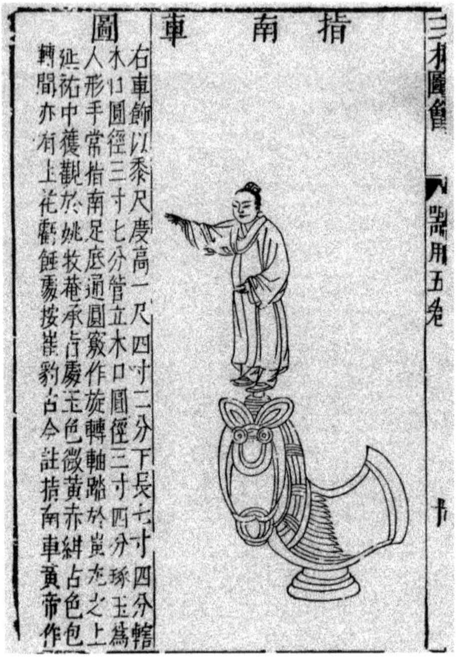

6.2 Imaginative rendering of the direction-indicating figure on a south-pointing carriage. The artist in this case obviously was content to convey his imaginary image but had no curiosity about how the mechanism worked.

and complicated machines were dreamed up that could never have possibly worked (Fig. 6.1). Finally, in yet another revolutionary development, we see a new interest in the *theory* of machines that started with Galileo (1564–1642) and led step-by-step to the discovery of laws that applied to the functioning of all machines and thus formed the basis for modern mechanics.[42]

In China, on the other hand, misgivings about mechanical devices and ingenuity are found in some of the earliest surviving writings and their echoes persisted among most educated people down to the nineteenth century.[43] This helps account for the absence in China of theoretical study of mechanics before the arrival of the Jesuits in the late sixteenth century. As Derk Bodde reminds us, there was in Chinese history only one group of thinkers who dealt analytically with mechanics, the later Mohists in the second half of the first millennium BCE. Significantly, their "utilitarian" views fell into obscurity after the founding of the empire by

42. Pacey, *Maze of Ingenuity*, 72–78.

43. Bodde (*Chinese Thought, Society, and Science*, 255–57) brings together a number of early stories that suggest some of the moral, economic and political concerns that undergirded this disdain for mechanical ingenuity. Blunden and Elvin provide in Cheng Tingzuo and his comments on the "excessive ingenuity" of Europeans a mid-eighteenth century example of how disdain for mechanical innovation was still alive and well among the literati. Caroline Blunden and Mark Elvin, *Cultural Atlas of China* (New York: Facts on File, 1983), 144–45. (See also Mark Elvin, "How Did the Cracks Open? The Origins of the Subversion of China's Late-Traditional Culture by the West," *Thesis Eleven* 57.1 (May 1999), 10 for an eloquent late nineteenth century expression of "[f]undamentalist anti-mechanism.") Mechanics and invention were never integrated into literati culture in China as they were in medieval high culture in Europe where God himself was often thought of as Artisan, Architect, Mechanic, Clockmaker. Francis C. Haber, "Time, Technology, Religion, and Productivity Values in Early Modern Europe," in J. T. Fraser, N. Lawrence, and F. C. Haber, *Time, Science, and Society in China and the West (Study of Time* V) (Amherst: The University of Massachusetts Press, 1986), 82–83. Haber also suggests that "the tendency in Christianity to think in terms of universal mankind and universal purpose provided a favorable environment for projecting machines and productivity [*even*] *to a cosmic role*, rather than just the everyday experience of making ends meet." Ibid., 82 (my emphasis).

the Qin emperor in the late third century BCE and had virtually no effect on later Chinese thought.[44]

The lack of interest in a theoretical understanding of mechanics, even when joined with suspicions about the usefulness of mechanical ingenuity, did not by any means preclude entirely any fascination with mechanical devices. As we have seen in Chapter 2, there is a long history of Chinese delight in automata or mechanical toys; this too continued down to modern times.[45] The Chinese also had, at least by the Sui and Tang dynasties (perhaps even as early as the Han[46]), vehicles that served as odometers and other vehicles ("south-pointing carriages") with mechanisms using differential gears that could indicate south regardless of the direction the cart faced (Figs. 6.2 and 6.3).[47] Moreover, European mechanical clocks, often with very complex jackwork, were highly prized items in China during the Ming and Qing and even proved an important tool in the Jesuit missionary efforts at the imperial court and among the literati.[48]

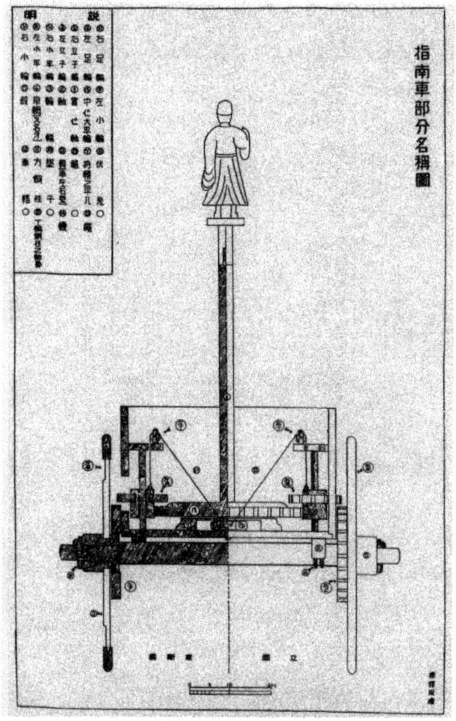

6.3 Modern attempted reconstruction of the mechanism for a south-pointing carriage. For details, see Needham and Wang, *SCC* 4:2, 293, Fig. 528 and discussion 288ff. No Chinese before the nineteenth or twentieth centuries ever came close to producing a working drawing with anything like this precision and complexity.

The Chinese, as we have seen repeatedly in our discussions above, *were* quite capable of developing productive machines of some complexity and ingenuity.[49] Yet there were very real limitations to the fascination mechanics held for the Chinese, above all when it came to theoretical understanding. Various explanations have sought to account for this. Joseph

44. Bodde, *Chinese Thought, Society and Science*, 168.
45. Needham and Wang, *SCC* 4:2, 156–65; Golas, "Emergence of Technical Drawing," 33–34.
46. Needham and Wang, *SCC* 4:2, 283.
47. Needham and Wang, *SCC* 4:2, 281–300, especially 298–300; Pomeranz, *The Great Diversion*, 64n142.
48. Catherine Pagani, *"Eastern Magnificence and European Ingenuity": Clocks of Late Imperial China* (Ann Arbor: The University of Michigan Press, 2001), Chap. 1. But it should be remembered that these clocks were usually much more appreciated as ornaments rather than as instruments; hence the emphasis on the mechanical accouterments. See Landes, *Revolution in Time*, 75ff.; Nicholas Standaert (ed.), *Handbook of Christianity in China*, vol. 1 (Brill: Leiden, 2001), 834–35.
49. For a handy overview of Chinese mechanical achievements, Needham's "Author's Note" in the "Mechanical Engineering" volume of *SCC* is still well worth reading. Needham and Wang, *SCC* 4:2, xliii–li.

Needham has suggested that the Chinese interest in actions at a distance (magnetism, tides, musical resonance) may have "handicapped" the Chinese in thinking theoretically about dynamics.[50] Other authors have suggested that the Chinese language was an impediment insofar as it did not place a strong emphasis on accuracy and the precise expression of ideas.[51] In any case, the Chinese were never tempted to duplicate the shift in Western thought from a view of the mechanical arts as an "artificial" human construction to a science of mechanics capable of contributing to a greater comprehension of nature and the cosmos.[52]

On the more practical level, it seems indisputable that most Chinese in the late empire were far less beguiled by intricate machinery for doing useful or productive things than were Europeans from the late Middle Ages on.[53] Mark Elvin has described the Chinese as too "severely pragmatic," without the European gift for "precisely visualized mechanical fantasy," a kind of "disciplined imagination."[54] Kenneth Stunkel suggests that, in the assessment of many Confucian officials, "a preoccupation with ingenious machines, like a fever for profit and material success, could eclipse the evaluating mind, stultify moral growth, and was best kept firmly in its place."[55] Moreover, if Needham is correct, the distrust of innovation in things mechanical was also widely shared by craftsmen.[56] This may help account for the fact that traditional Chinese drawings of machines usually focused on their overall, exterior appearance, seeing them mainly as complete entities rather than emphasizing their component parts. In direct contrast to a preference in medieval Europe for illustrating

50. Joseph Needham, Kenneth Girdwood Robinson and Ray Huang, "General Conclusions and Reflections," in *Science and Civilisation in China* (*SCC*), ed. Joseph Needham, vol. 7, part 2 (Cambridge: Cambridge University Press, 2004), 227.
51. Derk Bodde has argued (in opposition to Needham) that there was a "prevailing spirit" in the written classical language universally used by the educated in China that was "too often not conducive to clarity, precision and simplicity." For example, the frequently wide range of (even seemingly unrelated) meanings that could attach to individual characters/words easily fostered ambiguity. For a magisterial discussion of this topic with full consideration of opposing views, see Bodde, *Chinese Thought, Society and Science*, 88–96.
52. Long, "Power, Patronage, and the Authorship of *Ars*," 3; Haber, "Time, Technology, Religion, and Productivity Values," 82.
53. This is especially striking in conservative technologies such as mining (Golas, *SCC* 5:13, 11; Hans Ulrich Vogel, "The Mining Industry in Traditional China," in Helga Nowotny (ed.), *Cultures of Technology and the Quest for Innovation* (New York: Berghahn Books, 2006), 170 and 172) but tends to be a characteristic feature of most Chinese productive technologies, except perhaps clothmaking.
54. Elvin, "China as a Counterfactual," 108; Elvin "Skills and Resources in Late Traditional China," in Dwight H. Perkins (ed.), *China's Modern Economy in Historical Perspective* (Stanford: Stanford University Press, 1975), 85–113, especially 111; Blunden and Elvin, *Cultural Atlas of China*, 144–45. See also Pacey, *Maze of Ingenuity*, 74. We begin to see evidence of the "gift" in Europe already in the first half of the fourteenth century in the treatises of Guido da Vigevano. Hall, "Production et diffusion," 105–7; Lefèvre, *Picturing Machines*, 14, 20–21, 60–67.
55. Stunkel, "Technology and Values," 69.
56. ". . . one can see how great the difficulties were with which inventive mechanical genius had to contend in ancient times. The artisanate was wedded to those customary paths which the experience of ages had created and it must have distrusted innovations almost as much as the Confucian scholars, though for different reasons." Needham and Wang, *SCC*, 4:2, 49.

details rather than complete machines,[57] Chinese illustrators seldom showed an interest in taking a machine apart pictorially or explicating its inner workings.[58]

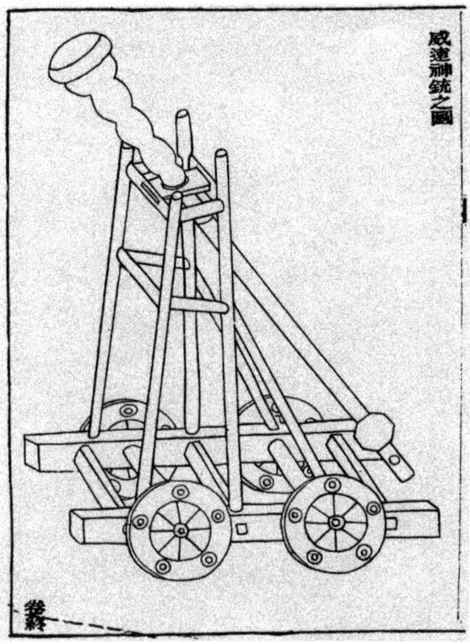

6.4 An unclear depiction of either a counterweighted trebuchet or an elevated bombard from the eighteenth-century *Gujin tushu jicheng* 古今圖書集成 (Compendium of books and illustrations past and present).

Making constant improvements to find the optimum design for a machine, developing better machines to do the same job, or experimenting consciously with different approaches to solve a given technological problem: these appear to have been rarely seen phenomena even in late imperial China.[59] Despite pioneering efforts to solve the problem of converting reciprocating to rotary motion,[60] no one in China ever remotely resembled a Leonardo da Vinci who designed *dozens* of ways just to solve this single problem.[61] Chinese over the centuries devised quite a variety of mills, but no Chinese figure ever came close to matching a Francesco di Georgio who alone illustrated fifty-eight different kinds of mills![62] In China, pretty good seems to have been all too often good enough when it came to mechanization.[63]

57. Hall "Production et diffusion," 153; Hall, "The New Leonardo," 467.

58. Zhang Yinlin, "Inventions and Inventors," 64.

59. Square-pallet chain pumps offer a good example. Elvin, "Skills and Resources," 111. Nor was this disinterest in experimentation limited only to mechanical devices: Agricola's plethora of fine alterations made to sluices to facilitate the concentration of different kinds of ore finds no parallel in China. Golas, *SCC* 5:13, 367. For two extreme examples of the European gift for precisely thinking through on paper machines that did not yet (and indeed would never) exist, see Elvin, "China as a Counterfactual," 110–11, Figs. 5a and 5b.

60. Needham and Wang, *SCC* 4:2, 380–90.

61. Bern Dibner, "Machines and Weaponry," in L. Reti (ed.), *The Unknown Leonardo* (London: Hutchinson, 1974), 167. In the same way, we have no illustrations of right-angle gearing aiming to show how it could be applied in two or more contexts or to solve similar but not identical problems. All Chinese illustrations of technology are "singular" in that they focus on a single, particular technological application.

62. Pamela O. Long, "Picturing the Machine: Francesco di Giorgio and Leonardo da Vinci in the 1490's," in Lefèvre, *Picturing Machines*," 125.

63. Or, in the words of Wang Chong of the Han (though he was not referring specifically to technology), "What can reasonably be assumed is not stated." Needham, Robinson and Huang, *SCC* 7:2, 98, fn. 14. Had he been referring to illustrations, he might well have said "depicted." In that form, his statement would reflect a characteristic all too frequently encountered in Chinese illustrations of technology, namely, the omission or fudging of (sometimes critical) details.

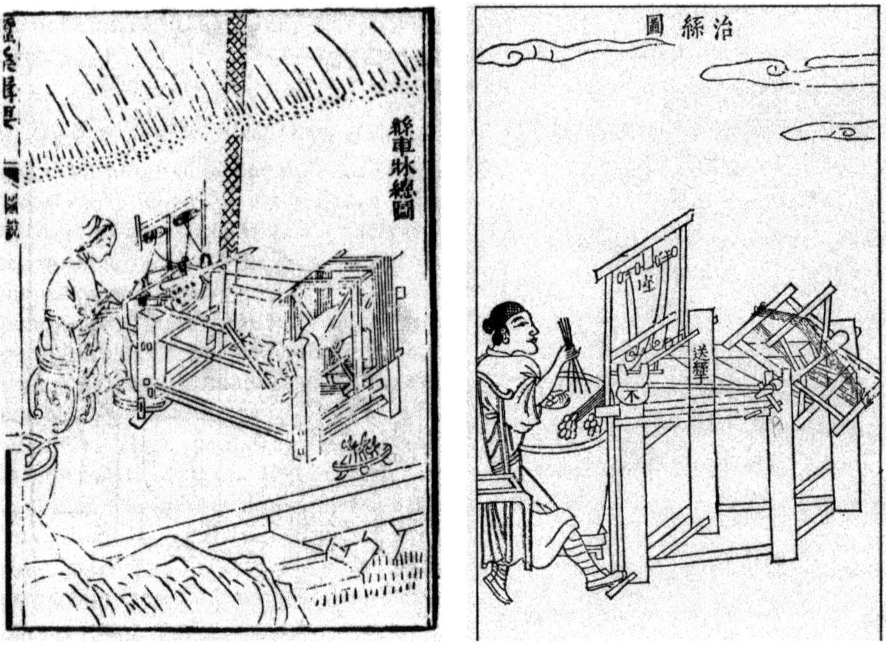

6.5 Silk-reeling frame, from (a) Kuhn, *SCC* 5:9, 392, Fig. 237; and (b) Sun and Sun, *T'ien-kung k'ai-wu*, 46, Fig. 2–6.

In their limited interest in the workings of complex mechanical devices as well as their frequent aversion to deliberate innovation, the Chinese differed not only from Europeans but also from the Japanese, especially in the last century of the Tokugawa period (1600–1868). The Japanese seem to represent an attitude toward machinery that was somewhere between that of the Europeans and that of the Chinese, inasmuch as they were apparently more interested in mechanical complexity in itself (as well as the imposing size of some of the mechanisms) and only secondarily in their effect on output. That such attitudes were widespread is suggested for example by that fact that city guidebooks of the period often point out complex machines that deserve a look by visitors.[64]

The frequently casual, not to say careless, approach to the depiction of mechanisms by Chinese artists who illustrated technical subjects reflects attitudes they held in common with their countrymen. Fig. 6.4 presents a particularly striking example: this drawing is such that it could be interpreted as either a counterweighted trebuchet or as an elevated bombard.[65] All too often illustrators were satisfied to provide visual *impressions* that may have shown how machines were used or what advantages they brought but seldom provided

64. Timon Screech, *The Western Scientific Gaze and Popular Imagery in Later Edo Japan: The Lens within the Heart* (Cambridge: Cambridge University Press, 1996), 63–64.

65. Needham, Ho, Lu and Wang, *SCC* 5:7, 286, Fig. 81. Only the caption, "wonder-working long-range awe-inspiring cannon" makes it clear that this is to be seen as a bombard. Indeed, in two other illustrations of a similar mobile weapon, we have in one case (p. 280, Fig. 76) clearly a trebuchet while in a later version (p. 281, Fig. 77), it has clearly transmogrified into a bombard.

precise information on a machine's exact appearance, or how it worked, or how it might be constructed. Another example in the realm of more complex machines is the 1871 illustration of a silk-reeling frame (Fig. 6.5 (a)), which is markedly better than the Song Yingxing version from the late Ming (Fig. 6.5 (b)), but which could never serve as more than an aide-memoire for someone who already knew how it worked or even knew how to build one.

The Role of Military Technology

Finally, it must be noted that China in the late Ming and in the Qing did not experience quite the same range of technological imperatives as one finds in Europe at this time. For example, in China, only a relatively limited number of productive technologies invited significant mechanization. We shall discuss below, for example, how the combination of cheap labor and costly resources inhibited the introduction of mechanization into farming which, as we have seen, was by many measures China's most important technology. In other cases such as mining, the economics of traditional production practices simply did not work to promote greater mechanization.[66]

However, one technology of prime importance in Europe, weaponry, might also have been expected to stimulate significant technological innovation in China in the late imperial period as it did earlier with the world's first development of gunpowder weapons. This was an area where cost considerations often gave way to more primal concerns of defense and, in some cases, even survival. Military force has always been an intrinsic component of political power and governments throughout human history have been willing to devote considerable resources to producing more and bigger and better weapons. This is exactly what one sees on a major scale among the increasingly rich and powerful states of early modern Europe. By the fifteenth century, an increasingly competitive political environment was giving rise to fierce competition in military technology. Even the very notion of rulership came to be intimately bound up with the possession of powerful, leading-edge technology.[67] Engineers with special skills in this area could count on steady work and high pay as they promoted their careers by offering themselves and their skills to the service of one or another political leader.[68] Workshops directed by these entrepreneurial innovators proliferated and became the source of much of the new Renaissance technology.[69]

66. Golas, *SCC* 5:13, 410–15. Similar constraints may have been operative in the making of paper, which over the centuries remained a handicraft industry using simple techniques. Despite China's very long lead in paper production, it was in Europe, not in China, that the mechanization of paper manufacture first occurred.
67. Long, "Power, Patronage and the Authorship of *Ars*," 3.
68. Paolo Galluzzi, "Leonardo da Vinci: From the "elementi macchinali" to the Man-Machine," in Claire Farago (ed.), *Leonardo's Science and Technology: Essential Readings for the Non-Scientist* (New York: Garland, 1999), 61; Edgerton, "Renaissance Development of Scientific Illustration," 172; White, "Flavor of Early Renaissance Technology," 41.
69. And even science. One thinks, for example, of the importance of ballistics.

In the initial stages, when gunpowder weapons were still something very new in Europe, there were few experienced and skilled artisans to make the weapons and few soldiers who knew how to use them. But in an environment of keen military competition, states had a strong incentive to increase those numbers as quickly as possible. In contrast to traditional methods of hands-on, one-on-one instruction and the haphazard experiences of soldiers on the battlefield, printed books offered an unprecedented opportunity to transmit information on the new technology rapidly and widely, and the illustrations in those books did much to make that information clear and precise. It is in these circumstances that we begin to see in Europe for the first time true technical illustration helping undermine established craft practices that, even at the beginning of the sixteenth century, continued to impede reliance on practical military and civil engineering books as conveyors of technological knowledge.[70]

Little in the way of similar developments occurred in China after the conquest by the Manchus and the establishment of the Qing dynasty in 1644. In large part, this may have been because much of the potential of premodern weaponry had already been realized in China in a great wave of military innovation in the eleventh to thirteenth centuries. Those innovations had brought the weapons at the disposal of the government to a point where they were unchallengeable by any enemies, at least until those enemies, above all the Mongols in the thirteenth century and the Manchus in the sixteenth and early seventeenth centuries, succeeded in borrowing the technology on a major scale for themselves. This, plus the fact that there were now simply fewer military advances possible, slowed still further the pace of change in the Qing up to the later nineteenth century.[71]

The story of how the Manchus caught up with the Chinese in the manufacture and use of firearms has been well told by Nicola Di Cosmo.[72] However, after their success in overthrowing the Ming and pacifying the country, efforts to improve weaponry largely came to an end. This arrested development probably also owed a great deal to a further factor, the successful Manchu adoption of long-proven Chinese policies and attitudes that not only disparaged war and violence[73] but also stressed tight government control over, and even a monopoly of, weapons.[74] The production of weapons for the army was largely confined

70. Edgerton, "Renaissance Development of Scientific Illustration," 184; Hall, "Production et diffusion," 167–68.

71. A potential breakthrough, more scientific than technological to be sure, but one that could have supported advances in gunpowder technology, was scientific ballistics. But although this new knowledge from the West was being explored by the Japanese at least as early as the beginning of the seventeenth century, in China, much like theoretical mechanics, it drew little or no interest until much later. Needham, Ho, Lu and Wang, *SCC* 5:7, 390–91.

72. Nicola Di Cosmo, "Did Guns Matter? Firearms and the Qing Formation," in Lynn A.Struve, *The Qing Formation in World-Historical Time* (Cambridge, MA: Harvard University Asia Center, 2004), 121–66.

73. For a discussion of Chinese ideas on warfare, see Bodde, *Chinese Thought, Society and Science*, 248–53.

74. Indeed, concern for their own security led the Manchus to limit Chinese participation in high-level military affairs and to ban many of the military works produced in the Ming. Mote, *Imperial China*, 927–28 and Needham and Yates, *SCC* 5:6, 28–29.

to massive government factories and arsenals mainly at the capital.[75] Out of necessity and habit, they were bureaucratically organized just like other government activities.[76] In contrast to the workshops of the Renaissance, this was a non-competitive environment where standardization and frugality were the dominating values, with little incentive to innovate.[77] There was nothing that even remotely resembled the "symbiotic relationship between armaments and money making" found in Europe after the fifteenth century.[78]

This kind of essentially stagnant weapons industry was something the government could live with in part because there was little to be gained and perhaps even something to be lost by attempting to produce improved weapons.[79] Ray Huang is surely at least partly correct when he emphasizes that most of the government's weapons would be used to maintain order within the empire. To maintain control over primitively armed agricultural populations, the government needed numbers of troops that it could control and maintain over long periods of time. Better weapons would be at most of marginal significance and thus there was little incentive to devise and produce them.[80] And without the stimulus of new and changing technology, there was little need to produce and distribute updated manuals with clear, informative illustrations.

That is not to say that writing about military matters, including weapons technology, came to a halt in the late imperial period. But it did suffer a significant decline in quantity if perhaps not quality: while we have the titles of at least 268 works on military topics written during the Ming, the rate of production of new titles falls by almost two-thirds during the Qing.[81] Moreover, the illustrations of weapons we find in the later works display little or no advance in techniques to render them more informative. There simply was no urgency to make either the weapons or the illustrations better. Apart from the likelihood that the

75. In large part for security reasons. Needham, Ho, Lu and Wang, *SCC* 5:7, 313–14. The official history of the Ming gives expression to these security concerns in regard to gun-founding: "The casting of guns and cannon is done in the Nei Fu 內府 palace compound, and it is forbidden to disclose any of the secrets of the *techniques and designs*." Needham, Ho, Lu and Wang, *SCC* 5:7, 341 (italics added).

76. This picture may have held less true in the extraordinary conditions of the late Ming when, according to Ray Huang, supply problems led to most of the weapons used by armies in the field being procured locally. In any case, these local arsenals were in no way centers of innovation and their products were, if anything, of poorer quality than the weapons produced in central government workshops. Ray Huang, *Taxation and Government Finance in Sixteenth-Century China* (Cambridge: Cambridge University Press, 1974), 320; Xu Dixin and Wu Chengming (eds.). *Chinese Capitalism, 1522–1840* (New York: St. Martin's Press, 2000), Chap. 3, especially 67 and 70.

77. You Zhanhong 游戰洪, "Lun junqi zeli" 論軍器則例 [A study of regulations and precedents on weapons and military equipment]," in Moll-Murata et al., *Chinese Handicraft Regulations of the Qing Dynasty*, 323.

78. Stunkel, "Technology and Values," 80.

79. As late as 1799, when the government melted down 160 old cannons, their newly manufactured replacements had a shorter range than the originals! You, "Regulations and Precedents," 323.

80. Ray Huang, *China: A Macro History* (Armonk, NY: M. E. Sharpe, 1990), 133.

81. Needham and Yates, *SCC* 5:6, 29. Some of this decline can probably be attributed to concerns arising from "the complicated quagmire of Manchu-Chinese relations" that could have seen Chinese interest in this technology as suspicious; Joanna Waley-Cohen, "China and Western Technology in the Late Eighteenth Century," *American Historical Review* 98.5 (1993), 1535.

government would presumably have been less worried about the circulation of military works that were none too exact in their treatment of the construction and use of weapons, one also gets the feeling that these works were meant mainly for two audiences, neither of which would have been overly concerned about the precision of the information in the illustrations. Many of the readers were surely military men who, like the viewers of illustrations in agricultural works, could rectify omissions and inaccuracies on the basis of their own experience. Most of the other readers were likely members of the literate elite who for one reason or another–perhaps because of some connection with their official duties–had an interest in military matters but hardly one that went so far as to desire or expect a high level of precision in pictures of this kind.[82]

The Jesuit Contribution and Its Limited Impact

As the Jesuits, led by Matteo Ricci, began their missionary work in China in the late sixteenth century, they relied greatly on information about European science and technology to spark among educated Chinese a broader interest in Western civilization.[83] In this way, they hoped to create receptivity to the teachings of Christianity as a product of the same civilization that had made such remarkable advances in understanding and exploiting nature. In the pursuit of this strategy, they had extensive recourse to books, drawings and paintings they had brought from Europe.

The drawings and paintings, with their realism and use of techniques such as linear perspective, chiaroscuro (light and shade effects) and hatching to depict volume attracted from the start considerable curiosity among many Chinese.[84] Thus a late Ming writer, Jiang Shaowen 姜紹聞, commented to the Wanli emperor on portraits of God and of the Virgin

82. There were also the pictures of weapons in works of fiction, if anything of even poorer quality, for readers whose interest was in the story being narrated, and who looked for comfortably familiar kinds of illustrations that had no reason to aim for accuracy even in the few cases where they pictured weapons other than ordinary swords, bows and arrows, and the like.

83. For an account of the scientific ideas the Jesuits introduced in China, Willard Peterson's "Western Natural Philosophy Published in Late Ming China" (*Proceedings of the American Philosophical Society* 117.4 (August 1973), 295–322), based on a careful and thorough reading of the Chinese books written by the Jesuits (with the help of Chinese collaborators), is still unsurpassed. The fascinating comparisons between Chinese and Western theories are the product of the author's thorough understanding of both Western and Chinese scientific thinking at that time. For an introduction to Chinese science stressing mathematics, astronomy and alchemy, see Ho Peng Yoke's *Li, Qi and Shu: An Introduction to Science and Civilization in China* (Mineola, NY: Dover Publications, 2000) which actually goes into significantly greater depth than one might expect from an "introduction."

84. Chiaroscuro was by no means entirely absent in the Chinese painting tradition (see Mary H. Fong, "Technique of 'Chiaroscuro'" and Sullivan, *Symbols of Eternity*, 64) but Chinese artists had done relatively little to exploit its possibilities. Linear perspective and the use of mechanical calculations to create a perspective view were entirely new to China and rapidly drew considerable interest. Michael Sullivan, "Some Possible Sources of European Influence on late Ming and Early Ch'ing Painting," in *Proceedings of the International Symposium on Chinese Painting* (Taipei: National Palace Museum, 1972), 596–97. (For early use of mathematical calculations to construct a perspective view, see Zhang and Tian, "Visual Presentation," 82–83.)

Mary presented by Ricci: "The facial features and the lines of clothing look like images of real things in a mirror, vividly alive. The dignity and elegance of the figures are beyond the technical capability of Chinese painters to produce."[85] Much later, in the mid-eighteenth century, the great flower painter Zou Yigui 鄒一桂 (1676–1782) who associated at the court with Jesuit painters such as Castiglione and Attiret and who wrote a treatise on Western painting entitled "Small Hill Painting Manual" (*Xiaoshan hua pu* 小山畫譜) extolled the skills of Western artists:

> Western artists excel in sketching and drawing; thus, when they depict light and shade and distance they are exact to the last detail. All human figures, houses and trees in their painting have shadows trailing behind them. The colors and brushes they use are different from those used in China. When they paint a scene, the perspective is presented as from broad to narrow, calculated in the dimensions of a triangle. Their mural pictures depicting palaces look so real that people are almost tempted to walk into them.[86]

Small wonder that the Jesuits hoped that admiration of these skills might also attract Chinese to the Christian message often present in these images.

Overall, however, even those Chinese painters who found it possible to appreciate the unique talents of Western artists tended to be qualified their admiration. One reaction to the new techniques was that, clever as such pictures might be, they simply did not measure up to the standards of fine painting as these were understood in literati circles.[87] This was exactly the assessment of Zou Yigui. While admitting that students could learn a thing or two from viewing Western pictures, he concluded that, because of the incompetence of their brushwork (judged by Chinese standards, of course), the Western paintings "have no place on our qualitative scale of painting."[88] More extreme reactions might even see the ability to create a perfect illusion of depth and distance on a flat surface as a kind of diabolical magic that was totally contrary to nature.[89]

Relatively few Chinese painters beyond imperial court circles (where the Qing emperors of the seventeenth and eighteenth centuries were keen sponsors of Jesuit painters) took

85. Cited in Hsiang Ta, "European Influences on Chinese Art in the Later Ming and Early Ch'ing Period," *Renditions* 6 (1976), 156.

86. Cited in Hsiang, "European Influences," 175 (slightly modified).

87. How widespread this and other attitudes were is hard to say, given the paucity of surviving evidence on the Chinese painters' reactions. Delahaye, "Du peu d'effet," 243 and 244n9. Moreover, Delahaye also notes (247–48) that there was more interest in the techniques among non-literati painters who used them, especially in the painting of portraits, a realm of painting not held in particularly high esteem by scholar painters. See also Clunas, *Art in China*, 194.

88. Cahill, *Compelling Image*, 72; compare Hsiang, "European Influences," 175–76 and Michael Sullivan, *The Arts of China* (Berkeley: University of California Press, 1973), 215. (Wherever possible, Hsiang's translations, particularly those from primary sources, should be checked especially against Cahill's which are consistently more accurate and often more elegant.) For a similar assessment of Western painting, with more specifics, see the comments by Wu Li 吳歷 (1632–1718), a prominent early Qing poet, painter and ordained Jesuit priest who was familiar with but totally resistant to Western painting techniques. Cahill, *Compelling Image*, 35.

89. F. Feuillet de Conches, "Les Peintres européens en Chine et les peintres chinois," *Revue Contemporaine* 25 (1856), 225; John Barrow, *Travels in China* (London: T. Cadell and W. Davies, 1804), 325.

their admiration of Western painting techniques to the further stage of incorporating them into their own painting. This was especially true for literati painters.[90] Their approach to painting was too different and too securely entrenched to be challenged by imports from well beyond the Chinese traditions. For example, indicating distance by varying the size of the elements making up a painting was a common and, presumably, satisfying convention that undercut experimenting with single point, linear perspective. The same was true for other Chinese techniques such as atmospheric or aerial perspective, that relied not only on placement ("up is back"[91]) but also on variations in color, tone and clarity, or parallel isometric projection which could provide an optically acceptable if not geometrically correct rendering of distance.[92] The elevated ("bird's-eye") view commonly adopted by Chinese painters, which dates back at least to the Later Han, could serve to separate and therefore make clearer activities occurring closer to or farther from the viewer, as in this relief showing two men on the left operating pedal tilt-hammers to hull grain and a man on the right working a winnowing fan (perhaps the earliest surviving comprehensible portrayal of this machine) (Plate 13).

Certain Western conventions could also at times make their works of art difficult for the Chinese to grasp. Edgerton offers a good example in a Chinese artist's effort in the early seventeenth century to reproduce an engraving of the nativity scene originally produced in Antwerp (Figs. 6.6 (a) and (b)). He rightly points out that the Chinese artist missed the idea of the Christ child as "symbolic source of illumination." (One might of course ask who, except in horror or science fiction films, ever saw a body throwing off illumination?) Being unfamiliar with chiaroscuro and perhaps misled by the marks of the engraver's burin, the Chinese artist also misinterpreted the black area above the figures, seeing it not as portraying the dark inside of a cave but as depicting reed matting attached to a cross beam below the roof (perhaps even thinking at the same time that the Western artist was not very successful here?)[93] This illustration also suggests another impediment to greater acceptance by the Chinese elite of Western painting and illustration techniques: many of the works they examined focused on didactic religious imagery, which not only was strange in itself to the Chinese but which was not even recognized as a suitable subject for any work that made a claim to artistic excellence. For the Chinese elite, "didactic art," especially with a religious theme, was an oxymoron. Craftsmanship, yes, perhaps even high-level craftsmanship, but not art.[94]

90. For some exceptions, see Cahill, *Compelling Image,* 13–26; Mayching Kao, "European Influences in Chinese Art, Sixteenth to Eighteenth Centuries," in Thomas H. C. Lee (ed.), *China and Europe: Images and Influences in Sixteenth to Eighteenth Centuries* (Hong Kong: The Chinese University Press, 1991), 259.

91. Silbergeld, *Chinese Painting Style,* 38.

92. Needham, Wang and Lu, *SCC* 4:3, 111–12; Sullivan, *Chinese Landscape Painting,* II, 91–92.

93. Edgerton, "Renaissance Development of Scientific Illustration," 188–92.

94. Clunas, *Art in China,* 130 and *Pictures and Visuality,* 181. Delahaye ("Du peu d'effet," 249) notes for an earlier time the small influence of Buddhism on Chinese painting and suggests that, given the equally large differences in worldview between Christianity and Chinese ideas, it may be unlikely that even a greater penetration of Christianity in China would have led to significantly greater acceptance of Western techniques by Chinese

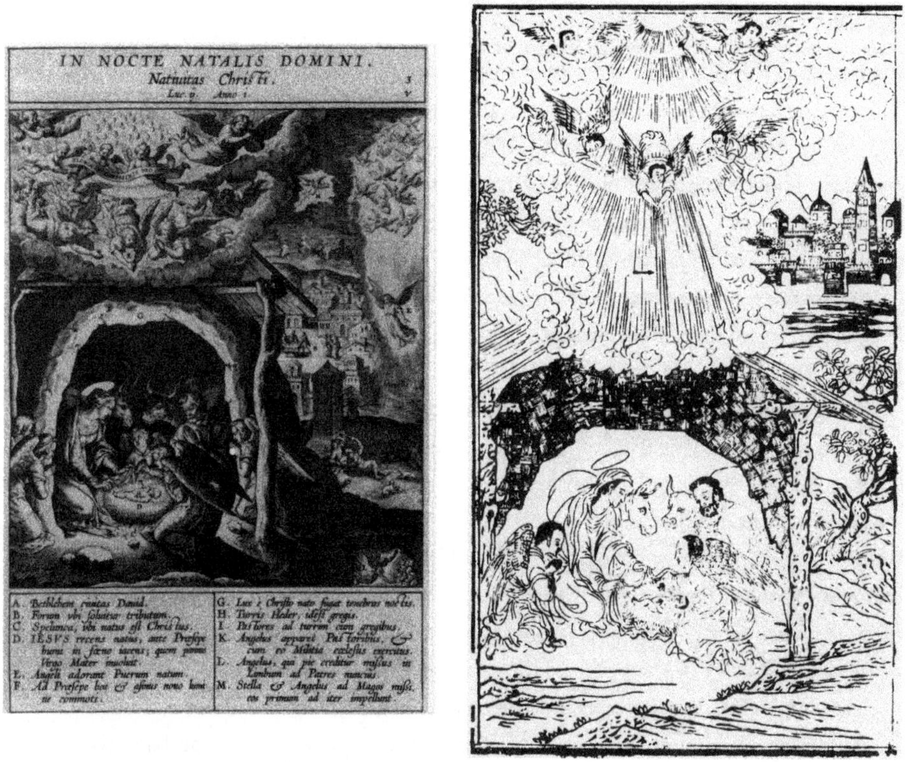

6.6 (a) *The Nativity* (1593) by Jerome Wierx; (b) Woodblock copy from a 1619/1620 Chinese edition of da Rocha's *Metodo de Rosario*.

All of which is not to say that Western artistic techniques could never have had significant influence on Chinese artists.[95] Two examples–and we must admit there seem to have been no others at this level of assimilation–will show us what effects these Western works could give rise to in rare cases and perhaps provide further hints why their influence was so limited. We shall first violate chronology and return to the end of the Ming to consider the scholar official Wang Zheng.[96] Wang's keen interest in machines as well as in Western engineering works brought by the Jesuits to China allowed him to achieve a not inconsiderable understanding of Western mechanics and technology. This even led Needham to characterize him, probably too optimistically, as "China's first modern engineer."[97] We then turn to the painter Jiao Bingzhen who at the end of the 17th century produced under imperial order a series of illustrations for a new edition of the *Pictures of Tilling and Weaving* that made extensive use of Western painting techniques, especially linear perspective.

painters.

95. For extensive bibliography of Western influences on Chinese art, see Cahill, "Compelling Image," 229–30n13 and 233–34nn1–13.

96. We are not the first to do so, since there is a biographical entry for Wang in *ECCP* but not in the *Dictionary of Ming Biography*!

97. Needham and Wang, *SCC*, 4:2, 171, fn. a.

China's First Modern Engineer?

Wang Zheng 王徵 was born in Jingyang, Shaanxi province in 1571, the son of a well-educated father who made his living as a teacher and was keenly interested in mathematics and nature studies.[98] His son's talent and education enabled him to pass the *juren* (second-highest) official examination in his early twenties (1594), but it would take him another twenty-eight years and nine failed attempts before he finally passed the (highest) *jinshi* exam in 1622, just after turning fifty.[99] Despite his many exam setbacks, Wang became one of the most famous scholars of his time, author of some forty-three works[100] and perhaps as well known as another intellectual giant of the period, the famous Jesuit patron Xu Guangqi 徐光啓.[101]

Wang's frequent visits to Beijing to take the exams enabled him to get to know the Jesuits there, an opportunity he used to the full. In the winter of 1615 or spring of 1616, he made the acquaintance of Diego de Pantoja[102] who probably baptized him in 1616.[103] Wang also studied Latin under Nicholas Trigault (1577–1628), perhaps the first Chinese to do so, though he seems to have developed at best a limited ability to read Latin texts.[104] He collaborated with Trigault on the *Xiru ermuzi* 西儒耳目資 (An aid to the ear and the eye of Western scholars), a Chinese-Latin phonetic dictionary of very high quality which was printed in 1626.[105] Wang supervised the printing of the book and signed as proof-reader.[106]

From his youth, Wang was keenly interested in agricultural machines and other devices for daily use.[107] He devoted considerable thought to improving them, perhaps even supervising construction of improved versions that were put to use on his own family's lands. These were recorded in a one chapter (*juan*) illustrated work, the *Zhu qi tushuo* (ZQTS) 諸器圖說 (Illustrations and explanations of various machines), completed in 1627.[108]

98. One of two short works he wrote was the *Suanshu gejue* 算數歌訣 or "Rhymed Rules for Calculating," *ECCP* 2, 807.

99. *ECCP* 2, 807.

100. Nineteen are still extant. The works are listed and some are described in Ren Dayuan, "Wang Zheng: Xixue yu xin sixiang de chuanbozhe yihuo yige tianzhujiaotu?" [Wang Zheng: A scientist, philosopher, and Catholic in Ming dynasty China], in Malek, *Western Learning and Christianity in China*, 341–43.

101. Ren, "Wang Zheng," 367.

102. Zhang Baichun, "Archimedian Mechanical Knowledge," 2nd of 16 pp.

103. de Pantoja may well have been responsible also for his new Christian name of Philip (Pei Libo 裴理伯), Standaert, *Handbook*, 406n5.

104. *ECCP* 2, 808; Zhang Baichun and Tian Miao, "Wang Zheng and the Transmission of Western Mechanical Knowledge to China," in Zhang and Renn, *Transformation and Transmission*, 83. In contrast to educated Japanese who often learned Latin or Portuguese, few Chinese officials or scholars chose to learn a foreign language. Nicholas Standaert, "The Transmission of Renaissance Culture in Seventeenth-Century China," *Renaissance Studies* 17.3 (2003), 370. Interestingly, the Japanese also proved much more adept at learning Western visual languages, a point to which we shall return below.

105. Standaert, "Transmission," 381.

106. Standaert, *Handbook*, 870; Zhang and Tian, "Transmission," 79–80.

107. *ECCP* 2, 808; Zhang and Tian, "Transmission," 79; Ren, "Wang Zheng," 344.

108. *ECCP* 2, 807–8. For the machines illustrated, besides the ZQTS itself, see Needham and Wang, *SCC* 4:2, 218

Needham characterizes it as a "companion book" to the slightly later *Qi qi tu shuo* (see below) and stresses that, whereas the latter work was a collaboration, this was a work by Wang alone that dealt with both machines developed in China as well as some purely Western mechanisms such as in the first account of the verge-and-foliot escapement in Chinese (Fig. 6.7).[109] It is in connection with the verge-and-foliot illustration that he makes a statement about how difficult it was for him to draw the mechanism.[110] Indeed, his portrayal resembles much more a highly simplified schematic diagram than a realistic portrayal, which is almost certainly what he would have liked to have drawn.[111]

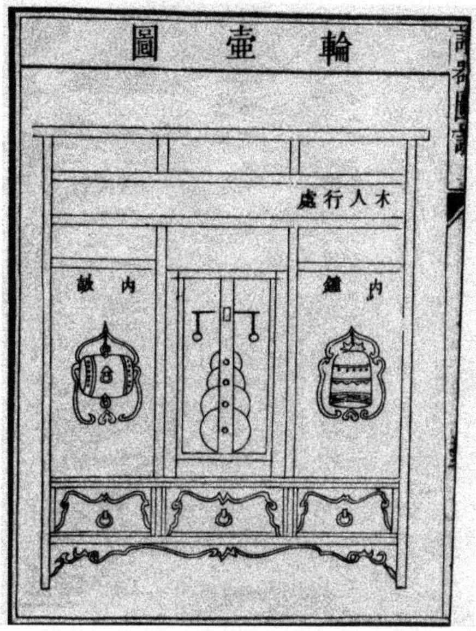

6.7 Wang Zheng's clock with its elegant but also primitive portrayal of a gear train and a verge and foliot escapement to slow down the release of the energy that drives the clock.

Wang's interest in machines, the fact that his native Shaanxi was home to a flourishing school of "Western studies,"[112] his opportunities to develop extensive contacts with the Jesuits in Beijing and perhaps also his reading of the1623 *Zhi fang waiji* 職方外紀 (Areas outside the concern of the Chinese imperial geographer)[113] led to a collaboration with Johann Schreck (Deng Yuhan 鄧玉函, 1576–1630), one of the most brilliant Jesuits of the China mission.[114] Together they produced in less than three months between November 1626 and February 1627[115] a major work introducing to the Chinese scholarly

and Zhang Baichun, "Wang Zheng *Xin zhi zhuqi tushuo* bianxi" [An inquiry into Wang Zheng's *Diagrams and Explanations of Various Newly Built Machines*], *Zhongguo keji shiliao* 17.1 (1996), 88–91.

109. ZQTS, 14a–16a; Needham and Wang, *SCC* 4:2, 218, 513–15; *HC*, 146.

110. HC, 146. Otto Mayr emphasizes the unprecedented complexity and ingenuity of the verge-and-foliot escapement: "Its characteristic feature, the carefully tuned dynamic interaction of several matching parts, is almost impossible to describe in words or to illustrate in a two-dimensional picture; to fully comprehend it, one should handle a working specimen." Otto Mayr, *Authority, Liberty & Automatic Machinery in Early Modern Europe* (Baltimore and London: The Johns Hopkins University Press, 1986), 5.

111. He relied on the same schematic convention in his illustrations for a weight-driven geared mill (ZQTS, 10a) and for a weight-propelled four-wheeled cart (ZQTS, 12a).

112. Spurred on in part by the discovery in 1625 in Xian (Chang'an) of a stele, dated 781, recounting in both Syriac and Chinese the Nestorian evangelization of China some nine centuries earlier. Ren, "Wang Zheng," 359.

113. Ren, "Wang Zheng," 349. This was a five-*juan* geographical work to accompany the world map produced for the Chinese by Ricci, which contained information on, among other things, mechanical devices used in European countries. *ECCP* 2, 808; Zhang, "Archimedian Mechanical Knowledge," 1st of 16 pp.

114. For his accomplishments, see Needham and Wang, *SCC* 4:2, 170–71, fn. i and 171, fn. a.

115. Zang and Tian, "Transmission," 79, 86.

world Renaissance mechanics and its applications. This [*Yuan xi*] *Qi qi tu shuo* [*lu zui*] [遠西] 奇器圖說 [錄最] ([A record of the best] Illustrations and descriptions of wonderful machines [from the Far West]) consisted essentially of translations from European works on mechanical theory and instruments. It included a series of fifty-four illustrations drawn by Wang, most of them copied from a variety of European originals.[116]

The *Wonderful Machines* was printed in 1627 in three chapters, the first two dealing with theory and the third serving as a collection of examples of applied mechanics.[117] It offered its readers a great deal of information on mechanics never previously available to the Chinese, but it was not without serious shortcomings. Although it devoted considerable discussion to mechanical concepts and propositions and gave examples of problem-solving procedures, it did not provide related propositions and proofs.[118] We are left to question just how well Wang really understood theoretical mechanics.[119]

Production problems also influenced the quality of the book. As in the case of Song Yingxing's *Exploitation*, the printing was a rush job. Because he had to leave for a new posting in Yangzhou, Wang had only a limited time to prepare the text and illustrations for printing. The same time constraints may also have precluded proofreading by Schreck who would surely have discovered many of the mistakes and misunderstandings in the text.[120] The quality of the illustrations was also affected by Wang's imperfect understanding of the complexities of some of the machines he was portraying as well as by the absence in China of many of the conventions and techniques for portraying mechanical devices that were integral to the original illustrations.[121]

Wang was surely aware of the limits of his drawing skills when it came to picturing complicated machines. We have seen this in his frustration over his drawing of the verge-and-foliot mechanism. Faced with that awareness as well as the very large number of illustrations and explanations available for inclusion,[122] Wang was forced to come up with some principles of selection that he outlines in the *Wonderful Machines* preface. The first was that the

116. For some of the European works drawn upon, see Standaert, "Transmission," 384. For sources of specific illustrations that can be identified, see Needham and Wang, *SCC* 4:2, 215–18, Table 58.

117. Zhang, "Archimedian Mechanical Knowledge," 2nd of 16 pp. The version commonly used today is that from the 1844 *Shou shan ge congshu* 守山閣叢書 [Mountain guardhouse collection], reprinted in 1889; see also *ECCP* 1, 36). The drawings in this edition are at least somewhat revised (Chinese stem and branch characters have been substituted for Latin identifying letters in the original, which apparently has not survived, and the almost immediate second edition, which has). See Fritz Jäger, "Das Buch von den wunderbaren Maschinen: Ein Kapitel aus der Geschichte der abendländisch-chinesischen Kulturbeziehungen," *Asia Major*, n.s., 1.1 (1944), 44 and 96n1 for further information on editions.

118. Zhang, "Archimedian Mechanical Knowledge," 10th of 16 pp.

119. Jäger, "Das Buch," 80, 93–94.

120. Jäger, "Das Buch," 93–94.

121. For two striking examples, see Needham and Wang, *SCC* 4:2, 212 and 214 and the originals on which they were based: Plates CLXX111 and CLXXIV. By contrast, the illustrations were highly praised by a certain Wang Yingkui 王應魁, a contemporary of Wang Zheng; see Jäger, "Das Buch," 95.

122. Ren, "Wang Zheng," 349 and 361. Wang speaks, probably with some exaggeration, of more than a thousand pictures and explanations.

devices included should have a practical use in daily life or should meet needs of the country (*guojia gongzuo* 國家工作). In practice, this meant mainly machines and implements that could be employed in rural production or for military defense. Secondly, he also chose to omit those machines that, while potentially useful, were so complicated that artisans would be incapable of constructing them. Cost of constructing the machines must also have been a consideration.[123]

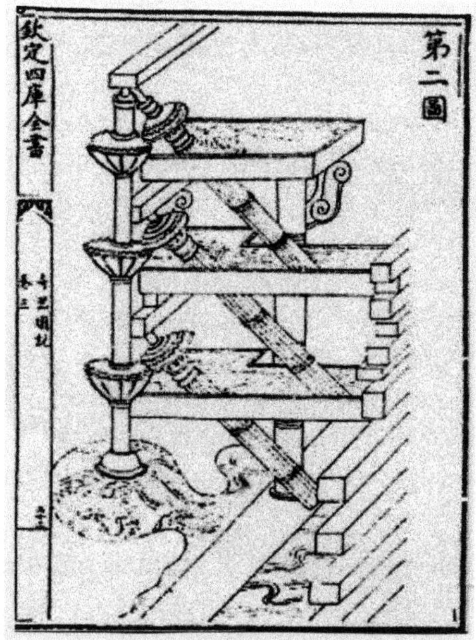

6.8 Three stacked Archimedian screw pumps from *Qi qi tu shuo* 奇器圖說 (Illustrations and explanations of wonderful machines)

In any case, the mechanical information from the third chapter of the *Wonderful Machines* was rarely put to practical use at the time or later. The inventory of machines already available to Chinese farmers and artisans had been adapted to Chinese conditions and resources over many centuries. The novelties presented in the *Wonderful Machines*, on the other hand, did not usually offer readily apparent advantages over machines currently in use, even less in those cases where the new machines would be costly and/or difficult to construct and maintain (Fig. 6.8). Moreover, as in most traditional works dealing with machines, the descriptions and the illustrations in the *Wonderful Machines* were often insufficiently complete, clear or accurate for construction of the machines by craftsmen who would not have been previously familiar with them. These screw pumps are a good example of an installation that would surely have been a great challenge to Chinese artisans of the time to construct, unfamiliar as they were with, among other things, worm gears.[124] Despite this, Wang gives not so much as a hint as to how to go about the construction, nor does he make any effort here or elsewhere to draw a clear screw mechanism.[125]

Wang's presentation of the theoretical side of the new Western mechanics drew even less attention in China than the machines he described. For example, when the *Wonderful Machines* was selected for inclusion in the great eighteenth-century compendium, the *Gujin tushu jicheng* 古今圖書集成 (Compendium of books and illustrations past and present),

123. Zhang and Tian, "Transmission," 84.

124. Needham and Wang, *SCC* 4:2, 220, 222.

125. *Wonderful Machines*, *juan* 3, 22a–b. By the end of the Ming, at least some Chinese also became familiar with the principle of the screw when they successfully made copies of Western clocks introduced at this time. Yet the extremely well-informed Rudolf Hommel in the first half of the twentieth century found no other use of screws in China. Hommel, *China at Work* (Cambridge, MA: MIT Press, 1969), 249.

it was only the third *juan* that was incorporated. The first two *juan* dealing with theoretical mechanics were simply omitted.[126] One is inclined to believe, along with Jäger,[127] that it was mainly because of the influence of Schreck that Wang even dealt with the theory of mechanics. Certainly, the dominant focus of the book was on everyday productive technologies.

Finally, it is rather remarkable how impervious Wang's drawing technique was to Western influences. He adopted much the same Chinese illustration style both for the *Various Machines* and for the *Wonderful Machines* even though it was often manifestly incapable of portraying the Western inspired machines precisely or, indeed, even comprehensibly. Here, too, time pressures probably precluded much effort to identify, to learn and to assimilate Western drawing techniques. But it is hard to suppress the suspicion that even someone as interested in mechanics *and* the portraying of machines as Wang found it impossible in his drawing to break through the limits imposed either by his natural talent, or by habits developed over many years, or perhaps by both.

Jiao Bingzhen's Pictures of Tilling and Weaving

Much of Chinese agricultural literature from the late Song on was marked by increasing reliance on stereotypes of literary description with lesser attention paid to actual practice.[128] This state of affairs was also reflected in a little-changing and limited repertoire of agricultural illustrations with copies of the same illustrations used over and over again.[129] Nevertheless, there were some striking exceptions to this overall trend. We have seen in Chapter 3 perhaps the most remarkable example in Wang Zhen's *Agricultural Treatise*. Among the Yuan, Ming and Qing official compilations of agricultural literature, the *Nong sang jiyao* 農桑輯要 (Fundamentals of agriculture and sericulture) of 1273 is especially notable for the considerable practical information it contains and has been considered to be one of the best of all traditional Chinese writings on agriculture.[130] An outstanding contribution at the end of the Ming was the *Nongzheng quanshu* 農政全書 (Complete treatise on agricultural administration) which was published in 1639 shortly after the death of its author-compiler, Xu Guangqi.[131] Xu incorporated Wang Zhen's entire *Agricultural Treatise* in his work, as well as most of its illustrations.[132] The generally conservative character of late

126. The extreme admiration of the editors of the annotated catalog to the Qianlong emperor's library for the machines portrayed in the third *juan* was matched by their disdain for the theoretical materials in the first two *juan*; Zhang, "Archimedian Mechanical Knowledge," 11th of 16 pp.

127. Jäger, "Das Buch," 93.

128. Franke, *Ackerbau und Seidengewinnung*, 56; Bray, *SCC* 6:2, 70–71.

129. Bray, *SCC* 6:2, 49.

130. Franke, *Ackerbau und Seidengewinnung*, 51; Bray, *SCC* 6:2, 71–72.

131. For the need to qualify Xu's authorship, see Francesca Bray and Georges Métailié, "Who Was the Author?" in Catherine Jami, Peter Engelfriet and Gregory Blue (eds.), *Statecraft and Intellectual Renewal in Late Ming China* (Leiden: Brill, 2001), 323.

132. Scholars have disagreed somewhat in their assessment of Xu's illustrations. For example, Franke held that Xu's illustrations surpassed those of all of his predecessors in their "Anschaulichkeit" or concreteness (Franke,

imperial agricultural illustration is nicely exemplified in Xu's work: significant technological changes and improvements between the time of Wang Zhen in the thirteenth century and Xu Guangqi in the early seventeenth century are mentioned in Xu's text but not reflected in the illustrations, which tend to be rough copies of the illustrations in the 1530 edition of Wang's work.[133]

Surviving evidence tells us little about how widely Lou Shu's *Pictures of Tilling and Weaving* circulated in the period from the end of the Song to the latter half of the Ming. After an original woodblock printing in the late 1220s or early 1230s, followed by a reprint with a preface dated 1237, we hear of no further printed editions of the *Pictures of Tilling and Weaving* for the next two and one-quarter centuries.[134] There was an apparently new edition by a Chinese official named Song Zonglu 宋宗魯 published in 1462, and it is a Japanese version of this edition that was reproduced by Otto Franke in 1913.[135] Earlier, in 1407, Lou's work, or a part of it, had been copied into the *Yongle Encyclopedia* (*Yongle dadian* 永樂大典), but those volumes no longer survive. In the flourishing book publishing environment of the late Ming, publishers of popular encyclopedias also turned to the *Pictures of Tilling and Weaving*: it found its way, usually in abbreviated versions, into household manuals such as the *Bianmin tuzuan* 便民圖纂 (Illustrated collection for the convenience of ordinary people) of 1593 and the *Bianyong xuehai qun yu* 便用學海群玉 (Seas of knowledge and mines of jade: Encyclopedia for convenient use) of 1607.[136] During these centuries, the term *geng zhi tu* 耕織圖 came often to be used as a general expression for all kinds of illustrations of agriculture and sericulture. Lou Shu was largely forgotten as the author of the original work and, as an independent work, his *Pictures of Tilling and Weaving* had at best a limited circulation.[137]

This would change significantly during the Qing. In 1689, on an inspection trip to south China, the Kangxi emperor was presented by the local literati with a number of rare books, among them the *Nongshu* 農書 (Book of agriculture) of Chen Fu 陳旉, the *Canshu* 蠶書 (Silkworm book) of Qin Guan 秦觀, and what was said to be a Song copy of Lou Shu's *Pictures of Tilling and Weaving*.[138] The emperor was so impressed by this last work and its

Ackerbau und Seidengewinnung, 52) but Bray assesses them much more negatively ("Introduction," 52–53 and "Agricultural Illustrations," 545).

133. Bray, "Agricultural Illustrations," 545.

134. Bray, "Agricultural Illustrations," 527; Philip K. Hu, *Visible Traces: Rare Books and Special Collections from the National Library of China* (New York: Queens Borough Public Library, 2000), 72–73; Kuhn, "Darstellungen," 338–39; Kuhn, "Marginalie," 143, fn. 2; Franke, *Ackerbau und Seidengewinnung*, 74; Pelliot, "A Propos du *Keng Tche T'ou*," 101. For a very handy chart of all the known versions of the *Pictures*, inside and outside of China, see Watabe, "Spread and Influence," 16, Table 3.

135. Franke, *Ackerbau und Seidengewinnung*. The number and the arrangement of the illustrations in the Japanese version as well as their content and linework suggest they may follow quite closely Lou Shu's original illustrations; Hu, *Visible Traces*, 73. Dieter Kuhn argues that the 1462 edition was the most important model for later versions of the *Pictures*; Kuhn, "Darstellungen," 336.

136. Kuhn, "Darstellungen."

137. Pelliot, "A Propos du *Keng Tche T'ou*," 101 and 107n1.

138. Franke, *Ackerbau und Seidengewinnung*, 78–80. Franke takes this to mean a copy made from the 1210 stone

illustrations that he ordered the painting of new illustrations. The task was assigned to one of the most eminent of his court painters, Jiao Bingzhen 焦秉貞 (active 1689–1726). When Jiao, probably with the help of his students,[139] completed the illustrations (this time, two sets of twenty-three scenes), a delighted emperor not only rewarded him generously but also ordered woodblocks to be prepared containing Jiao's illustrations, the old Lou Shu lyrics, a new poem composed by the emperor himself for each illustration, and a preface by the emperor dated the third month of 1696. These blocks were used to print copies of the work for distribution to government officials.[140]

Kangxi admired this new version of Lou Shu's work not mainly for aesthetic reasons or for the technical information it conveyed but, above all, for the effectiveness with which it reiterated the crucial role of the political authorities in encouraging agriculture and the need for the elite to recognize and remember the hardships of the common people on which their comfortable way of life rested.[141] The enthusiasm of the emperor and his promotion of a new edition of the *Pictures of Tilling and Weaving* brought the work a prestige greater than anything it had enjoyed up to that time.[142] It would be followed by three more imperial editions (one under Yongzheng and two under Qianlong) and was incorporated into other compilations that gave it a broad readership. The illustrations also became a rich source of decorative motifs employed in many different crafts such as ceramics, fans and lacquerware. In this way, Lou Shu's illustrations in various versions pervaded Chinese visual culture and were even to be found on export objects (including wallpaper) that made their way to the west in the eighteenth and nineteenth centuries.[143]

Kangxi's choice of Jiao Bingzhen, whom Anita Chung characterizes as "the most influential *jiehua* artist in the Kangxi court,"[144] to do the new illustrations rested on the emperor's high admiration for Jiao's unrivaled artistic abilities. Interestingly, one of the few comments one can find in this period directly linking mathematical and artistic ability (not at all uncommon in Europe since the Renaissance) comes from Kangxi's explanation of what he found to be distinguishing about Jiao: "As he has long been versed in the measurement of the latitudes of the heavenly bodies [i.e., has worked in the imperial observatory] and the topographical differences of the earth, he can show, within the space of a single foot in his paintings, layer upon layer of mountains and high peaks which represent distances as far

engraving of the work as assembled by his grandsons or from the woodblock edition of 1237.

139. Delahaye, "Du peu d'effet," 245.

140. Franke, *Ackerbau und Seidengewinnung*, 80–81.

141. These themes dominate the preface Kangxi wrote for the work. For a full annotated translation into German, see Franke, *Ackerbau und Seidengewinnung*, 101–5; for an English translation, see Hu, *Visible Traces*, 74–75.

142. This edition is considered to be one of the finest woodblock editions produced by the Qing court; Wang, *Farming and Weaving Pictures*, 81.

143. Franke, *Ackerbau und Seidengewinnung*, 85; "Zur Geschichte des Kêng Tschi T'u," 202; Bray, "Agricultural Illustrations," 530–31. Actually, there is some evidence in the form of a painting found on a round silk fan from the late Song that Lou's illustrations had already begun to enter into the decorative repertoire of craftsmen. Watabe, "Spread and Influence," 8.

144. Chung, *Drawing Boundaries*, 49.

as ten thousands of *li* [a *li* equaled about a third of a mile]."[145] It apparently did not occur to the emperor that these particular talents might not be of great relevance for depicting close-up scenes of agriculture and silkmaking. Hu Jing's 胡敬 *Guochao yuanhua lu* 國朝院畫錄 (Record of Qing court painting, 1816), citing this remark of Kangxi, adds that Jiao "had worked in the Imperial Observatory and was versed in the science of survey[ing] and mathematics. This helped him appreciate the Western method and adapt it to his own painting. His Late Majesty Kang Xi [Kangxi]'s commendation of his painting is accordingly also a commendation of his scholarship in mathematics."[146] Based on Jiao's background in the imperial observatory, Hsiang Ta suggests reasonably that he learned these techniques from missionaries with whom he worked there.[147] If so, we do not know who exactly these teachers were; the famous missionary painters Castiglione (1688–1766) and Jean Denis Attiret (1702–68) did not arrive at the court until 1715 and 1739 respectively. This was well after Jiao had completed his illustrations for the 1696 edition of the *Pictures of Tilling and Weaving*, at a point when his reputation as a painter skilled in Western painting techniques was already firmly established.[148]

The original paintings made by Lou Shu, any rubbings from the engraving of the work on stone arranged by his grandsons in 1210, or any illustrations made for the first two woodblock editions from the 1220s and 1237 seem all to have disappeared in the centuries following the Song. In the second half of the thirteenth century, however, an otherwise little known painter named Cheng Qi 程棨 painted new versions that seem to have followed Lou's originals quite closely.[149] Line drawings of Cheng's versions were also carved on stones from which an undetermined number of rubbings were made. In the eighteenth century, Cheng Qi's paintings came into the hands of the Qianlong emperor who also had line drawing copies made of them which were then inscribed on stone slabs.[150] A set of rubbings from these inscriptions was obtained by the leading French sinologist at the turn of the twentieth century, Paul Pelliot, who included them in a major study devoted to Lou's

145. Hsiang, "European Influences," 167.

146. Hsiang, "European Influences," 167 (slightly modified). Kangxi was so fascinated with the techniques of perspective that he asked the Jesuits to send to China an expert in perspective. They sent Giovanni Gherardini who stayed in China only for about five years or less but left behind impressive illusionistic frescos in the new Jesuit church in Beijing as well as students he had trained in perspective and oil painting. Sullivan, *Meeting of Eastern and Western Art*, 54.

147. Hsiang, "European Influences," 167. Though Jiao is universally credited with the *Pictures of Tilling and Weaving* illustrations, there is at least one piece of evidence suggesting that he was helped by a certain Leng Mei 冷枚 who himself achieved a considerable reputation for both his figural and architectural paintings. Ibid., fn. 34; Chung, *Drawing Boundaries*, 49, 50.

148. Thorp and Vinograd (*Chinese Art and Culture*, 360) and others have suggested the Jesuit Ferdinand Verbiest as the person who probably taught Jiao Bingzhen Western mathematical perspective techniques.

149. Pelliot, "A Propos du *Keng Tche T'ou*," 94. Two scrolls that many scholars would accept as the original Cheng paintings (Bray, "Agricultural Illustrations," and Watabe, "Spread and Influence," 9, 11, and 16, Table 3) are held by the Freer Gallery in Washington, D.C., which generously allowed me to examine them carefully and provided me with complete copies for my research.

150. Bray, "Agricultural Illustrations," 529.

work.[151] In the same period, the leading German sinologist, Otto Franke, produced the results of his own research in which he relied on the Japanese version of Song Zonglu's 1462 edition as well as the Jiao Bingzhen illustrations which had been included in the 1742 agricultural compendium *Comprehensive Study of the Farming Year* (*Shoushi tongkao* 授時通考).[152]

Comparison of Jiao Bingzhen's approach to the portrayal of these farming and silkmaking activities with earlier versions deriving ultimately from Lou Shu's originals six hundred years earlier and Cheng Qi's early copies turns up some striking differences.[153] In a very influential 1930 article dealing with European influences on late Ming and early Qing painting, the noted Chinese scholar Hsiang Ta (Xiang Da 向達) summed up what he saw as the contrast between Jiao's illustrations and those of a Japanese version closer to Lou's originals.[154] Taking as an example the two versions of the "second weeding" scene (Figs. 6.9 and 6.10), Hsiang comments:

> Lou's picture presents mainly a view of toil and is true to life, while Chiao[Jiao]'s picture, exquisitely and decoratively executed, presents a view of peacefulness and leisure, as if the artist, idling in the calmness of nature and envisioning the life of the countryside, took Lou's work as a blueprint, and used his painter's imagination to limn his own picture, without realizing that he has rendered the scene much more delicately and, in doing so, made it fall short of reality.[155]

151. Pelliot, "A Propos du *Keng Tche T'ou*," esp. 80–86.

152. Franke, *Ackerbau und Seidengewinnung*. See also his later "Zur Geschichte des Kêng Tschi T'u," esp. 169–77. (Anyone seeking examples of late nineteenth and early twentieth century European sinology at its breathtaking best could do worse than look at these prodigious studies by Pelliot and Franke.)

153. Without the existence of Lou's originals, of course, we can never say for sure how closely the later copies resembled Lou's paintings. Even leaving aside the question of the liberties Cheng himself may have taken in making his copies, there is also the question of changes introduced by the very processes (i.e., engraving on stone from which rubbings were taken) by which Cheng's versions were reproduced. We can assume that inevitably many of the finer details of Cheng Qi's paintings were lost, for example facial expressions and coloring. Nevertheless, since we are viewing the illustrations primarily for their technological content, we can be fairly comfortable that Cheng Qi probably captured most of the essential elements of Lou's paintings such as their overall design, what was included and what was excluded, and the actions of the people involved in using the technology (Pelliot, "A Propos du *Keng Tche T'ou*," 91). Moreover, insofar as the illustrations show technological details, it seems likely that the level of accuracy in Lou's paintings was at least the equal of what we see in the Cheng versions. Given the strong tendency in the treatment of such subjects, it would be likely for the copies to be inferior rather than superior to the originals.

154. The article is available in (a not always satisfying) English translation by Wang Teh-chao. Hsiang, "European Influences." Unfortunately, some of what Hsiang has to say is vitiated by his taking a fifteenth-century Japanese version of the *Pictures* as representative of Lou's illustrations instead of, for example, the Semallé scroll illustrations reproduced in Pelliot, "A Propos du *Keng Tche T'ou*" or the (probably) Cheng Qi copy which are clearly closer to the originals.

155. Hsiang, "European Influences," 170. See also Berthold Laufer's use of "forced mannerism" to characterize the paintings of Jiao vs. the "grave simplicity" of the Japanese Kanô edition. Pelliot, "A Propos du *Keng Tche T'ou*," 91. On the other hand, Francesca Bray rightly points out that Lou's illustrations paid considerable attention to details little connected with actual farming or sericulture (scenic vistas, children playing etc.). Bray, "Agricultural Illustrations," 531–32. In any case, the two versions of this scene, in their very layout, certainly show a decidedly different importance assigned to the farmers actually doing the weeding.

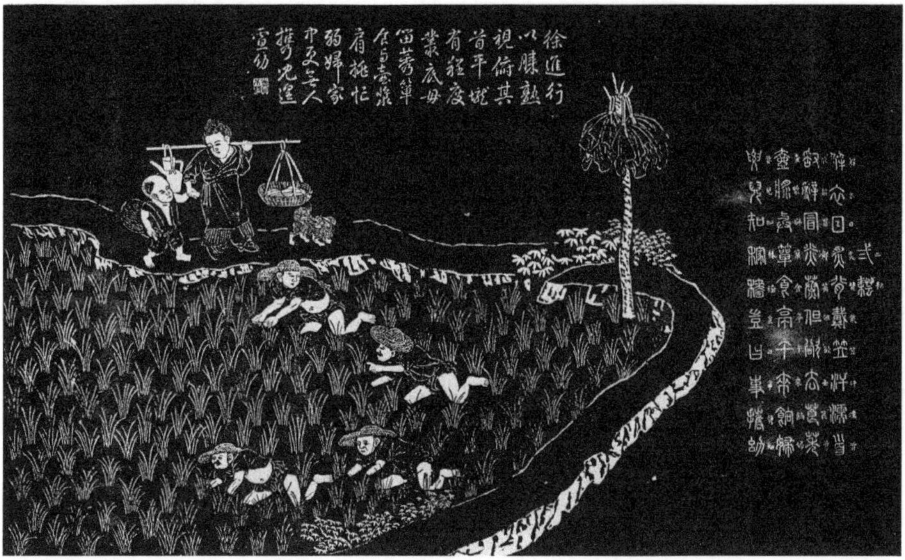

6.9 "Second Weeding" scene from the *Pictures of Tilling and Weaving,* Qianlong stone-carved copy of Cheng Qi's version.

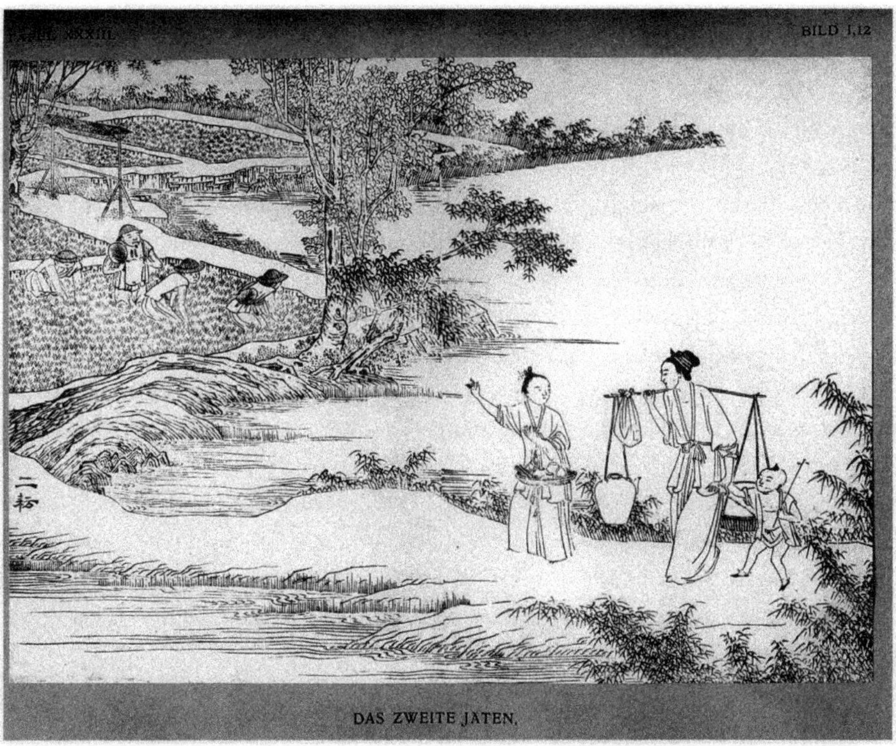

6.10 "Second Weeding" scene from the *Pictures of Tilling and Weaving,* Jiao Bingzhen version.

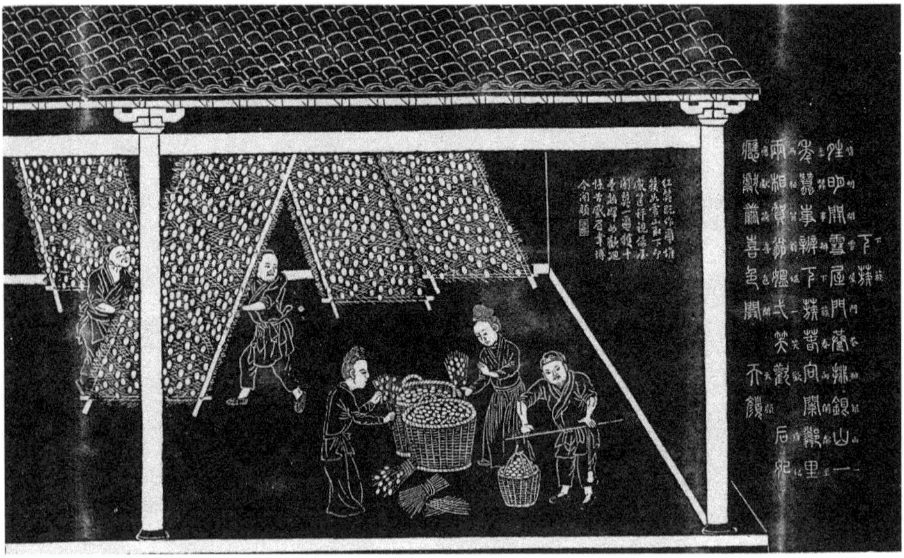

6.11 "Removing silkworms from the spinning trellises" scene from the *Pictures of Tilling and Weaving*, Qianlong stone-carved copy of Cheng Qi's version.

As a painter of unusual skill with a highly personal style, Jiao gave greater priority to producing a distinctive rendering of Lou's illustrations than to copying them closely. Moreover, Jiao had not acquired his reputation without developing a keen awareness of the tastes of his audience, which included above all the emperor. Thus, in addition to producing pictures marked by considerable delicacy, he also sought in small ways to make the scenes less strange for his mostly urban viewers. For example, he did not hesitate to alter clothing or hairstyles to make them better reflect familiar contemporary fashions.[156]

Along with subtly changing the feeling of the illustrations, many of Jiao's modifications also tended to deemphasize even more than Lou's versions the technological elements of the scenes being portrayed. Lou's original pictures in their straightforward simplicity at least keep the focus clearly on the activities portrayed. Very few figures beyond those engaged in productive tasks are included. Extraneous details are generally omitted. By contrast, Jiao seems to miss few opportunities to add details that, one must admit, add vitality to his drawings: children play and fight; mothers and grandmothers try to control them; someone serves tea; chickens, dogs and birds cavort to no discernable purpose except lively decoration. Sometimes, Jiao's focus on less important or even extraneous details leads to the disappearance or near-disappearance of the activity to which the illustration is supposedly dedicated. We see this in the two versions of the removal of the silkworms from the spinning trellises prior to extracting the silk filaments: the actual handling of the boards has disappeared entirely in Jiao's illustration (Figs. 6.11 and 6.12).

156. Interestingly, Lou's men always wear a headcovering of some sort while the women never do. Jiao usually follows this rule, but not always. Compare Franke, Tafel LXXIV where the seated figure preparing the cocoons for the unwinding of the silk is a woman but has a headcovering, with Pelliot, Planche L.

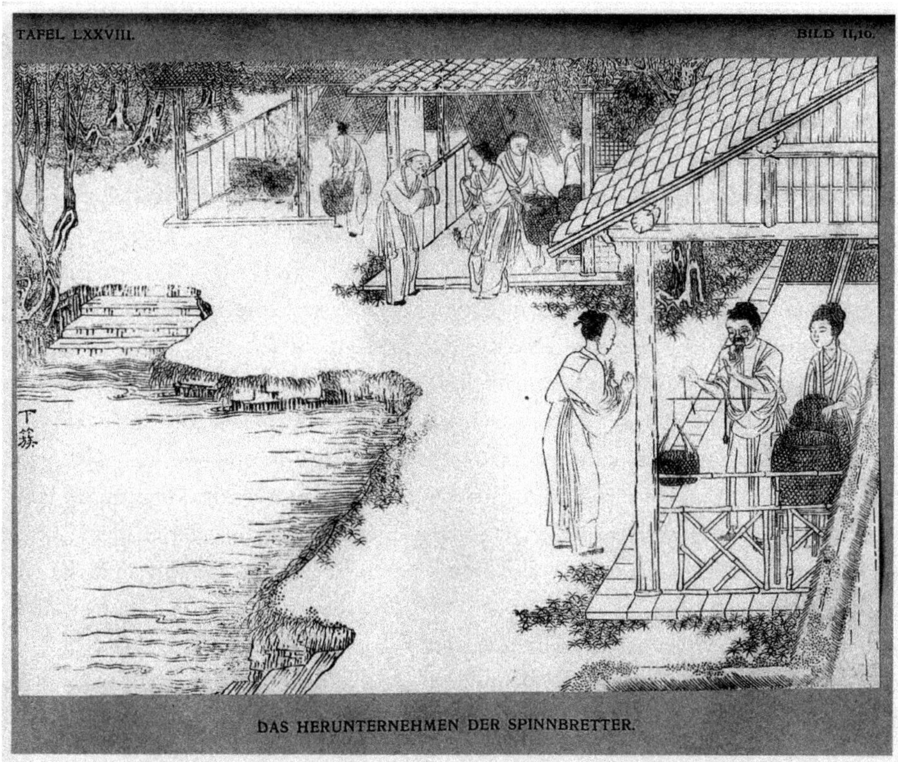

6.12 "Removing silkworms from the spinning trellises" scene from the *Pictures of Tilling and Weaving*, Jiao Bingzhen version.

Earlier, in discussing Lou's original *Pictures of Tilling and Weaving*, we suggested that his decision to break down rice farming and silkmaking to twenty-one and twenty-four stages respectively and illustrate each of those stages made it possible for viewers to gain a more complete picture of what were very elaborate production processes extending over months. That this technique came to be widely used in later centuries for the portrayal of other extended productive activities suggests at the least its appeal for Chinese viewers. Whether it served to give a better sense of the technology in use — which of course was a lower Chinese priority than it tends to be for us — is open to question. We have already considered important ancillary concerns in our discussion of Song Yingxing's *Exploitation of the Works of Nature*, questions such as the effectiveness with which the illustrations and their accompanying texts work together, how the illustrations do or do not complement the verbal descriptions. In the case of the *Pictures of Tilling and Weaving*, the limited technological content of both the illustrations and, even more, the accompanying lyrics (whether those of Lou or of the Kangxi and Qianlong emperors) greatly inhibits the possibility of the viewer gaining anything more than a superficial comprehension of the technology from either the individual illustrations or the whole series. Moreover, the ability of the illustrations to provide accurate information rather than misinformation requires that, at a minimum, the processes illustrated be presented in the proper order. It is therefore somewhat disconcerting to note that Jiao sometimes got the order of his illustrations mixed up (or, if not Jiao, someone else dealing with the carving and printing of the illustrations). This happened most egregiously in steps of rice processing from the completion of the harvesting to the storing of the grain: threshing, winnowing, hulling and milling, and sieving. And not only did Jiao get the order wrong but it appears that, in all or virtually all the editions of *Pictures of Tilling and Weaving* that followed, these processes never corresponded to the correct order of Lou's original.[157]

Jiao's limited interest in portraying technological processes correctly is further suggested by the two entirely new illustrations he added to the series on farming, one portraying the first rice shoots and the other depicting thanksgiving offerings to the gods after the harvest was in. Whatever Jiao's reasons for including these new illustrations, it is clear that they are devoid of technological information. That was not what the emperor was looking for. Francesca Bray is likely correct in seeing the new thanksgiving offerings image as one more indication that the enthusiasm of the Qing emperors for the *Gengzhi tu* rested on its "iconic value" as a portrayal of a harmonious and well-functioning society rather than on its portrayal of rural technology.[158]

More than for their portrayal of technology, Jiao's illustrations have been admired especially for their effective use of Western linear perspective techniques that gave them a spatial sense new to Chinese painting.[159] Linear or single point perspective is also sometimes referred to as "convergent" perspective since "lines and points on each side of the observer's position, though in fact parallel, appear to meet at a vanishing point upon the horizon."[160] With these techniques, the painter has a powerful tool not only for introducing an accurate, mathematically-based sense of distance but also for relating elements of a scene one to another in convincing fashion. Neither of these concerns is of primary importance in most technological illustrations which usually are relatively close-up views of a single subject. However, use of appropriate perspective techniques, including foreshortening, may be essential to make clear the structure and functioning of a more or less complicated machine. This was a problem that most earlier drawing systems, east or west, found difficult

157. Or at least so it would seem from an examination of four editions, three printed and one manuscript, that followed Jiao's version. See Franke, *Ackerbau und Seidengewinnung*, 157–59 and 94; Walter Fuchs, "Zum Keng-chih-t'u der Mandju-Zeit und die japanische Ausgabe von 1808," *Ostasiatische Studien* 48 (1959), 67 and 80. The same problem had already made its appearance earlier in popular encyclopedias of the Ming and may date back even to the 1462 edition that, except for its preface, has not survived. Kuhn, "Darstellungen," 343, 345 and 350; Hu, *Visible Traces*, 73.

158. Bray, "Agricultural Illustrations," 535. We might add that it can also be seen to reflect a ritual aspect that was part and parcel of agricultural work; Brian Pfaffenberger, "Social Anthropology of Technology," *Annual Review of Anthropology* 21 (1992), 501.

159. As we have noted earlier, the Chinese had their own methods (especially atmospheric, aerial and parallel projection) for representing three-dimensional space on a flat surface but these had none of the mathematical exactitude that governed single-point perspective methods of the Renaissance.

160. Needham, Wang and Lu, *SCC* 4:3, 111.

to handle. From the Western tradition, an excellent example is Villard de Honnecourt's effort to portray a water-powered saw (Fig. 6.13). The paucity of any form of coherent perspective assures that the drawing will be incomprehensible to any but those who happen to have seen the actual machine, preferably in operation. For a Chinese example of how the lack of control over perspective can thoroughly muddy the workings of a relatively complex machine, we can compare two illustrations of the armillary sphere of Su Song's clocktower, one probably very close to the original Song version where the perspective is handled quite well (Fig. 6.14 (a)) and the other from an eighteenth-century manuscript of the *System Essentials* where it clearly is not (Fig. 6.14 (b)).

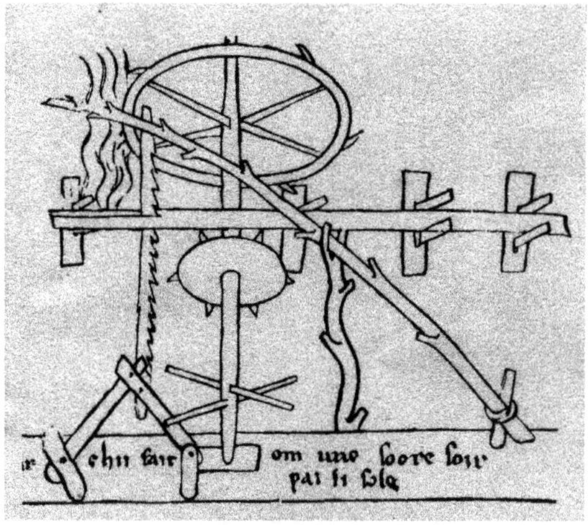

6.13 Villard de Honnecourt's water-powered saw

Despite the borrowing of Western techniques, Jiao felt no need to abandon traditional Chinese painting techniques. Many elements of his paintings such as houses, trees and figures were rendered in the traditional manner and gave his illustrations a pronounced Chinese flavor. Moreover, with all of his skill in handling the new perspectival techniques, Jiao could still be at a complete loss when it came to portraying a complicated machine. One of the most striking examples is the attempt to portray a loom operated by a single weaver (Fig 6.15). His incomprehensible portrayal is very disappointing, all the more when compared with Lou Shu's original (Fig. 6.16 (a)) or the Japanese copy of 1676 (Fig. 6.16 (b)).

Jiao's portrayal of a pattern-loom operated by two people, with just a hint of convergent perspective, Fig. 6.17, is a better overall rendition but hardly any more informative in depicting the way the machine worked. Here again, Lou's version was apparently far superior, perhaps the finest depiction of a "jacquard" loom ever produced in China before the nineteenth century (Fig. 6.18).

The 1696 edition of the *Pictures of Tilling and Weaving* has been widely regarded as one of the finest woodblock editions produced at Kangxi's court.[161] It benefited also from the carving of the blocks by Zhu Gui 朱圭 (c. 1644–1717) and Mei Yufeng 梅裕鳳 (fl. c. 1696) who were famous as engravers as well as being painters.[162] Its lavish production and

161. Wang Chaosheng, *Farming and Weaving Pictures*, 81 (79).

162. Wang Chaosheng, *Farming and Weaving Pictures*, 80 (78); Marcia Reed and Paola Demattè (eds.), *China on Paper: European and Chinese Works from the Late Sixteenth to the Early Nineteenth Century* (Los Angeles: Getty Research Institute, 2007), 196. It is an interesting question (unfortunately with too little surviving evidence to answer it in any but a speculative way) whether illustrations both drawn and carved by the same person(s) were generally superior to the normal pattern where an illustrator made the drawing and a

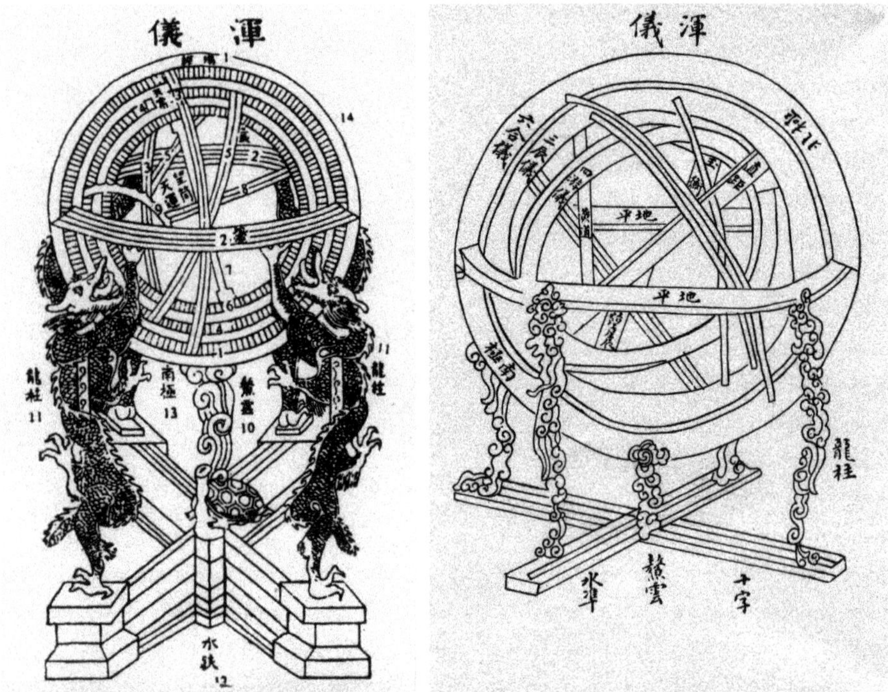

6.14 (a) The armillary sphere of Su Song's clocktower; (b) Eighteenth-century drawing of Su Song's armillary sphere.

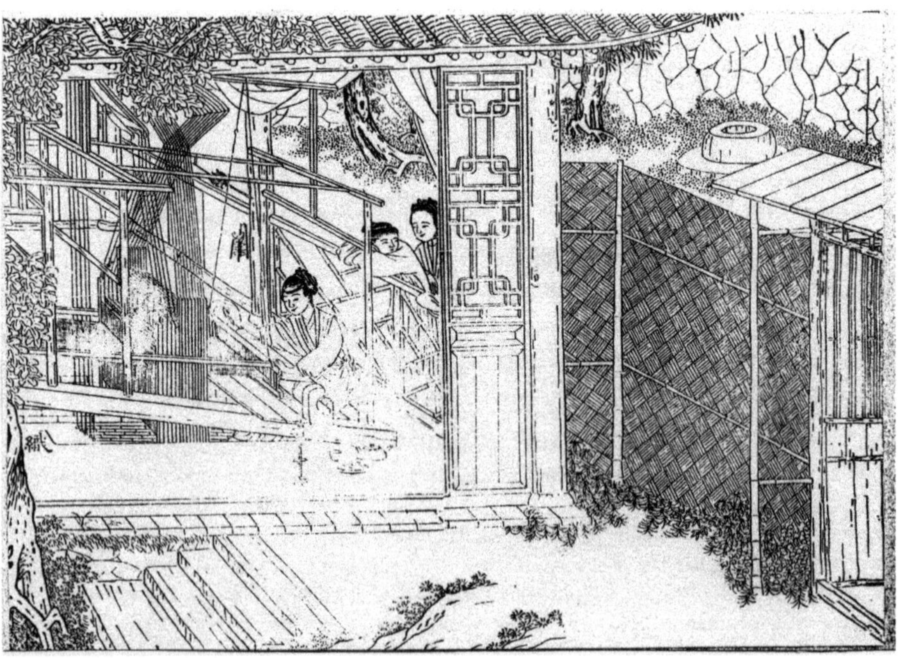

6.15 Single-operator loom from the *Pictures of Tilling and Weaving*, Jiao Bingzhen version.

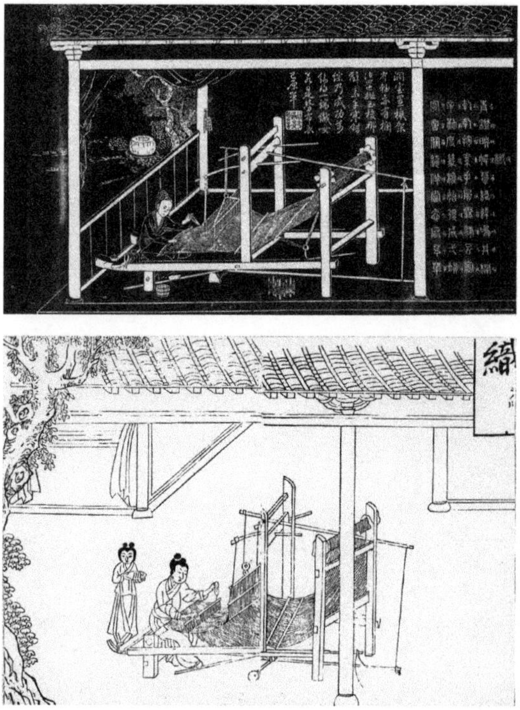

6.16 (a) Single-operator loom from the *Pictures of Tilling and Weaving*, close to Lou Shu version; (b) Japanese copy of Lou Shu's single-operator loom.

wide distribution[163] was made possible by strong support from the emperor. Despite his support, however, and despite the fact that another imperial edition, sponsored by Qianlong, was printed and distributed in 1739, Jiao's *Pictures of Tilling and Weaving* as well as other paintings incorporating more advanced drawing techniques never came to exercise an important influence on Chinese scholar-painting in general or Chinese illustrations of technology in particular. Although the eighteenth century was notable for a proliferation of schools and tendencies in scholar-painting (in part in reaction against the Court painting establishment), none of them drew inspiration from Western approaches and techniques.[164] The early to mid-eighteenth-century critic Zhang Geng 張庚 probably expressed the prevailing view among scholar-painters when he wrote that Jiao's painting was "not worthy of refined appreciation, and lovers of antiquity will not adopt it."[165]

Most Chinese literati painters simply found uncongenial a system of painting so reliant on geometry and measurement. The almost total absence of systematic geometric thinking in China must have made it especially difficult for them to learn how to use this system. As Benjamin March incisively notes, the Chinese scholar-artists "had no taste for the mathematical theorizing necessary to produce a single exact geometrical formula for the

professional blockcarver did the carving. Lucille Chia suggests quite reasonably that the latter practice would require "some general understanding of how to translate text and image on paper into a standardized linear system to facilitate and routinize the carving." The result could be "less opportunity for technical and artistic innovation . . . " Chia, "Text and *Tu*," 243.

163. Franke, "Zur Geschichte des Kêng Tschi T'u," 202.

164. Delahaye, "Du peu d'effet," 247, 249–50. Jiao's paintings did have at least a modest influence on some folk and professional painters, especially as we have noted above in decorative contexts. Indeed, the influence may have been greater that has been recognized, given the fact that so little of this painting production has survived. Kao, "European Influences," 272; Delahaye, "Du peu d'effet," 248.

165. Cahill, *Compelling Image*, 72.

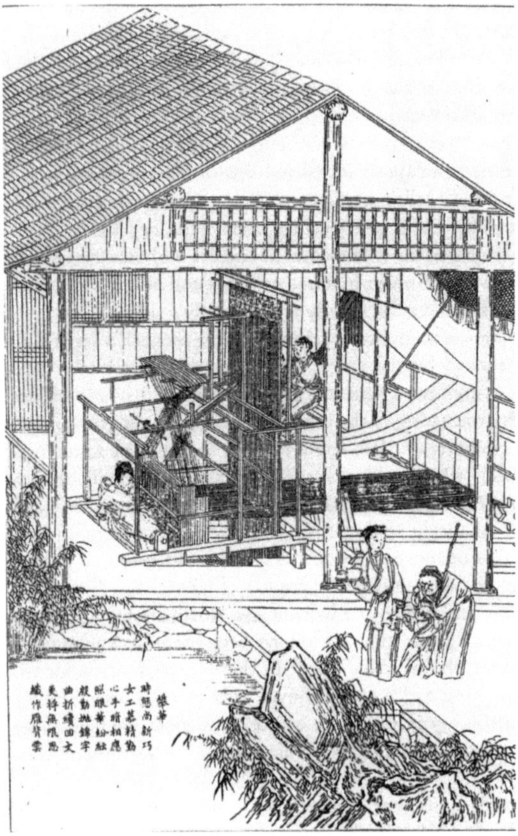

6.17 Pattern-loom operated by two people, from the *Pictures of Tilling and Weaving*, Jiao Bingzhen version.

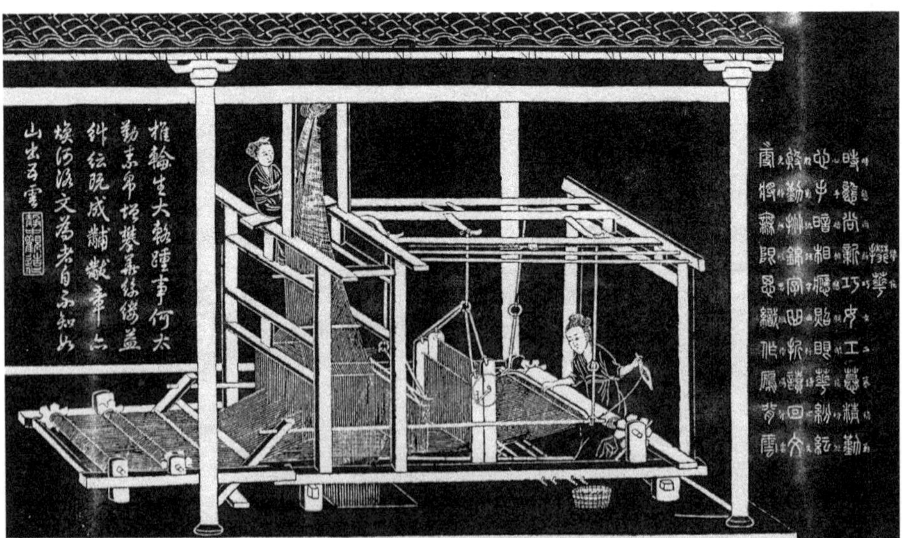

6.18 Pattern-loom after Lou Shu

universal solution of a general problem, even had such a solution been desirable, which, according to the psychology of the artists, it was not."[166] Moreover, we have many examples such as Lou Shu's looms showing that traditional conventions and techniques could, in the hands of a skillful artist, serve quite well to portray even complicated machines, provided the artist understood the machine and how it functioned and provided he wanted to get it right. Moreover, potentially useful Western techniques such as modeling strokes, hatching, and chiaroscuro often were not easy for Chinese artists to learn to use convincingly.[167] Thus, when imperial patronage ended, so too did the prestige of the kind of painting associated with Jiao.[168]

A Qualified Last Hurrah for Realism: Architectural Painting in the Qing

As this account comes to a close around the time of Qianlong's reign (1736–95), we find the Chinese environment probably less conducive to further developments in the picturing of technology than in most earlier periods. Yet, even during this period of relatively strong resistance to change, change did continue, though it tended to be of an accommodative sort that modified rather than posing serious challenges to existing patterns of thought and action. In the case of painting, for example, one can detect something of a revived interest in more descriptive or realistic painting that harked back to the great achievements of the Northern Song, but which also could reflect Chinese painters' acquaintance with similar characteristics in European pictures that were now available in China.[169] Among works portraying one or another technology, this realism is perhaps best seen in the *jiehua* or architectural paintings of which the Qianlong emperor was so fond and which he promoted enthusiastically (Plate 14).[170]

The beginnings of such paintings go back at least to the First Emperor of Qin who, as a part of his efforts to unify China, decided to duplicate at his capital city the palaces of the feudal lords he had defeated. The elaborate drawings that would have been necessary as a prelude to construction must have been among the earliest drawings created in China for a clear technological purpose. In later times, much of Chinese painting owed a great deal to various kinds of imperial patronage, but that patronage rarely had much to do with technological considerations even when technology was portrayed. Rather, as we have seen for example in the many imperial-sponsored versions of the *Pictures of Tilling and Weaving*, including those of Jiao Bingzhen, these paintings were meant above all to convey a symbolic or ideological message that would bolster imperial legitimacy.

Architectural painting in China is generally considered to have reached a peak of realistic verisimilitude in the Yuan, though this by no means meant that the paintings were intended

166. March, "Linear Perspective," 137, 139.

167. Compare Reed and Demattè, *China on Paper*, 77, Fig. 33 with Fig. 32 on the previous page.

168. Hsiang, "European Influences," 171–72.

169. Cahill, *Compelling Image*, 76.

170. Chung, *Drawing Boundaries*, 48, 54, 60.

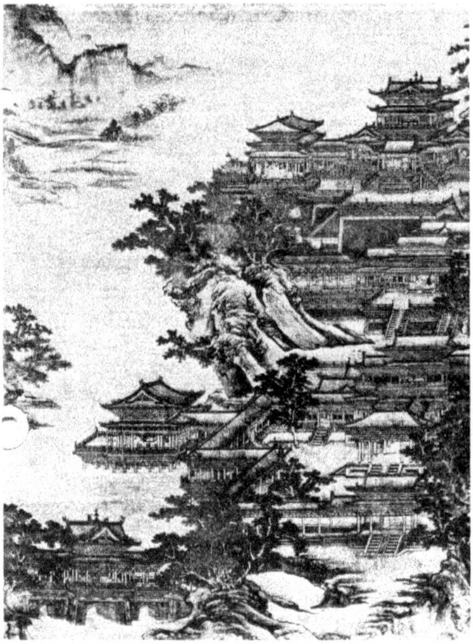

6.19 Li Rongjin, *The Han Palace*

to convey technological information or succeeded in doing so. An excellent example is the remarkable painting of an imaginary Han palace by Li Rongjin 李容瑾, a little-known painter active in the mid-fourteenth century (Fig. 6.19). It contains a remarkable amount of accurate detail but tells us very little about building technology.

In the following Ming dynasty, a dramatically different approach to architectural paintings came to prevail, one which historians of Chinese painting have generally associated with a decline in quality. Architectural painting joined in the general retreat from realism and verisimilitude that characterized so much of later painting (Fig. 6.20). By comparison with earlier surviving paintings, the architectural elements in most Ming paintings tend to be dwarfed by an all-dominating landscape; the images become smaller, delicate instead of monumental, and less precise in their presentation of details. In some cases, this was because of the tendency in Ming architecture to decrease the size of brackets relative to the size of the buildings, but to increase their numbers.[171] Moreover, in many Ming architectural renderings, the individual buildings are scrunched together, further precluding any presentation of details except in the exterior elements and leading even to a loss of "structural clarity."[172] In general the buildings in Ming paintings seem meant to convey more a sense of elegance and luxury than a realistic picture of their appearance (Fig. 6.21).[173] As Trousdale comments, "[m]any Ming architectural landscapes would have severely tried the credulity of sober observers had realism any longer been a matter of concern."[174]

In the Qing, architectural and *jiehua* painting experienced an impressive resurgence not only at the Manchu court under the enthusiastic patronage of the emperors, but also in prosperous regional urban centers such as Yangzhou, whose commercial vitality and mercantile patronage created a vibrant art scene that could compare with that of Beijing.[175] At

171. Pan Guxi, "Yuan and Ming Dynasties," in Nancy Shatzman Steinhard, *Chinese Traditional Architecture* (New Haven and London: Yale University Press, 2002), 204.

172. Chung, *Drawing Boundaries*, 36–39.

173. Even in Ming writings on painting, rarely are there references to particular skill in measurement or getting proportions right. Chung, *Drawing Boundaries*, 33.

174. Trousdale, "Architectural Landscapes," 312.

175. Chung, *Drawing Boundaries*, 3.

the court, as suggested above, most of the finest work was produced in the time of Qianlong and under his vigorous and consistent support.[176] In spite of the widespread disparagement of the genre by literati artists and critics who did not share the emperor's enthusiasm, there were some scholar-officials, even holders of high bureaucratic appointments, who specialized in painting architectural subjects.[177]

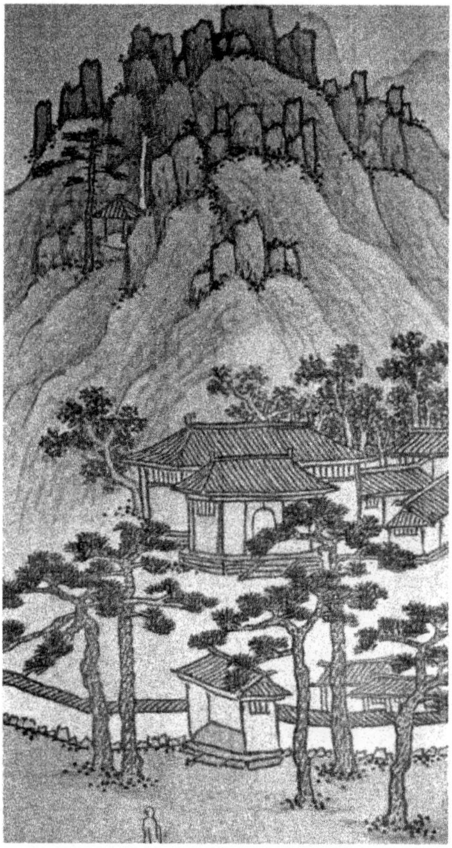

6.20 "Tianping Mountain" from Shen Zhou's *Ten Views of Two Rivers*

Despite the realistic achievements of architectural painting in the Qing, these paintings are ultimately no more technologically informative than their great predecessors of the Song and Yuan. As we noted regarding Li Rongjin's *The Han Palace*, it is packed with fine detail but tells us virtually nothing about the technology used to create this breathtaking complex.[178] Indeed, to a large extent, what we are looking at is a very skillful creation of the illusion of detail. This becomes clear the moment one tries to determine with any accuracy the size of one element compared with another. The scale alone makes this impossible. One meets with the same problem in the great architectural paintings of the Qing. The virtually complete dominance of broad vistas and the absence of any kind of close-ups precluded inclusion of truly informative detail.[179]

---

176. Qianlong was even known to order a scholar-official painter and a professional (*jiehua*) painter to collaborate on a single work. Chung, *Drawing Boundaries*, 58. Chung may overstate the ideological motivations behind the great Qing architectural paintings, above all the masterpieces produced at Qianlong's court. See Cary Y. Liu, "Review of Anita Chung, *Drawing Boundaries*," *JAS* 65.2 (May 2006), 42. We certainly do not have textual evidence like, for example, Kangxi's preface to Jiao Bingzhen's version of the *Pictures of Tilling and Weaving* that unequivocally establishes such motives. Nevertheless, that these works were meant to convey the splendor and magnificence of the emperor and of the imperial establishment as well as to bolster acceptance of Manchu rule seems impossible to deny; see especially the discussion in Chung, *Drawing Boundaries*, Chapter 3.

177. Chung, *Drawing Boundaries*, 57, 58.

178. See previous page, Fig. 6.19.

179. For a number of examples, see the excellent selection of color plates in Chung, *Drawing Boundaries*, between pages 118 and 119.

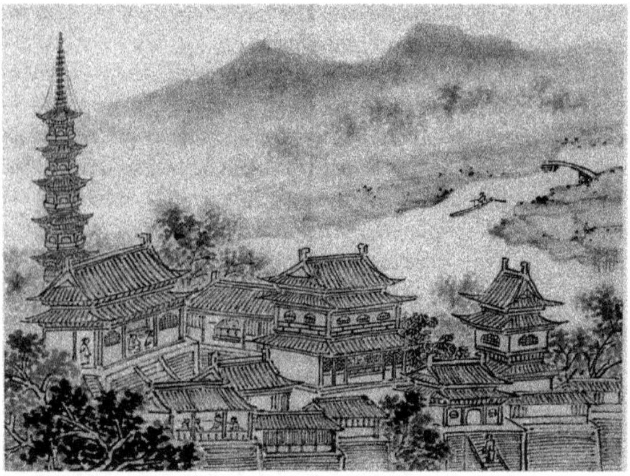

6.21 Shen Zhou, leaf from *Twelve Views of Tiger Hill*

So too did the fact that, since so much of Chinese large-scale building was highly ornamental, artists often found themselves focusing on reproducing those decorative details. For example, it is clear that once the artist had learned to draw or paint, say, a certain kind of bracket set, that technique could serve again and again since one of the characteristics of mature architectural technology in China was the repetition of a single kind or just a few kinds of bracket sets in a given building. What was probably most difficult for many artists was the patience necessary to repeat these often detailed elements over and over. But what they were portraying in their paintings represented only visible outer surfaces that often only hinted at the very complex but largely invisible building techniques employed. It is therefore at least open to serious doubt, contrary to the comment by Guo Ruoxu cited above,[180] how much understanding the technology that lay behind those surfaces would lead to a "better" painting. One can be equally doubtful of the idea of people like Nian Xiyao 年希堯 (d. 1738), translator of a Western classic on perspective, that Western methods, above all perspective techniques, were the key to more accurate and informative architectural depictions.[181] More important in practice was the assembling of a multitude of individual elements into a coherent whole that would *appear* to be simultaneously accurate and aesthetically right. As Anita Chung points out, "[on] the whole, it is the achievement of verisimilitude in architectural representation that makes an image look faultless."[182]

Since Chinese architectural paintings by their very nature could do little to convey information on building techniques, we should not be surprised that those who depicted other technologies found little or nothing in these paintings to help them (again, always insofar as that was their goal) to present their own technologies with greater precision or accuracy.

180. See Chapter 3, especially note 24.

181. Standaert, *Handbook*, 741 and 794; Chung, *Drawing Boundaries*, 62. The translated work was Andrea Pozzo's *Perspectiva Pictorum et Architectorum*.

182. Chung, *Drawing Boundaries*, 20.

Closing Comments

By way of closing, we focus below on a number of themes that are of paramount importance for understanding the role played by portrayals of technology in traditional Chinese culture.

Non-technological Aims in Portrayals of Technology

The introduction to this study emphasized that one of the most important assumptions to jettison when considering premodern Chinese depictions of technology is that they responded primarily to what we would identify as "technological" needs or concerns. Whether in the form of paintings or the book illustrations that are the source of most of our surviving portrayals, pictures of technological subjects typically sought to appeal to a viewership the overwhelming majority of whom would have little or no direct experience with or special interest in the technologies portrayed. Most of these viewers would be drawn above all to pictures that were aesthetically pleasing or entertaining. It is true that these portrayals also often embodied a didactic element. But if they did so, it was one that focused not so much on providing technical information and understanding[1] as on conveying moral or ideological instruction or inspiration. Such themes might range from the importance of hard work in order to produce the necessities of daily life to the obligation of the ruler to see that living conditions throughout the land supported a harmonious social order. Well-drawn agricultural illustrations could be admired for a certain kind of "ideological efficacy" since agriculture, as we have noted, was often seen to be at the heart of a well-functioning socio-political order and even a well-ordered cosmos.[2] The dominance of non-technological aims in the illustrating of technological subjects often meant that it was left to written texts rather than to illustrations to present technological information for practical use.

There were of course exceptions. We have seen Wang Zhen's *Agricultural Treatise* as one of the most remarkable. Another example, dating possibly from the nineteenth century, involves the case of the official who tried to revive silkmaking in Shaanxi but found the accounts in standard works such as the *Can sang jiyao* 蠶桑輯要 (A summary of sericulture)

1. The kind Bert Hall discusses in "The Didactic and the Elegant."
2. As an imperial edict summarized it in BCE 111: "Agriculture is the basis of the whole world." Needham, Wang and Lu, *SCC* 4:3, 264.

and the *Bin feng guang yi* 豳風廣義 (An overview of the customs of Bin) to be too difficult for ordinary people to understand. He therefore had straightforward pictures printed that would be "understandable at a glance" (*yi mu liao ran* 一目了然).[3] Nevertheless, even in attempting to understand the illustrations of a self-consciously "technological" work such as Song Yingxing's *The Exploitation of the Works of Nature,* we do well to keep in mind that it is *our* tendency to judge the illustrations according to the accuracy and clarity with which the technologies (and especially the tools and machines) are portrayed. Most Chinese readers would as likely have judged the success of the illustrations on their ability to generate in a pleasing way a feeling for what performing this work was like, or what they would like to think it was like.[4]

Most of the portrayals of technology therefore could be and were accomplished quite adequately by artists with a less than thorough understanding of the technology involved. So long as there was little stress on precision, accuracy and verisimilitude, the portrayals did not require the talents and skills of people like the hyphenated artist-physicians, artist-engineers etc. so prominent in Renaissance Europe. Communication between scholar-artists or scholar-official artists (useful hyphenations in the Chinese context) or even professional artists and the users of technology was rare in China.[5] Thus the illustrating of technology continued to "improve" so long as such improvement was consonant with mainly aesthetic values. When they diverged, aesthetic values — very often inimical to better technological portrayals — tended to prevail. For advances in technological illustration to have continued, China would have had to produce a new kind of illustrator/designer/painter with a keen interest in technology. Such figures or even Chinese variations of them were rare to non-existent in the Chinese context.

Pre-eminence of Agriculture and Human Inputs

Agriculture as we have seen was the uniquely important Chinese productive technology. The amount of food needed to feed a large, dense population, the numbers of people involved in producing it, the sophistication of Chinese agricultural practice (drawing as it did on a vast reservoir of knowledge deriving from millennia of experience across a highly varied environment),[6] and the focus on agriculture in so much of the thinking of both the people and the government: all of this put agriculture right at the heart of Chinese life. It is

3. Wang Chaosheng, *Farming and Weaving Pictures*, 170 (169).
4. Recall the discussion of the idealizing character of many technical illustrations in Jiao Bingzhen's version of the *Pictures of Tilling and Weaving* in Chapter 6.
5. As it was generally between scholars of any kind and craftsmen. Needham, Robinson and Huang, *SCC* 7:2, 230.
6. A point that particularly struck the Jesuits in the seventeenth and eighteenth centuries was the range of considerations for determining just what kind of manure to use. Mark Elvin, "The Technology of Farming in Late-Traditional China," in Randolph Barker and Radha Sinha (eds.), *The Chinese Agricultural Economy* (Boulder: Westview Press, 1982), 13–14.

therefore hardly surprising that agriculture played a crucial role in the early development of painting in China (as discussed in Chapter 1). In later times, it also became the subject of a vast literature as well as enjoying at least quantitatively considerable preeminence in visual portrayals of technology in China.[7]

Agriculture was what one might call an organic technology. Like China's other great rural technology, the making of cloth, it dealt with the manipulation of living matter. Rather unlike clothmaking, however, its tools tended to be simple and easy to illustrate, its tasks often not amenable to the use of machinery of any complexity.[8] More important than tools and machines were care and dedication in carrying out the tasks of farming.[9] Farmers drew upon a considerable experience-based knowledge relating to soil, climate, seeds, fertilizer and water needs, much of which was intuitive and did not easily lend itself either to verbal or to visual elucidation.[10] A particularly explicit and striking example of the essential role of intuitive knowledge is found in the poem that accompanies the illustration of a draw loom in the *Pictures of Tilling and Weaving*.[11] The third line of the poem speaks of "hand and mind [working] in *obscure resonance*" to produce an intricate design; it neglects to mention the complex loom without which the task would have been impossible.

This emphasis on human skills and, often enough, strength too, may provide at least part of the explanation for less reliance than one might expect on mechanical devices even in the final centuries of the empire. The relative paucity of complicated machinery in Ming and Qing China, especially complex machines with concealed workings, shows the Chinese as distinctly less inclined to seek out mechanical solutions for technological tasks than either the Europeans or the Japanese, as we noted in Chapter 6. This tendency may have been reinforced by economic conditions in the last two dynasties. China at that time had a vast body of cheap but skilled rural labor that, despite a growing shortage of resources including wood and iron for constructing machines, was able to maintain rises in agricultural production necessary for an ever-growing population. In these conditions, using machines to substitute for labor risked increasing costs as well as lowering yields when it was precisely maximization of those yields that was needed to feed China's dense and growing population.[12]

7. There is a striking contrast here with what we find, say, in the late Middle Ages in Europe, where technical illustrations typically focused on tools and machines used in surveying, building and warfare; Knobloch, "Technische Zeichnungen," 50.
8. Elvin, "How did the Cracks Open?," 11. See also Francesca Bray's distinction between "mechanical technologies" which were typical of European farming and the "skill-oriented" technology that characterized wet-rice cultivation. Bray, *The Rice Economies: Technology and Development in Asian Societies* (Oxford: Blackwell, 1986), 7.
9. Thus, in north China, the small plots of poor peasant farmers could regularly be made to produce far better harvests through garden farming than the more extensive holdings in large estates where the same level of diligence and skill was not applied. Elvin, "Technology of Farming," 14.
10. Elvin, "Technology of Farming," 13. We should keep in mind, however, that there were some cases, as in the weeding illustrations of the *Pictures of Tilling and Weaving*, where verbal explanations could convey knowledge extremely difficult or even impossible to depict in a visual image.
11. Bray, *Technology and Gender*, 201, Fig. 14; compare Hammers, *Pictures of Tilling and Weaving*, 211.
12. For these points, the works of Mark Elvin, especially "The High Level Equilibrium Trap," "China as a

Thus we see, for example, that Chinese farmers devised a way of hulling grain by means of a water-powered stamp but commonly found it more cost-effective to do the hulling by hand-pounding or by using a foot-powered stamp.[13]

Given the de-emphasis of mechanization in so much of Chinese productive technology, it followed quite naturally that Chinese illustrations of technology, tending to reflect the actual technology in use, regularly present us with a picture of technology even in the nineteenth century in which practitioners applied knowledge, skills and muscles in place of more or less complex machines and inanimate sources of power.

Underrepresented Technologies

We can gain further insights into the strengths and weaknesses of Chinese portrayals of technology by considering a number of quite important productive activities in traditional China whose technologies are not well represented in the surviving illustrations, especially those dating from before the nineteenth century. Mining is a good example.[14] That mining did not generate more images may at first be somewhat surprising since mining was, after agriculture and clothmaking, the productive technology that engaged the largest number of laborers in traditional China.[15] But mining as practiced in premodern China was also a technology that relied much more on experiential knowledge not easily pictured than on a well-developed toolkit. Moreover, it did not, in the Chinese circumstances, invite significant mechanization: the excavation and transport of its product relied overwhelmingly on the availability of abundant supplies of cheap labor supplemented by very basic equipment that changed little over the centuries. In addition, a perennial shortage of investment capital for mining[16] further inhibited the introduction of machinery into the mines and the stimulus that might have given to mining illustrations.

Two further reasons may also help explain the paucity of mining illustrations. Chinese illustrators seem to have had special difficulty portraying an underground environment convincingly. We can compare this nineteenth-century illustration of a large underground copper mine in Yunnan (Fig. 7.1) with two examples of the far better illustrations of mine interiors Japanese painters and illustrators were capable of in the same period (Plates 15 and

---

Counterfactual," and "The Technology of Farming," are indispensable.

13. Franke, *Ackerbau*, 123; 155–56n41; 157–59n49.
14. As so often, Song Yingxing's late Ming *The Exploitation of the Works of Nature* is the main exception, providing illustrations of coal mining (Sun and Sun, *T'ien-kung k'ai-wu*, 204, Fig. 11–3), silver mining (239, Fig. 14–2), digging a shaft (Sun and Sun, 243, Fig. 14–5), surface mining of iron ore (Sun and Sun, 246, Fig. 14–8), concentrating ore by washing (Sun and Sun, 249, Fig. 14–9), washing tin ore (Sun and Sun, 253, Fig. 14–11 and 254, Fig. 14–12), and gem mining (Sun and Sun, 301, Fig. 18–3 and 302, Fig. 18–4).
15. Golas, *SCC* 5:13, 1–3, 14–16. Also, in total value of its product, mining likely ranked directly behind agriculture. It is perhaps not without significance that Hommel begins his volume *China at Work* with a discussion of mining, especially coal mining.
16. Golas, *SCC* 5:13, 410–15.

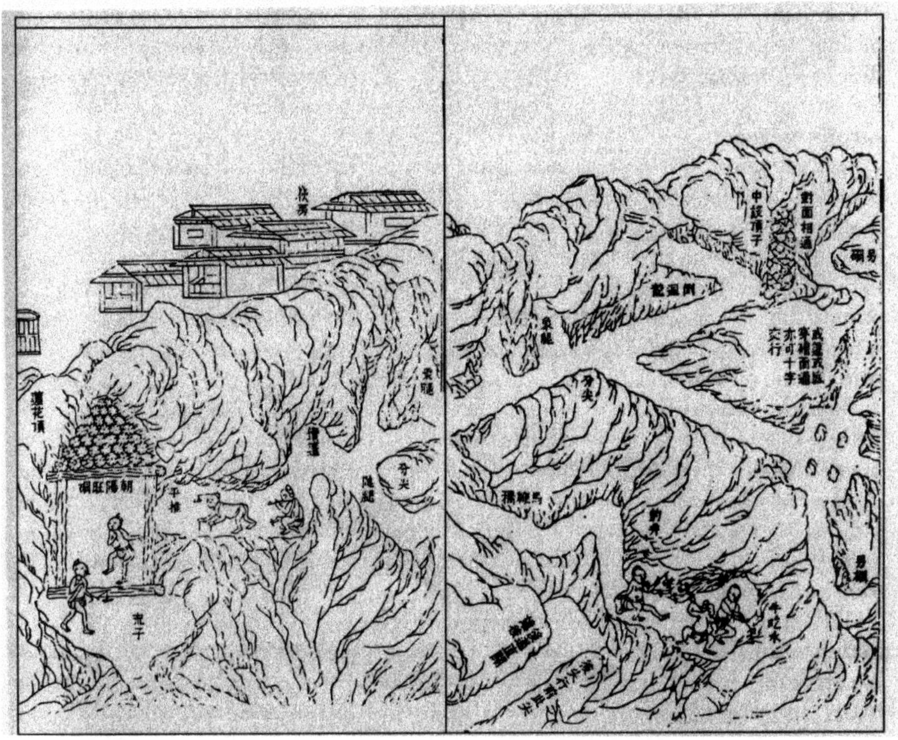

7.1 Large underground copper mine in nineteenth-century Yunnan

16).[17] Finally, an instinctive aversion to mining and miners, by no means unique to China, also may well have discouraged the writing about and therefore also the picturing of mining technology.

Control of water was another activity that generated much less in the way of interesting technological illustrations than might have been expected given the rich history of water control projects such as irrigation works,[18] flood control, polders, river conservancy and transport canals[19] that did so much over the centuries to modify the Chinese countryside. Such projects were often sponsored, directed and funded either by the central government or by local officials and were seen as an important part of the government's responsibility to provide for the needs and welfare of the population. They thus benefitted from the considerable resources the government had at its disposal. That we do not have more in the way of enlightening illustrations undoubtedly reflects the fact that the implements and machines

17. For a sample of European portrayals of mining from the sixteenth century, together with a succinct but superb discussion, see Lefèvre, "*Picturing Machines*."

18. Elvin provides an excellent overview and classification of irrigation techniques in "Technology of Farming," 21–24.

19. For a breathtakingly thorough overview of these activities, crowning achievements of Chinese technology that are often insufficiently recognized, see Needham and Wang, *SCC* 4:2, 330–62 and Needham, Wang and Lu, *SCC* 4:3, 211–378; a large proportion of the relevant surviving illustrations, especially those from Song Yingxing's *Tian gong kai wu*, are reproduced in these pages.

for excavation, building embankments and such tasks were generally simple and similar to those long used in other activities such as agriculture and mining. As for other knowledge and techniques more specific to hydraulic engineering such as the behavior of water flows in different conditions, these did not lend themselves well to visual portrayals. Occasionally we come across an illustration of what, right down into modern times, was more responsible than anything else for the greatest achievements of the Chinese in this technology: the unparalleled ability of Chinese officials to organize vast numbers of laborers to accomplish large, sometimes even massive, projects (Fig. 7.2).[20] This mid-nineteenth century illustration from the autobiography of an official skilled in hydraulic engineering shows the cutting of a canal by large numbers of workers using for the most part simple digging instruments, wheelbarrows, hand-swung buckets and square-pallet chain pumps. It powerfully makes the point that the best available "technology" in projects of this kind consisted mainly in the organizing of a massive labor force, which, in itself, effectively became a giant machine.

Shipbuilding was another under-reported and under-illustrated technology in traditional China. Despite the ubiquity of water transport especially in south China but also along the coast in the north, the plethora of river, lake and sea battles in China's military history, and the seven great voyages of Zheng He 鄭和 (1371–1433) in the Ming,[21] no systematic manuals on shipbuilding seem to have been published in China before nineteenth century.[22] Even one important manuscript that clearly deserves such a characterization was a product of the first half of the nineteenth century, or just possibly the late eighteenth.[23] Illustrations of nautical technology from other sources are also quite rare. Although Song Yingxing's *Exploitation of the Works of Nature* includes a fairly extensive discussion of ships, it contains only two illustrations despite the fact that many more types of ships are described in the text.[24] And just as Song's illustrations are disappointing from a technological point of view, the same tends to be true for most other surviving depictions of ships.[25] We have seen in Chapter 3 one arresting example in the warship from the important *Collection of Most Important Military Techniques* which sports three decks but no masts or sails![26] Indeed, one frequently has the feeling that portrayals of ships were done by draftsmen or artists who had little familiarity with things nautical. Partly this reflected the fact that, as in Europe, Chinese

20. Needham, Wang and Lu, *SCC* 4:3, 261–63.
21. In the years 1403–1419 at the beginning of Zheng's voyages, Chinese maritime shipyards built 2149 sea-going vessels. SCC 4:3 (Needham, Wang and Lu), 479, fn. f.
22. Needham, Wang and Lu, *SCC* 4:3, 380. This volume of SCC, pp. 379–699, is the invaluable first reference for all topics having to do with Chinese nautical technology.
23. Needham, Wang and Lu, *SCC* 4:3, 406–8, Figs. 941 and 942.
24. Sun and Sun, *T'ien-kung k'ai-wu*, 171–80. (This translation should be compared with the extensive translated segments in Needham, Wang and Lu, *SCC* 4:3, 411–16.)
25. Needham, Wang and Lu, *SCC* 4:3, 424–29. The great exceptions are, of course, the magnificent Song paintings of ships by Guo Zhongshu and Zhang Zeduan, examples of which we saw in Chapter 3. It is just our good fortune, however, that they are so revealing since the authors apparently did not create their paintings with any specifically technological purpose in mind.
26. Compare also Needham, Wang and Lu, *SCC* 4:3, 426, Fig. 949 and 691, Fig. 1051.

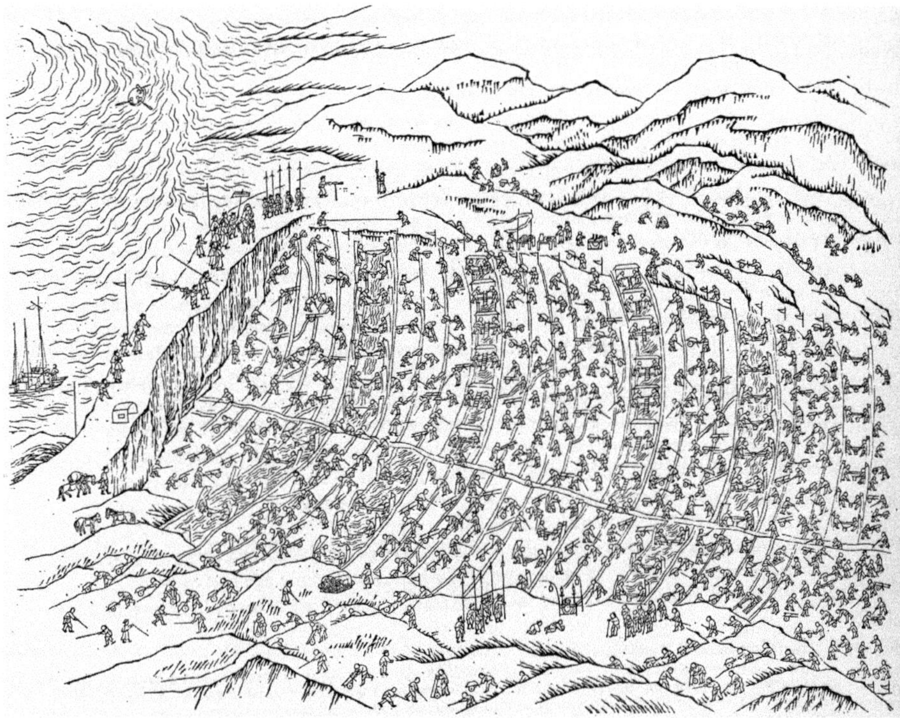

7.2 The cutting of a canal at Zhongmou in Jiangsu "doubtless between 1833 and 1842." Needham, Wang and Lu, *SCC* 4:3, 262, Fig. 876. In recent times such as during the Great Leap Forward in the 1950s, projects of this kind might mobilize hundreds of thousands of laborers.

shipbuilders made little use of drawings of any kind in their work. As Needham points out, even in the twentieth century, "the Chinese traditional shipwrights used no templates or blueprints, depending rather upon the skill or sureness of eye of the oldest and most experienced craftsmen."[27] Thus, although there were undoubtedly some working drawings that have not survived, perhaps even more than we might expect, it is likely that the general paucity of shipbuilding illustrations derives above all from the fact that they were not made in the first place.

Finally, there is the fascinating case of ceramics. We have here the curious phenomenon of a highly developed technology that made the Chinese for centuries world leaders for quality and quantity of ceramic production, but which has also left behind only a small body of rather undistinguished illustrations of ceramic production processes. Here, too, a number of reasons suggest themselves. To begin with, we are dealing with not only a paucity of visual materials but also of written accounts.[28] In part, this is attributable to the widespread

27. Needham, Wang and Lu, *SCC* 4:3, 413.

28. Rose Kerr and Nigel Wood, "Chemistry and Chemical Technology: Ceramic Technology," in *Science and Civilisation in China (SCC)*, ed. Rose Kerr, vol. 5, part 12 (Cambridge: Cambridge University Press, 2004), 20.

disinterest of educated Chinese in production processes.[29] Even connoisseurs deeply drawn to ceramics focused overwhelmingly on the finished product. They took little or no interest in the dirty and dusty workshops where those products were made.[30] Without accounts of how ceramic objects were produced, there was no demand for illustrations to accompany such descriptions. Before the eighteenth century, it is only in the *Exploitation of the Works of Nature* with its twenty pages of text and twelve accompanying full-page illustrations[31] that we get a serious examination of the ceramics production process.[32]

Moreover, just as in the case of agriculture, many of the most interesting and sophisticated aspects of ceramic production — selection of the best clays, mixing of glazes, firing schedules, etc. — did not lend themselves very well to illustration. Nor did the simple and largely unchanging equipment inspire artistic depictions. As we have seen so often, the quality of the objects produced owed much less to the rather simple equipment used than to the skills and knowledge, often intuitive, that the potters brought to their task. These also lay beyond visual portrayal.

The Absence of Standards for Technical Drawing

As we have more than once noted above, another obstacle to the emergence of improved portrayals of technological subjects in China was the continuing absence down to the Ming and Qing periods of any generally accepted norms for what constituted appropriate technical illustrations. This situation contrasts dramatically with the more or less well-defined schools of painting and general book illustration.[33] Of course, one should not exaggerate. Dieter Kuhn has argued that the illustrations of the successive editions of Wang Zhen's *Agricultural Treatise* and of Lou Shu's *Pictures of Tilling and Weaving* display a "pictorial continuity" (*bildnerische Kontinuität*) and that errors transmitted over succeeding editions are "most convincing proof of a tradition."[34] I would suggest two caveats, however: first, most of the continuity in the illustrations resides in the subjects treated and much less in any drawing styles and techniques specifically identified with technological portrayals; second, apparent stylistic continuities surely derived as much from skilled copying (at which many

29. Recall the comments of Song Yingxing; Sun and Sun, *T'ien-kung k'ai-wu*, xiii–xiv.
30. Kerr and Wood, *SCC* 5:12, 31 discusses this attitude in the late Ming. There were only a few notable exceptions, including the occasional literatus who actually turned his hand to working clay (31–33).
31. In the original 1637 edition; see the excellent 1959 Zhonghua shuju reprint. For the three surviving copies of the original edition, see Golas, "Like Obtaining a Great Treasure," 569 and 594n109.
32. It is worth noting that *all* the traditional illustrations of ceramic production in the magnificent Kerr/Wood volume on Chinese ceramics come from Song Yingxing's work.
33. I do not find persuasive the contention of Liu Keming and his colleagues that "engineering drawing" became a separate discipline in China in the Song-Yuan period. See Liu Keming et al., "A Brief Survey of Engineering Drawing in the Song Dynasty." Persuasive evidence to support this is simply lacking. Moreover, it is hard even to imagine that there could have been enough demand for technical illustrations to encourage artists to specialize their efforts in this area.
34. Kuhn, "Marginalie," 144; Kuhn, "Some Notes Concerning the Textile Technology," 409n5.

Chinese artisans excelled and which frequently provided the illustrations for successive printed works) as from any agreed upon standards for portraying technical subjects (for which there is no explicit evidence).

This absence of commonly accepted drafting standards was in part due also to the complete absence of professionalization, not to speak of academization, of technical knowledge in traditional China.[35] It was a situation pithily captured by Francesca Bray in her characterization of China as "a society without engineers."[36] As a result, what we might in hindsight recognize as important breakthroughs in illustration techniques tended to occur haphazardly and did not lead, especially after the Song, to substantial cumulative improvements over time. Plan drawings of the sort found in the *Building Standards* and in non-technological catalogs of archaeological objects such as the *Kaogutu* 考古圖 (Researches on archaeology with drawings) of Lü Dalin 呂大臨 (1044–93)[37] are rarely to be found in other works dealing with technological subjects.[38] No institutional mechanisms encouraged the use of improved techniques by other artists at the time or later. In the worst case, advanced drawing techniques such as the component parts drawings in the *System Essentials* were not only rarely used in later centuries but even largely forgotten.

What the artists who did illustrations of technology had in common was an approach to what constituted a good or at least a proper drawing of any kind. An important element of this approach was the use of *baimiao* or outline illustration techniques that were at best only imperfectly capable of naturalistic and precise representation of objects.[39] In those rare cases where an artist or illustrator sought not to copy but to draw an unfamiliar, complex piece of machinery using the *baimiao* technique, he inevitably tended to "generalize" the drawing, a process in which details tended to fall by the wayside.[40] This tendency was further reinforced when the artist was drawing from memory, making use of "observations stored in the mind's eye" with all their inevitable imperfections.[41]

35. The failure of the Chinese to "institutionalize" technology is a major theme of Stunkel, "Technology and Values" and is also dealt with in Vogel, "Mining Industry." One might even hypothesize that the relatively autonomous development of various technologies may be yet another reflection of the modular mode of thinking among the Chinese that Lothar Ledderose has so brilliantly brought to our attention in *Ten Thousand Things*. See also Mark Elvin, "The Man Who Saw Dragons: Science and Styles of Thinking in Xie Zhaozhe's *Fivefold Miscellany*," *The Journal of the Oriental Society of Australia* 25 & 26 (1993–94), 22: ". . . the Chinese, in science, seem to have been loners in comparison with the Europeans." Feuillet de Conches makes the same point, emphasizing that this was also true for painters. F. Feuillet de Conches, "Les peintures européens," 234–35.

36. Bray, *Technology and Gender*, 210.

37. E.g., Wu Jiming, *History of Chinese Drawing*, 40, Fig. 29.

38. One exception is a number of wheels in the *System Essentials* that might be viewed as plan drawings; Needham, Wang and de Solla Price, *Heavenly Clockwork*, 35, 38–39.

39. By their nature, *baimiao* drawings left out details as well as color, texture and techniques for modelling objects that contributed to making them appear lifelike. It was left to the reader's imagination to supply these. Hegel, *Reading Illustrated Fiction*, 325; Maeda, "*Chieh-hua*: Ruled-line Painting," 123n3.

40. See Matthies, "Medieval Treadwheels," 512, for a nice example (treadwheels) of the same thing happening in Europe.

41. Needham points out two excellent examples in his caption for an anonymous illustration of a thirteenth or

The Role of Government Workshops

When we examine the impact of the major written works dealing with technology that were encouraged by the government and even written by officials, it seems clear that works like the *Building Standards* or Wang Zhen's *Agricultural Treatise* served far more to codify existing technological practice, albeit perhaps best practice, than to encourage improvements and advances. On the other hand, the "bureaucratic" concerns that dominated these works were probably conducive on the whole to the production of illustrations that were more functional and correspondingly less influenced by aesthetic considerations.[42] Meant to help officials and others in the performance of their duties, they tend to display an unprecedented emphasis on tools, machines, instruments and components. We see this especially in the *System Essentials*, the *Building Standards* and the *Agricultural Treatise* where human beings are largely absent from the drawings instead of being right at their heart as was so often the case in drawings or paintings that pictured practical technology.

Thus the important role played by political authorities in the promotion of illustrations having to do with technology has been a recurrent theme in our story. Partly this was because government officials from very early times saw the maintenance of a harmonious society as crucially dependent on an adequately clothed, fed and housed populace. They therefore sought to encourage the production especially of essentials and, consequently, the skills and technology that contributed to it. Thus, even though most of traditional China's agricultural and textile production, for example, was in private hands, officials still frequently composed or sponsored writings that would encourage more effective production techniques.

From Shang times onward, the political authorities were also engaged in producing in government-run factories and workshops items necessary or desirable for their own personal use.[43] Mainly these were luxury products such as bronze ritual vessels, ornamented jades, lacquered utensils and silk cloth. But they also included military weapons and accoutrements, ceramic items for daily use, as well as the tools and implements needed to

---

fourteenth century water-powered milling plant. Needham and Wang, *SCC* 4:2, Plate CCXLI, Fig. 627b; see also Fig. 50 above. The painter, probably painting from memory, not only confused paddle-wheels with gear-wheels but also was able to provide only a botched version of the crank, connecting-rod and piston rod combination that made up the reciprocator that worked a flour-sifter.

42. To be sure, there are many examples where one gets the impression that learning the terminology was often more important than understanding the technology. Vogel, "Important Sources of the History of Premodern Chinese Salt Production Techniques," in Hua Jueming et al. (eds.), *Study on Ancient Chinese Books and Records of Science and Technology* (Zhengzhou: Daxiang chubanshe [Elephant Press], 1998), 173. Again, this is not surprising among a thoroughly book-oriented officialdom.

43. For a very negative, but probably mostly accurate picture of at least many of the government workshops and other production facilities, focusing on the Ming period, see Xu and Wu, *Chinese Capitalism*, Chapter 3. Lothar Ledderose (*Ten Thousand Things*, 75–76) draws a distinction between "workshops" (small establishments run by a master craftsman) and "factories" (larger establishments run by managers who might or might not themselves be craftsmen). The distinction can be useful provided it is not applied too rigidly, all the more since we frequently lack the evidence to make it with much confidence.

manufacture these and other goods.[44] The Shang workshops were the predecessors of an extensive system of government production facilities that one finds in China over most of the following three thousand years and which constituted the closest thing to a "technological sector" in premodern China. Needham offers as a "provisional conclusion" that "a considerable proportion of the most advanced technologists in all ages in China were either directly employed by, or under the close supervision of, administrative authorities forming part of the central bureaucratic government."[45]

This statement, however, invites certain qualifications. First of all, it may well apply better to the situation in earlier times when, in addition to actual governmental production facilities, nobles and powerful officials acting as patrons often attracted to their entourage people with special skills in scientific/technological matters and even sometimes operated quite large production units in their residential compounds.[46] After the huge economic expansion in late Tang and Song, however, technological expertise was much more widely spread throughout the society, proportionally less monopolized by those who held political power, and more responsive to broader consumer demand. However, even in this later period, the role of the government could be decisive for broad technological advance. For example, if Kerr and Wood are correct, even in the Ming and Qing period, "massive support that central government procurement of ceramics guaranteed gave technological superiority to the whole Chinese ceramics industry."[47]

Second, even if one might generally expect a higher level of quality in the productions of government factories, workshops, arsenals etc. than was typical across Chinese society, that does not necessarily mean that these facilities were a significant force for technological innovation. Apart from certain important exceptions such as the imperial ceramics works at Jingdezhen, there is little evidence either to support or to counter such a contention. We have few surviving examples of written descriptions and even fewer drawings arising directly out of the government production environment that might alert us to innovations; insofar as they were produced, these ephemeral jottings apparently were not deemed worthy of wider distribution, either in manuscript or printed form. Only occasionally can we tell from tantalizingly laconic statements or from the products produced that innovation had taken place. Thus we learn that, at Jingdezhen, "technicians at the factory were subjected to repeated demands for innovatory products, many of them stimulated by receipt of tributary artifacts from beyond the borders of China."[48]

---

44. On Shang workshops in general, see David M. Keightley, "Public Work in Ancient China: A Study of Forced Labor in the Shang and Western Chou" (PhD Dissertation, Columbia University, 1969), 39–65. For silk workshops, see Dieter Kuhn, "The Silk-Workshops of the Shang Dynasty," in Li Guohao et al., *Explorations*, 397–405, which, however, includes some speculations that are very much open to question.
45. Needham and Wang, *SCC* 4:2, 20. A striking early example is the Three Seasons Tailoring Workshops of the Later Han where the workforce consisted of "several thousand male and female workers, craftsmen and artisans who were trained and skilled in all branches of the textile trade." Kuhn, "Silk Weaving," 103.
46. Needham and Wang, *SCC* 4:2, 10, 17, 26, 32, 33 and 400.
47. Kerr and Wood, *SCC* 5:12, xlvi–xlvii.
48. Kerr and Wood, *SCC* 5:12, xlvi. These demands could sometimes be so onerous that they led the potters

Finally, embedded in the very organization of government productive facilities were practices and ideas that must often have inhibited invention and innovation, even as they promoted the production of large quantities of consistently higher quality goods than may have been typical of most private producers. Many of these facilities consisted of a highly bureaucratic organization with modular production techniques that made wide use of ultra-specialized artisans who had at best very limited ability or opportunity to apply to their tasks creativity that might have led to improvements and innovations.[49] Moreover, government workshops often put considerable emphasis on standardization and holding down costs (though corruption and waste appear to have been at least as prevalent here as in other government operations[50]). These were not values likely to encourage innovation.[51]

On balance, then, government factories and workshops very possibly impeded more than they promoted technological innovation, and hence the design sketches, working drawings and the like that might have been a part of such efforts. The extent to which they encouraged the spread of technological knowledge is also open to question. The rotational workers at government works (insofar as they were not soldiers or convicts) were certainly capable of bringing home with them at the end of their tours any interesting technical knowledge that they might have picked up in their work.[52] On the other hand, in special cases such as the major government arsenals at the capital where most of the huge number of weapons to supply Chinese armies were manufactured, secrecy must have regularly played its inhibiting role both in the advancement of the technology itself and, perhaps even more, in the production of written materials that might lead to the leaking of "dangerous" knowledge to the populace.[53] Such emphasis on secrecy may also partially account for the fact that, even in eras of important advances in weaponry such as the Song, those advances seem to have occurred and spread quite slowly.

---

to riot and/or destroy what they had produced. Ledderose, *Ten Thousand Things*, 86. On the other hand, abundant government resources made it possible to meet the costs of experiments and of the high rates of failure that often accompanied the development of new clays, glazes, kilns, firing techniques and the like.

49. Ledderose, *Ten Thousand Things*, 4–6, 25, 37, 84 and, especially, 48–49; 70 and 75–76. Ledderose contends that figures in the Qin First Emperor's terra-cotta army made by workers from the state factories display more stylistic uniformity — as well as a more consistent, higher quality workmanship — than those made by workers from private local workshops. See also the contrast drawn by Keightley between Chinese workshops with their modular production under bureaucratic supervision and the more individualistic ancient Greek workshops organized "around a series of acts performed by single craftsmen." Keightley, "Early Civilization in China," 19. For a fascinating example of how modular production (in this case, nose shapes) can make possible identification of different workers or work teams on a single project, see Barbieri-Low, *Artisans*, 87–88 and 89, Fig. 3–13.
50. Xu and Wu, *Chinese Capitalism*, 67.
51. Moll-Murata et al., *Chinese Handicraft Regulations*, 17 and 323.
52. Bray points out that, especially in the period up to the Song, technological skills in silk-weaving were constantly enriched throughout the country by weavers returning home from their yearly stints in the imperial workshops; Bray, *Technology and Society*, 47.
53. Or to non-Chinese on the borders who might threaten the empire. As late as the 1620s, the Manchus seem not yet to have developed the ability to manufacture gunpowder weapons. di Cosmo, "Did Guns Matter?" 141.

Overall, it may well be that the greatest positive impact of the government on the illustration of technology came not from the government workshops or factories but rather from the actual printing or subventions for printing of illustrated books that dealt with technological or related subjects. Most portrayals of technological subjects that have survived, as we have noted, are found in printed books. The political authorities — central government as well as local officials — continued active here right down to the very end of the imperial period.

Maturity or Stagnation?

It has been suggested that late imperial Chinese institutions and culture, including science, technology and art, had reached by the end of the eighteenth century "a magnificent dead end."[54] Without necessarily subscribing to quite so stark a characterization, most scholars familiar with this period of Chinese history would today agree that China's technological inventiveness and innovation declined substantially in Ming and Qing China even as the economy from the sixteenth to the eighteenth centuries underwent vigorous growth. Particularly relevant for our purposes is the fact that Chinese printing, right down to the nineteenth and twentieth centuries, remained an unmechanized handicraft industry using techniques little changed for over a millennium.[55] This could hardly be more different than the situation in Europe where, by the late sixteenth century, copper engravings had largely replaced woodblock prints for illustrations. Copper engravings not only were capable of better expression of light and shade but, even more important for technological subjects, they could convey finer details. In China, book illustrations continued overwhelmingly to be produced by the woodblock process, which of course had its own advantages in that, for example, copies of illustrations for new editions could be produced easily and cheaply by artisans of average skills using well-established techniques. Chinese book illustrations thus regularly display, as noted above, a continuity that finds no counterpart in Europe.[56]

The reasons for the paucity of new developments in technology overlap to some extent with the reasons, often economic, that we have previously discussed to account for the limited spread of even quite readily available improvements, especially but not only in the area of mechanization. As in the workshops and factories, government efforts to promote the most important technology, agriculture, present a mixed record. Even when the population/land crisis of the eighteenth century had quite clearly outrun local solutions, the court and its officials, according to Francesca Bray, did not "press for sustained innovation in agriculture."[57] The operative phrase here is "press for." According to Pierre-Étienne Will, and accepted by Bray, there was a clear disconnect between government rhetoric encouraging

54. Blunden and Elvin (*Cultural Atlas of China*, 144–47) provide a superb summary discussion of the near-stasis that had come by this time to mark so much of Chinese life and thought.

55. Tsien, *SCC* 5:1, 382.

56. Chia, *Printing for Profit*, 12.

57. Bray, *Technology and Gender*, 27, fn. 45.

agriculture and efforts actually to bring about improvements. The latter failed in some measure because the government ruled out compulsion or coercion and instead tried to get its policies adopted through education and persuasion.[58] Moreover, if Bray is correct, many of the agricultural reform schemes put forth by officials in the late Ming and early Qing sought conflicting goals that could not easily be harmonized. For most officials at that time, farming and weaving were still the very basis not only of essential economic production but also of a proper social order based on household (mainly rural) productive activities that were often divided on a gender basis whereby men farmed and women produced cloth. In fact, such standards had been greatly undercut as commercialization spread and production came to be shaped increasingly by market forces. Nevertheless, the ruling elite continued to believe that it was possible to devise policies that would reform work in such a way as to restore Confucian values.[59]

Technological advance, especially in machinery that would have invited better illustration techniques, was also impeded by the fact that the Chinese, even by the beginning of the Ming, seem to have accomplished many of the most difficult innovations in the development of sophisticated mechanical applications. Or so Elvin has argued, basing himself partly on textile machinery and pointing to such breakthroughs as mechanizing hand movements, replacing human and animal with inanimate power, and devising linkages that made possible control of two or more different motions by a single power source, animate or inanimate.[60] The very success of these developments must have lessened the inclination of the Chinese to carry on further experiments, with all the costs and risks involved. For example, the Archimedian screw pump, introduced in China by the Jesuits in the early seventeenth century, had the capacity to raise as much water as up to five square-pallet chain pumps. But the difficulty and costs of constructing and maintaining it usually made it an unattractive choice for Chinese farmers who instead stayed with the familiar, relatively cheaper and either more reliable or more easily repaired pallet pumps.[61]

Of course, it is not easy to make the case that further technological advances would inevitably have led to better illustrations. As we noted earlier, there were some significant

58. Bray, *Technology and Gender*, 31, fn. 53. Sustained efforts to increase agricultural and other production must also have been undercut to some extent by the short periods local officials served in a given post.

59. Bray, "Towards a Critical History of Non-Western Technology," in Brook and Blue, *China and Historical Capitalism*, 168.

60. Elvin, "China as a Counterfactual," 108. One illustration of how complex certain machines had become by the end of the Ming is the loom described by Song Yingxing that, if he is correct, was made up of more than 1800 parts; Bray, "Towards a Critical History," 178–79, esp. fn. 57; Sun and Sun, *T'ien-kung k'ai-wu*, 55. See also Needham and Wang, *SCC* 4:2, 225.

61. Zhang Baichun, "Archimedian Mechanical Knowledge," 11–12 of 16; Brook, "The Spread of Rice Cultivation and Rice Technology," 686. Gary Hamilton and Wei-an Chang have made a further argument that the merchant-or distribution-driven economy they see in late imperial times increasingly pushed production down into rural households, leading to the *simplifying* of the technology used. Gary G. Hamilton and Wei-an Chang, "The Importance of Commerce in the Organization of China's Late Imperial Economy," in Giovanni Arrighi, Takeshi Hamashita and Mark Selden (eds.), *The Resurgence of East Asia: 500, 150 and 50 Year Perspectives* (London and New York: Routledge, 2003), 179, 180–81, 186, 203.

technological changes and improvements in agriculture between the time of Wang Zhen in the thirteenth century and that of Xu Guangqi in the early seventeenth century. Xu mentions them in his *Nongzheng quanshu* (Complete treatise on agricultural administration) but *none* are reflected in the illustrations.[62] Even the mechanical innovations that were made in Ming and Qing farming and for which we have textual evidence almost never appear in technological illustrations.[63] Nevertheless, there is no denying that a largely static technology was not likely to create an environment that encouraged improved technological depictions.

When we reverse our focus to examine the effect that stagnation in the portrayal of technology may have had on advances in technology itself, it seems highly plausible that the shortcomings so pervasive in Chinese portrayals of technology must have impeded technological inventiveness at least to some degree. A considerable body of scholarship over the last several decades has argued that images play a much greater role in our thinking processes than was once recognized. In other words, visual thinking, the manipulation of images, can often be as important or even more important than verbal thinking, the manipulation of words. Originally, it was scholars interested in the psychology of art who pioneered these investigations.[64] But the case can be made equally well for thinking about technology. Eugene Ferguson, in his remarkable study *Engineering and the Mind's Eye*, summarizes the argument:

> Many features and qualities of the objects that a technologist thinks about cannot be reduced to unambiguous verbal descriptions; therefore they are dealt with in the mind by a visual, nonverbal process. . . . It has been nonverbal thinking, by and large, that has fixed the outlines and filled in the details of our material surroundings.[65]

When one realizes, however, that the images we carry in our minds, whether or not we use them to think with, are experience-based (derived from sense impressions, mainly from our eyes but also from our ears, nose, fingers, etc.), it is obvious that different sets of experiences will lead to very different sets of mental images.[66] In the case of technology, our repertoire of images derives not only from the tools and machines we have seen and perhaps even

62. Bray, "Agricultural Illustrations," 545.

63. As Bray notes: "When it came to practical, technical matters like farming, its seems that those educated Chinese who recorded changes or improvements were satisfied with the power of words to convey material processes, and felt no need or desire to resort to graphics." Bray, "Agricultural Illustrations," 552. An example of another kind is provided by Georges Métailié who notes that the maize (corn) plant does not appear in accurate depictions until the mid-19th century, three centuries after its introduction to China! Georges Métailié, "The Representation of Plants: Engravings and Paintings," in Bray, Dorofeeva-Lichtmann and Métailié (eds.), *Graphics and Text in the Production of Technical Knowledge in China*, 493–94.

64. I am thinking, for example, of E. H. Gombrich, Rudolph Arnheim and William M. Ivins, Jr.

65. Ferguson, *Engineering and the Mind's Eye*, xi. This book is the fuller development of ideas first presented much earlier in the author's pathbreaking article, "The Mind's Eye: Nonverbal Thought in Technology," *Science* 197 (August 26, 1977), 827–36. See also Chapter 2, "Visual Thinking" in Arnold Pacey, *Meaning in Technology* (Cambridge, MA: MIT Press, 2001) and Pfaffenberger, "Social Anthropology of Technology," 507–9.

66. As in Aesop's fable of the seven blind men examining an elephant.

used, but also from illustrations, including illustrations of technological artifacts that we may or may not have experienced in reality. In the technologically explosive Renaissance, for example, the drawings of a Georgio Martini or a Leonardo da Vinci threaten to overwhelm us with their profusion of carefully drawn details. By contrast, we have seen that Chinese illustrations even in works that focused on technology regularly underplayed precision and attention to detail.[67] They therefore did little to encourage that increasingly precise kind of technological thinking that alone could promote technological advance once machines had reached a certain level of complexity. William Ivins may not be entirely correct when he argues that creating "a culture of technologies requires much harder and more accurate thinking [than building up a culture of art and philosophy]," but it seems quite clear that, in later imperial times at least, the Chinese preferred to put their best thinking into philosophy, art and other aesthetic and academic studies, to the relative neglect of technology. To this, their depictions bear witness.

67. And also precise measurement. Bodde makes the interesting suggestion that traditional Chinese instruments and machines may overall have required less accurate measurement for their construction than those of the West; Bodde, *Chinese Thought, Society and Science*, 140.

Bibliography

Acker, W. R. B. *Some T'ang and Pre-T'ang Texts on Chinese Painting*. 2 vols. Leiden: E. J. Brill, 1954, 1974.

Ackerman, James. "The Involvement of Artists in Renaissance Science." In Shirley and Hoeniger, *Science and the Arts in the Renaissance*, 94–129.

Ames-Lewis, Francis. *Drawing in Early Renaissance Italy*. New Haven and London: Yale University Press, 1981.

Atwell, William S. "The T'ai-ch'ang, T'ien-ch'i, and Ch'ung-chen Reigns." *CHC*, vol. 7, part 1, 587–88.

Baigrie, Brian S. *Picturing Knowledge: Historical and Philosophical Problems Concerning the Use of Art in Science*. Toronto: University of Toronto Press, 1996.

Bailey, Gauvin Alexander. *Art on the Jesuit Missions in Asia and Latin America, 1542–1773*. Toronto: University of Toronto Press, 1999.

Barbieri-Low, Anthony J. *Artisans in Early Imperial China*. Seattle: University of Washington Press, 2007.

Barker, David. *Traditional Techniques in Contemporary Chinese Printmaking*. Honolulu: University of Hawai'i Press, 2005.

Barnhart, Richard M. *Painters of the Great Ming: The Imperial Court and the Zhe School*. Dallas: Dallas Museum of Art, 1993.

Barnhart, Richard M. et al. *Three Thousand Years of Chinese Painting*, 1997. (See under Yang Xin.)

Barrow, John. *Travels in China*. London: T. Cadell and W. Davies, 1804.

Basalla, George. *The Evolution of Technology*. Cambridge: Cambridge University Press, 1988.

Baynes, Ken and Francis Pugh. *The Art of the Engineer*. Woodstock, NY: The Overlook Press, 1981.

Behr, Wolfgang. "Placed into the Right Position - Etymological Notes on Tú 圖 and Congeners." In Bray et al., *Graphics and Text*, 110–34.

Bickford, Maggie. "Stirring the Pot of State: The Sung Picture-Book *Mei-Hua Hsi-Shen P'u* and Its Implications for Yüan Scholar-Painting." *Asia Major* (Third Series) 6.2 (1993), 169–226 and Figs. 1–5.

Billeter, Jean François. *The Chinese Art of Writing*. New York: Skira/Rizzoli, 1990.

Biot, Edouard. "Researches into the Manners of the Ancient Chinese, according to the She-king." In Legge, *The She King*, 142–71.

Blunden, Caroline and Mark Elvin. *Cultural Atlas of China*. New York: Facts on File, 1983.

Bodde, Derk. *Chinese Thought, Society, and Science: The Intellectual and Social Background of Science and Technology in Pre-modern China*. Honolulu: University of Hawai'i Press, 1991.

Boltz, William G. *The Origin and Early Development of the Chinese Writing System*. New Haven: American Oriental Society, 1994.

Booker, Peter Jeffrey. *A History of Engineering Drawing*. London: Chatto & Windus, 1963.

Bray, Francesca. "Biology and Biological Technology: Agriculture." In *Science and Civilisation in China (SCC)*, ed. Joseph Needham, vol. 6, part 2. Cambridge: Cambridge University Press, 1984.

———. *The Rice Economies: Technology and Development in Asian Societies*. Oxford: Blackwell, 1986.

———. *Technology and Gender: Fabrics of Power in Later Imperial China*. Berkeley: University of California Press, 1997.

———. "Towards a Critical History of Non-Western Technology." In Timothy Brook and Gregory Blue, *China and Historical Capitalism: Genealogies of Sinological Knowledge*, 158–209. Cambridge: Cambridge University Press, 1999.

———. *Technology and Society in Ming China (1368–1644)*. Baltimore: Society for the History of Technology, 2000.

———. "Agricultural Illustrations: Blueprint or Icon?" In Bray et al., *Graphics and Text*, 521–67.

———. "Chinese Literati and the Transmission of Technological Knowledge: the Case of Agriculture." In Schäfer, *Cultures of Knowledge*, 299–325.

———. "Introduction: the Powers of *Tu*." In Bray et al., *Graphics and Text*, 1–6.

Bray, Francesca, Vera Dorofeeva-Lichtmann, and George Métailié (eds.). *Graphics and Text in the Production of Technical Knowledge in China: The Warp and the Weft*. Leiden and Boston: Brill, 2007.

Bray, Francesca and Georges Métailié. "Who Was the Author of the *Nongzheng quanshu*?" In Jami, Engelfriet and Blue, *Statecraft and Intellectual Renewal in Late Ming China*, 322–59.

Brokaw, Cynthia J. "Commercial Publishing in Late Imperial China: The Zou and Ma Family Businesses of Sibao, Fujian." *Late Imperial China* 17.1 (June 1996), 49–92.

———. "On the History of the Book in China." In Brokaw and Chow, *Printing and Book Culture*, 3–54.

Brokaw, Cynthia J. and Kai-wing Chow (eds.). *Printing and Book Culture in Late Imperial China*. Berkeley: University of California Press, 2005.

Brook, Timothy. "The Spread of Rice Cultivation and Rice Technology into the Hebei Region in the Ming and Qing." In Li Guohao et al., *Explorations*, 659–90.

———. "Mapping Knowledge in the Sixteenth Century: The Gazetteer Cartography of Ye Chunji." *East Asian Library Journal* 7.2 (Winter 1994), 5–32.

———. "Weber, Mencius, and the History of Chinese Capitalism." *Asian Perspective* 19.1 (1995), 79–97.

———. *The Confusions of Pleasure: Commerce and Culture in Ming China*. Berkeley: University of California Press, 1998.

———. "Communications and Commerce." *CHC*, vol. 8, part 2, 636–37, 649, 927.

Bulliet, Richard W. et al. *The Earth and Its Peoples*. Boston: Houghton Mifflin, 1997.

Burstall, Aubrey F. *A History of Mechanical Engineering*. Cambridge, MA: MIT Press, 1965.

Bush, Susan. *The Chinese Literati on Painting: Su Shih (1037–1101) to Tung Ch'i-ch'ang (1555–1636)*. Cambridge, MA: Harvard University Press (Harvard-Yenching Institute), 1971.

Bush, Susan and Hsio-yen Shih (eds.). *Early Chinese Texts on Painting*. Cambridge, MA: Harvard University Press (Harvard-Yenching Institute), 1985.

Bussagli, Mario (ed.). *Cotton and Silk Making in Manchu China*. New York: Rissoli, 1980.

Bussotti, Michela. "Woodcut Illustration: A General Outline." In Bray et al., *Graphics and Text*, 461–83.

Cahill, James. "Wu Chen: A Chinese Landscapist and Bamboo Painter of the Fourteenth Century." Ph.D. Dissertation. University of Michigan, 1958.

———. "Confucian Elements in the Theory of Painting." In Arthur Wright (ed.). *The Confucian Persuasion*, 115–40. Stanford: Stanford University Press, 1960.

———. *The Distant Mountains; Chinese Painting of the Late Ming Dynasty, 1570–1644*. New York and Tokyo: Weatherhill, 1982.

———. *The Compelling Image. Nature and Style in Seventeenth-Century Chinese Painting*. Cambridge and London: Harvard University Press, 1982.

———. *Three Alternative Histories of Chinese Painting*. Kansas City, KS: Spencer Museum of Art, University of Kansas, 1988.

———. *The Painter's Practice; How Artists Lived and Worked in Traditional China*. New York: Columbia University Press, 1994.

———. *Chinese Painting*. Geneva: Skira, 1995.

———. "Approaches to Chinese Painting, Part II." In Yang, Barnhart et al., *Three Thousand Years*, 5–12.

———. "Chinese Painting: Innovation after 'Progress' Ends." In Lee, *China, 5000 Years*, 174–292.

———. *Pictures for Use and Pleasure*. Berkeley: University of California Press, 2010.

Camerota, Filippo. "Renaissance Descriptive Geometry: The Codification of Drawing Methods." In Lefèvre, *Picturing Machines*, 175–208.

Cardwell, Donald. *Wheels, Clock, and Rockets: A History of Technology*. New York: W. W. Norton, 1995.

Carter, Thomas Francis. *The Invention of Printing in China and Its Spread Westward*. 2nd ed., revised by L. Carrington Goodrich. New York: Ronald Press, 1955.

Cartier, Roger. "Gutenberg Revisited from the East." *Late Imperial China* 17.1 (June 1996), 1–9.

Chang Po[Pai]-ch'un 張柏春. See Zhang Baichun.

CHC: Twitchett, Denis, John K. Fairbank et al. (eds.). *The Cambridge History of China*. Cambridge: Cambridge University Press, 1978–.

Chen Mingda 陳明達. *Yingzao fashi da mu zuo yanjiu* 營造法式大木做研究 [Studies on structural carpentry in the *Building Standards*]. Beijing: Wenwu, 1981.

Chen Tongbin 陳同濱 et al. (eds.). *Zhongguo gudai jianzhu da tudian* 中國古代建築大圖典 [Illustrations of ancient Chinese architecture]. Beijing: Jinri Zhongguo chubanshe, 1996.

Chen Wenhua 陳文華. *Zhongguo gudai nongye kejishi tushuo* 中國古代農業科技史圖說 [Illustrations with commentary on the history of ancient Chinese agricultural science and technology]. Beijing: Nongye chubanshe, 1991.

Cheng Weiji. *History of Textile Technology of Ancient China*. Rego Park, NY: Science Press New York, 1992.

Cherniack, Susan. "Book Culture and Textual Transmission in Sung China." *Harvard Journal of Asiatic Studies* 54 (1994), 5–125.

Chia, Lucille. "*Mashaben*: Commercial Publishing in Jianyang from the Song to the Ming." In Smith and von Glahn, *The Song-Yuan-Ming Transition*, 284–328.

———. "Text and *tu* in Context; Reading the Illustrated Page in Chinese Blockprinted Books." In Drège, *Dossier*, 241–76.

———. *Printing for Profit; the Commercial Publishers of Jianyang, Fujian (11th to 17th Centuries)*. Cambridge, MA: Harvard University Press, 2002.

Chiang Yee. *Chinese Calligraphy: An Introduction to Its Aesthetic and Technique*, 2nd ed. London: Methuen, 1954.

———. *The Chinese Eye: An Interpretation of Chinese Painting*. Bloomington and London: Indiana University Press, 1970.

Chou, Ju-hsi. "In Defense of Qing Orthodoxy." In *The Jade Studio: Masterpieces of Ming and Qing Painting and Calligraphy from the Wong Nan-P'ing Collection*. Introductory Essays by Richard M. Barnhart et al., 35–42. New Haven, CT: Yale University Art Gallery, 1994.

———. "Painting Theory in Eighteenth-Century China." In Peterson et al., *The Power of Culture*, 321–43.

Chow, Kai-wing. "Writing for Success: Printing, Examinations, and Intellectual Change in Late Ming China." *Late Imperial China*, 17.1 (June 1996), 120–157.

Chung, Anita. *Drawing Boundaries: Architectural Images in Qing China*. Honolulu: University of Hawai'i Press, 2004.

Cipolla, Carlo M. *European Culture and Overseas Expansion*. Harmondsworth: Penguin, 1970.

———. *Clocks and Culture 1300–1700*. New York: W. W. Norton, 1977.

Clunas, Craig. *Superfluous Things: Material Culture and Social Status in Early Modern China*. Cambridge: Polity Press, 1991.

———. *Pictures and Visuality in Early Modern China*. Princeton: Princeton University Press, 1997.

———. *Art in China*. Oxford and New York: Oxford University Press, 1997.

———. "Luxury Knowledge: the *Xiushilu* (Records of Lacquering) of 1625." *Techniques et cultures* 29 (1997), 27–40.

———. "Modernity Global and Local: Consumption and the Rise of the West." AHR 104.5 (Dec. 1999), 1497–1511.

———. "Text, Representation and Technique in Early Modern China." In Karine Chemla (ed.). *History of Science, History of Text*, 107–21. Dordrecht, The Netherlands: Springer, 2004.

Combaz, Gisbert. "La peinture chinoise vue par un peintre occidental." *Mélanges chinois et bouddhiques* 6 (1938–1939), 42–141.

Combridge, J. H. "The Astronomical Clocktowers of Chang Ssu-Hsun and his Successors." *Antiquarian Horology* 9.3 (June 1975), 288–301.

Cooper, Michael, S. J. (ed.). *The Southern Barbarians: the First Europeans in Japan*. Tokyo: Kodansha, 1971.

Crosby, Alfred W. *The Measure of Reality: Quantification and Western Society, 1250–1600*. Cambridge: Cambridge University Press, 1997.

Cullen, Christopher. "The Science/Technology Interface in Seventeenth-Century China: Song Yingxing 宋應星 on *Qi* 氣 and the *Wu Xing* 五行." *Bulletin of the School of Oriental and African Studies* 53.2 (1990), 295–318.

Damerow, Peter and Urs Schoefplin. "Western Sources of the *Qiqi tushuo*." In Zhang and Renn, *Transformation and Transmission*, 89–93.

de Bary, Wm. Theodore. "Some Common Tendencies in Neo-Confucianism." In Nivison and Wright, *Confucianism in Action*, 25–49.

Delahaye, Hubert. "Du peu d'effet de la peinture occidentale en Chine aux XVIIe et XVIIIe siecles." In Jami and Delahaye, *L'Europe en Chine*, 111–35.

Demiéville, Paul. "[Review of the] Photolithographic edition of Li Ming-tchong's *Ying tsao fa che*." *Bulletin de l'Ecole Francaise d'Extreme-Orient (BEFEO)* 25 (1925): 213–64.

Dibner, Bern. "Machines and Weaponry." In Reti, *The Unknown Leonardo*, 166–89.

Di Cosmo, Nicola. "Did Guns Matter? Firearms and the Qing Formation." In Struve, *Qing Formation*, 121–66.

Dolza, Luisa M. "Reframing the Language of Inventions: The First Theatre of Machines." In Lefèvre, Renn and Schoepflin, *Power of Images*, 89–104.

Dong Wen 董文 (ed.). *Jiaozheng Tian gong kai wu* 校正天工開物 [A corrected *Tian gong kai wu*]. Taipei: Shijie, 1962.

Drège, Jean-Pierre (ed.). *Dossier: Texte et image dans le livre illustré chinois*. Special section in *Bulletin de l'École française d'Extreme-Orient* 89 (2002), 237–326.

Drège, Jean-Pierre et al. *Images de Dunhuang; Dessins et peintures sur papier des fond Pelliot et Stein.* Paris: École française d'Extrême-Orient (Mémoires archéologiques 24), 1999.

Dubery, Fred and John Willats. *Perspective and Other Drawing Systems.* New York: Van Nostrand and Reinhold, 1972.

Ebrey, Patricia Buckley. *China.* Cambridge: Cambridge University Press, 1996.

ECCP: Hummel, Arthur W. (ed.). *Eminent Chinese of the Ch'ing Period.* 2 vols. Washington D.C.: Library of Congress, 1943, 1944.

Ecke, Tseng Yu-ho. "A Reconsideration of *Ch'uan-mo-i-hsieh,* the Sixth Principle of Hsieh Ho." In *Proceedings of the International Symposium on Chinese Painting.* Taipei: National Palace Museum, 1972.

Edgerton, Samuel Y., Jr. "The Renaissance Development of Scientific Illustration." In Shirley and Hoeniger, *Science and the Arts in the Renaissance,* 168–97.

———. *The Heritage of Giotto's Geometry; Art and Science on the Eve of the Scientific Revolution.* Ithaca and London: Cornell University Press, 1991.

Edgren, Sören et al. *Chinese Rare Books in American Collections.* New York: China Institute in America, 1984.

Edwards, Richard. *The World around the Chinese Artist: Aspects of Realism in Chinese Painting.* Ann Arbor: LSA Checkpoint College of Literature, Science, and the Arts, University of Michigan, 1987.

Elman, Benjamin A. *On Their Own Terms: Science in China, 1550–1900.* Cambridge, MA: Harvard University Press, 2005.

———. *A Cultural History of Modern Science in China.* Cambridge, MA: Harvard University Press, 2006.

Elvin, Mark. "The High-Level Equilibrium Trap: The Causes of the Decline of Invention in the Traditional Chinese Textile Industries." In W. E. Willmott (ed.), *Economic Organization in Chinese Society,* 137–72. Stanford: Stanford University Press, 1972. Reprinted with an important "Postscript" in Mark Elvin, *Another History: Essays on China from a European Perspective,* 20–63. Honolulu: Wild Peony/University of Hawai'i Press, 1996.

———. *The Pattern of the Chinese Past.* Stanford: Stanford University Press, 1973.

———. "Skills and Resources in Late Traditional China." In Dwight H. Perkins (ed.). *China's Modern Economy in Historical Perspective,* 85–113. Stanford: Stanford University Press, 1975.

———. "The Technology of Farming in Late-Traditional China." In Randolph Barker and Radha Sinha (eds.), *The Chinese Agricultural Economy,* 13–35. Boulder: Westview Press, 1982.

———. "China as a Counterfactual." In Jean Baechler, John A. Hall and Michael Mann (eds.). *Europe and the Rise of Capitalism,* 101–12. Oxford and New York: Basil Blackwell, 1988.

———. "The Man Who Saw Dragons: Science and Styles of Thinking in Xie Zhaozhe's *Fivefold Miscellany.*" *The Journal of the Oriental Society of Australia* 25 & 26 (1993–94), 1–41.

———. "How Did the Cracks Open? The Origins of the Subversion of China's Late-Traditional Culture by the West." *Thesis Eleven* 57.1 (May 1999), 1–16.

Escande, Yolande. "Perspectives et limites des recherches récentes sur la calligraphie et la peinture." *Revue Bibliographique de Sinologie* 14(1996): 219–42.

Feng, Jiren. *Chinese Architecture and Metaphor.* Honolulu and Hong Kong: University of Hawai'i Press and Hong Kong University Press, 2012.

Feuillet de Conches, F. "Les peintres européens en Chine et les peintres chinois." *Revue Contemporaine* 25 (1856), 216–60.

Ferguson, Eugene S. "The Mind's Eye: Nonverbal Thought in Technology." *Science* 197 (August 26, 1977), 827–36.

———. *Engineering and the Mind's Eye*. Cambridge, MA: MIT Press, 1992.

Finsterbusch, Käte. *Verzeichnis und Motivindex der Han-Darstellungen*. Wiesbaden: Otto Harrassowitz, 1971.

Fong, Mary H. "The Technique of 'Chiaroscuro' in Chinese Painting from Han through T'ang." *Artibus Asiae* XXXVIII.2/3 (1976), 91–127.

Fong, Wen. "The Study of Chinese Bronze Age Arts: Methods and Approaches." In Fong, *The Great Bronze Age of China*, 20–34.

———. *The Great Bronze Age of China*. New York: Metropolitan Museum of Art, c. 1980.

Fong, Wen C. *Beyond Representation: Chinese Painting and Calligraphy, 8th to 14th Century*. New Haven: Yale University Press, 1992.

Fong, Wen C. et al. *Images of the Mind*. Princeton: Princeton University Press, 1984.

Fontein, Jan and Wu Tung. *Han and T'ang Murals Discovered in Tombs in the People's Republic of China and Copied by Contemporary Chinese Painters*. Boston: Museum of Fine Arts, 1976.

Franke, Herbert. "Siege and Defense of Towns in Medieval China." In Frank A. Kierman, Jr. and John K. Fairbank, *Chinese Ways in Warfare*, 151–201. Cambridge, MA: Harvard University Press, 1974.

———. (ed.). *Sung Biographies*. Wiesbaden: Franz Steiner, 1976.

Franke, Otto. *Kêng Tschi T'u: Ackerbau and Seidengewinning in China*. Hamburg: L Friederichsen & Co., 1913. Digital version available at Biblioteca Sinica 2.0 (www.univie.ac.at/Geschichte/China-Bibliographie/blog/2011/05/23/franke-keng-tschi-tu/).

———. "Zur Geschichte des Kêng Tschi T'u. Ein Beitrag zur Chinesischen Kunstgeschichte und Kunstkritik." *Ostasiatische Zeitschrift* 3.2 (Juli–Sept. 1914), 169–208.

Fraser, Sarah E. *Performing the Visual: The Practice of Buddhist Wall Painting in China and Central Asia, 618–960*. Stanford: Stanford University Press, 2004.

Fuchs, Walter. "Zum Keng-chih-t'u der Mandju-Zeit und die japanische Ausgabe von 1808." *Ostasiatische Studien* 48 (1959), 67–80 and Pls. A1–A23; B1–B23.

Galluzzi, Paolo. "Leonardo da Vinci: From the 'elementi macchinali' to the Man-Machine." In Claire Farago (ed.), *Leonardo's Science and Technology: Essential Readings for the Non-Scientist*, 395–408. New York: Garland, 1999.

——— (ed.). *Leonardo da Vinci; Engineer and Architect*. Montreal: Montreal Museum of Fine Arts, 1987.

Gernet, Jacques. "La société chinoise a la fin des Ming." *Recherches de science religieuse* 72.1 (1984), 27–36.

———. *China and the Christian Impact: A Conflict of Cultures*. Trans. Janet Lloyd. Cambridge: Cambridge University Press, 1985.

———. *A History of Chinese Civilization*. Trans. J. R. Foster and Charles Hartman. 2nd ed. Cambridge: Cambridge University Press, 1996.

Girmond, Sybille. "Chinesische Bilderalben zur Papierherstellung. Historische und stilistische Entwicklung der Illustrationen von Produktionsprozessen in China von den Anfängen bis ins 19. Jahrhundert." In Wolfgang Schlieder et al., *Chinesische Bambuspapier*, 18–38. Berlin: Academie Verlag, 1993.

Glahn, Else. "On the Transmission of the *Ying-tsao fa-shih*." *T'oung Pao* 61 (1975), 232–65.

———. "Li Chieh." In Herbert Franke (ed.), *Sung Biographies*, Vol. 2, 523–29.

———. "*Ying-tsao fa-shih*." In Hervouet, *A Sung Bibliography*, 186–88.

———. "Unfolding the Chinese Building Standards: Research in the *Yingzao fashi*." In Nancy Shatzman Steinhardt, *Chinese Traditional Architecture*, 48–57. New York: China Institute, 1984.

Golas, Peter J. "Rural China in the Song." *Journal of Asian Studies* 39.2 (February, 1980), 291–325.

———. "Agricola in China: A Little Problem of Translation." In Hashimoto Keizô, Catherine Jami and Lowell Skar (eds.), *East Asian Science: Tradition and Beyond*, 91–96. Osaka: Kansai University Press, 1995.

———. "Chemistry and Chemical Technology; Mining." In *Science and Civilisation in China (SCC)*, ed. Joseph Needham, vol. 5, part 13. Cambridge: Cambridge University Press, 1999.

———. "Apples and Oranges; Coal and Gold: Some Reflections on Comparisons in the History of Technology." In Kim Yung Sik and Francesca Bray (eds.). *Current Perspectives in the History of Science in East Asia*, 74–85. Seoul: Seoul National University Press, 1999.

———. "The Emergence of Technical Drawing in China: The *Xin Yi Xiang Fa Yao* and Its Antecedents." *History of Technology* 21 (1999), 29–63.

———. "Technology and Science." In Edward L. Shaughnessy (ed.), *China; Empire and Civilization*, 166–83. Oxford: Oxford University Press, 2000.

———. "Technological Illustration in China: A post-Needham Perspective." In Alain Arroult and Catherine Jami (eds.). *Science and Technology in East Asia: The Legacy of Joseph Needham*, 43–58. Turnhout: Brepols, 2001.

———. "'Like Obtaining a Great Treasure': The Illustrations of Song Yingxing's *The Exploitation of the Works of Nature*." In Bray et al., *Graphics and Text*, 569–614.

———. "Technical Representation in China: Tools and Techniques of the Trade." *EASTM* 20 (2003), 11–44.

———. "Technical Drawing in Ming/Qing China and in Renaissance Europe." In Jiang Xiaoyuan (ed.). *History of Science in the Multiculture*, 199–208. Shanghai: Shanghai Jiao Tong University Press, 2005.

Gombrich, E. H. *Art and Illusion: A Study in the Psychology of Pictorial Representation*. New York: Bollingen Foundation, 1965.

Haber, Francis C. "Time, Technology, Religion, and Productivity Values in Early Modern Europe." In J. T. Fraser, N. Lawrence and F. C. Haber, *Time, Science, and Society in China and the West (The Study of Time V)*, 79–92. Amherst: The University of Massachusetts Press, 1986.

Hall, A. R. "Guido's *Texaurus*, 1335." In B. S. Hall and D. C. West (eds.). *On Pre-Modern Technology and Science: A Volume of Studies in Honor of Lynn White, Jr.*, 11–51. Malibu, CA: Undena Publications, 1976.

Hall, Bert S. "The New Leonardo." Review of *Leonardo da Vinci: The Madrid Codices*, ed. L. Reti; and *The Unknown Leonardo*, by L. Reti. *Isis* 67 (1976), 463–75.

———. "Giovanni de'Dondi and Guido da Vigevano: Notes Toward a Typology of Medieval Technological Writings." *Annals of the New York Academy of Sciences* 314 (1978), 127–42.

———. "Production et diffusion de certains traités de techniques au moyen âge." In G. H. Allard and S. Lusignan (eds.). *Cahiers d'études médiévals 7; Les arts mechaniques au moyen age*, 146–70. Montreal: Bellarmin, 1982.

———. "The Didactic and the Elegant: Some Thoughts on Scientific and Technological Illustrations in the Middle Ages and Renaissance." In Baigrie (ed.). *Picturing Knowledge*, 3–38.

Hamilton, Gary G. and Wei-an Chang. "The Importance of Commerce in the Organization of China's Late Imperial Economy." In Giovanni Arrighi, Takeshi Hamashita and Mark Selden, *The Resurgence of East Asia; 500, 150 and 50 Year Perspectives*, 173–213. London and New York: Routledge, 2003.

Harley, J. B. and David Woodward (eds.). *The History of Cartography*. vol. 2, book 2. *Cartography in the Traditional East and Southeast Asian Societies*. Chicago and London: The University of Chicago Press, 1994.

Hay, John. "'Along the River during Winter's First Snow': A Tenth-Century Handscroll and Early Chinese Narrative." *Burlington Magazine* 114 (May 1972), 294–303.

HC: see Needham et al., *Heavenly Clockwork*.

Hegel, Robert E. *Reading Illustrated Fiction in Late Imperial China*. Stanford: Stanford University Press, 1998.

Henderson, John B. *The Development and Decline of Chinese Cosmology*. New York: Columbia University Press, 1984.

Herrmann, Konrad (trans.). *Erschliessung der himmlischen Schätze*. Bremerhaven: Wirtschaftsverlag NW, Verlag für neue Wissenschaft, 2004.

Hervouet, Yves. *A Sung Bibliography*. Hong Kong: The Chinese University Press, 1978.

Ho Peng Yoke. *Li, Qi and Shu: An Introduction to Science and Civilization in China*. Mineola, NY: Dover Publications, 2000.

Ho, Wai-kam et al., *Eight Dynasties of Chinese Painting*. Cleveland: Cleveland Museum of Art, 1980.

Hollister-Short, Graham. "On the Origins of the Suction Lift Pump." *History of Technology* 15 (1993), 57–75.

Hommel, Rudolf P. *China at Work*. Cambridge, MA: MIT Press, 1969.

Hsiang Ta. "European Influences on Chinese Art in the Later Ming and Early Ch'ing Period." Trans. Wang Teh-chao. *Renditions* 6 (1976), 152–78.

Hsiao Chen-shih: see Xiao Zhenshi.

Hu, Philip K. (comp. and ed.). *Visible Traces; Rare Books and Special Collections from the National Library of China*. New York: Queens Borough Public Library, 2000.

Huang, H. T. "Biology and Biological Technology: Fermentations and Food Science." In *Science and Civilisation in China (SCC)*, ed. Joseph Needham, vol. 6, part 9. Cambridge: Cambridge University Press, 2000.

Huang, Ray. *Taxation and Government Finance in Sixteenth-Century China*. Cambridge: Cambridge University Press, 1974.

———. *China: A Macro History*. Armonk, NY: M. E. Sharpe, Inc., 1990.

Hummel, Arthur W. (ed.). *Eminent Chinese of the Ch'ing Period*. 2 vols. Washington: United States Government Printing Office, 1944.

Hutt, Julia. *Understanding Far Eastern Art: A Complete Guide to the Arts of China, Japan and Korea–Ceramics, Sculpture, Painting, Prints, Lacquer, Textiles and Metalwork*. New York: E. P. Dutton, 1987.

Ivins, William M., Jr. *Art and Geometry: A Study in Space Intuitions*. Cambridge, MA: Harvard University Press, 1946.

———. *Prints and Visual Communication*. Cambridge, MA: Harvard University Press, 1953.

Jäger, Fritz. "Das Buch von den wunderbaren Maschinen; Ein Kapitel aus der Geschichte der abendländisch-chinesischen Kulturbeziehungen." *Asia Major*, n.s., 1.1 (1944), 78–96.

Jami, Catherine. "Learning Mathematical Sciences during the Early and Mid-Ching." In Benjamin A. Elman and Alexander Woodside (eds.). *Education and Society in Later Imperial China, 1600–1900*, 223–56. Berkeley: University of California Press, 1994.

———. "'European Science in China' or 'Western Learning.'" *Science in Context* 12.3 (1999), 413–34.

Jami, Catherine, Peter Engelfriet, and Gregory Blue (eds.). *Statecraft and Intellectual Renewal in Late Ming China*. Leiden: Brill, 2001.

Jami, Catherine and Hubert Delahaye (eds.). *L'Europe en Chine. Interactions scientifiques, religieuses et culturelles aux XVIIe et XVIIIe siècles*. Paris: College de France, Institut des Hautes Études Chinoises, 1993.

Kao, Mayching. "European Influences in Chinese Art, Sixteenth to Eighteenth Centuries." In Thomas H.C. Lee (ed.). *China and Europe: Images and Influences in Sixteenth to Eighteenth Centuries*, 251–303. Hong Kong: The Chinese University Press, 1991.

Karlgren, Bernhard. *The Book of Documents*. Göteborg: Elanders Boktryckeri Aktiebolag, 1950. (Reprint from *The Museum of Far Eastern Antiquities, Bulletin* 22 (1950), 1–81.)

Keightley, David N. "Public Work in Ancient China: A Study of Forced Labor in the Shang and Western Chou." PhD Dissertation. Columbia University, 1969.

———. "Early Civilization in China: Reflections on How It Became Chinese." In Paul S. Ropp (ed.). *Heritage of China; Contemporary Perspectives on Chinese Civilization*, 15–54. Berkeley: University of California Press, 1990.

———. "Archaeology and Mentality: The Making of China." *Representations* 18 (1987), 91–128.

———. "The Quest for Eternity in Ancient China: The Dead, Their Gifts, Their Names." In George Kuwayama (ed.). *Ancient Mortuary Traditions of China: Papers on Chinese Ceramic Funerary Sculptures*, 12–24. Los Angeles: Far Eastern Art Council, Los Angeles County Museum of Art, 1991.

———. "A Measure of Man in Early China: In Search of the Neolithic Inch." *Chinese Science* 12 (1995), 18–40.

Kerr, Rose and Nigel Wood. "Chemistry and Chemical Technology: Ceramic Technology." In Rose Kerr (ed.). *Science and Civilisation in China (SCC)*, vol. 5 part 12. Cambridge: Cambridge University Press, 2004.

Kesner, Ladislav. "Likeness of No One: (Re)presenting the First Emperor's Army." *Art Bulletin* 77.1 (March 1995), 115–32.

Kim, Yung Sik. "Chu Hsi (1130–1200) on Calendar Specialists and their Knowledge: A Scholar's Attitude toward Technical Scientific Knowledge in Traditional China." *T'oung Pao* 78 (1992), 94–115.

Klemm, Friedrich. *A History of Western Technology*. Trans. Dorothea Waley Singer. London: Allen & Unwin, 1959.

Kline, Morris. *Mathematics in Western Culture*. New York: Oxford University Press, 1953.

Knobloch, Eberhard. "Technische Zeichnungen." In Uta Lindgren (ed.). *Europäische Technik im Mittelalter 800 bis 1400: Tradition und Innovation*, 45–72. Berlin: Gebr. Mann Verlag, 1996.

Ko, Dorothy. *Teachers of the Inner Chambers; Women and Culture in Seventeenth-Century China*. Stanford: Stanford University Press, 1994.

Kuhn, Dieter. "Die Darstellungen des *Keng-chih-t'u* und ihre Wiedergabe in populär-enzyklopädischen Werken der Ming-Zeit." *Zeitschrift der Deutschen Morgenländischen Gesellschaft* 126.2 (1976), 336–67.

———. "Marginalie zu einigen Illustrationen im Nung-shu 農書 des Wang Chen 王禎." In R. Goepper, D. Kuhn and U. Wiesner (eds.). *Zur Kunstgeschichte Asiens: 50 Jahre Lehre und Forschung an der Universität Köln*, 143–52. Wiesbaden: Franz Steiner, 1977.

———. "Some Notes Concerning the Textile Technology Pictured in the *Keng-chih-t'u*." *Zeitschrift der Deutschen Morgenländischen Gesellschaft* (1980), 2, 408–16.

———. "Silk Technology in the Sung Period (960–1278 A.D.)." *T'oung Pao* (1981), 1–2, 48–90.

———. "The Silk-Workshops of the Shang Dynasty (16th–11th Century B.C.)." In Li Guohao et al. *Explorations in the History of Science and Technology in China*, 367–408.

———. *Die Song-Dynastie (960–1279): Eine neue Gesellschaft im Spiegel ihrer Kultur*. Weinheim: Acta Humaniora, VCH, 1987.

———. *Science and Civilisation in China*. vol. 5, *Chemistry and Chemical Technology*, part IX, *Textile Technology: Spinning and Reeling*. Cambridge: Cambridge University Press, 1985.

———. "Silk Weaving in Ancient China: From Geometric Figures to Patterns of Pictorial Likeness." *Chinese Science* 12 (1995), 77–114.

———. *The Age of Confucian Rule: The Song Transformation of China*. Cambridge, MA: Harvard University Press, 2009.

Landes, David. *Revolution in Time: Clocks and the Making of the Modern World*. Cambridge, MA: Belknap Press, 1983.

Lawton, Thomas. *Freer Gallery of Art Fiftieth Anniversary Exhibition. II Chinese Figure Painting*. Washington: Smithsonian Institution, 1973.

Ledderose, Lothar. "Chinese Calligraphy: Its Aesthetic Dimension and Social Function." *Orientations* 17.10 (Oct. 1986), 35–50.

———. *Ten Thousand Things: Module and Mass Production in Chinese Art*. Princeton: Princeton University Press, 2000.

Lee, Sherman. *China, 5000 Years: Innovation and Transformation in the Arts*. New York: Solomon R. Guggenheim Museum, 1998.

Lefèvre, Wolfgang. "The Limits of Pictures: Cognitive Functions of Images in Practical Mechanics–1400 to 1600." In Lefèvre et al., *Power of Images*, 69–88.

——— (ed.). *Picturing Machines*. Cambridge, MA: MIT Press, 2004.

Lefèvre, Wolfgang, Jürgen Renn and Urs Schoepflin (eds.). *The Power of Images in Early Modern Science*. Basel: Birkhäuser Verlag, 2003.

Legge, James. *The Chinese Classics*. Volume III: *The Shoo King*. Hong Kong: Lane Crawford, 1865.

———. *The Chinese Classics*. Volume IV: *The She King*. Hong Kong: Lane Crawford, 1871.

Levarie, Norma. *The Art and History of Books*. New Castle, DE and London: Oak Knoll Press and The British Library, 1995.

Lewis, E. E. *Masterworks of Technology: The Story of Creative Engineering, Architecture, and Design*. Amherst, NY: Prometheus Books, 2004.

Lewis, Mark Edward. *Writing and Authority in Early China*. Albany: State University of New York Press, 1999.

Li Guohao, Zhang Mengwen and Cao Tianqin, *Explorations in the History of Science and Technology in China*. Shanghai: Ku-Chi, 1982.

Li Yan and Du Shiran. *Chinese Mathematics; A Concise History*. Trans. John N. Crossley and Anthony W.-C. Lun. Oxford: Clarendon Press, 1987.

Liang Sicheng (Liang Ssucheng) 梁思成. *Yingzao fashi zhushi* 營造法式注釋 [An annotated translation of the *Building Standards*]. 2 vols. Beijing: Zhongguo jianzhu gongye chubanshe, 1983.

Lim, Lucy. "Form and Content in Early Chinese Pictorial Art." In Lim, *Stories from China's Past*, 51–53.

——— (ed.). *Stories from China's Past: Han Dynasty Pictorial Tomb Reliefs and Archaeological Objects from Sichuan Province, People's Republic of China*. San Francisco: The Chinese Culture Foundation of San Francisco, 1987.

Lindqvist, Cecilia. *China: Empire of Living Symbols*. Trans. Joan Tate. Reading, MA: Addison-Wesley, 1991.

Liscomb, Kathlyn Maurean. *Learning from Mount Hua: A Chinese Physician's Illustrated Travel Record and Painting Theory*. Cambridge: Cambridge University Press, 1993.

Liu, Cary Y. "Review of Anita Chung, *Drawing Boundaries: Architectural Images in Qing China*." JAS 65.2 (May 2006), 400–402.

Liu Heping. "Painting and Commerce in Northern Song Dynasty China, 960–1126." Ph.D. Dissertation. Yale University, 1997.

———. "*The Water Mill* and Northern Song Imperial Patronage of Art, Commerce, and Science." *The Art Bulletin* 84.4 (December 2002), 566–95.

Liu Keming 劉克明. "*Songdai gongcheng tuxue de chengjiu*" 宋代工程圖學的成就 [Achievements of engineering drawing in the Song], *Wenxian* 1991.4, 238–46.

———. *Zhongguo gongcheng tuxue shi* 中國工程圖學史 [A history of Chinese engineering graphics]. Wuhan: Huazhong kejixue chubanshe, 2003.

Liu Keming 劉克明 and Gong Shixing 龔世星. "*Zhongguo gudai gongcheng tuxue de ruogan chengjiu*" 中國古代工程圖學的若干成就 [Some achievements of ancient Chinese engineering graphics]. *Gongcheng tuxue ji jisuanji tuxue* [Journal of engineering graphics and computer drafting]. (1992.1), 75–81, 68.

Liu Keming 劉克明, Hu Xianzhang 胡顯章, and Fan Weiwei 樊葳萎. "A Brief Survey of Engineering Drawing in the Song Dynasty." Paper presented at the Second International Conference on Graphic Education Research in China and Japan. Chengdu, 1995.

Liu Wu-chi and Irving Yucheng Lo (eds.). *Sunflower Splendor: Three Thousand Years of Chinese Poetry*. Garden City, New York: Anchor Press/Doubleday, 1975.

Loehr, Max. "Some Fundamental Issues in the History of Chinese Painting." *Journal of Asian Studies* 23.2 (1964), 185–93.

Long, Pamela O. "Power, Patronage, and the Authorship of *Ars*: From Mechanical Know-how to Mechanical Knowledge in the last Scribal Age." *Isis* 88.1 (March 1997): 1–41.

———. "Picturing the Machine: Francesco di Giorgio and Leonardo da Vinci in the 1490s." In Lefèvre, *Picturing Machines*, 117–41.

———."Technological Transmission in China and Europe: A Comparative View." In Schäfer, *Cultures of Knowledge*, 75–84.

Lu Jingyan 路敬嚴 and Hua Jueming 華覺明 (eds.). *Zhongguo kexue jishu shi* 中國科學技術史 [A history of Chinese science and technology]. Beijing: Kexue, 2000.

Lust, John. *Chinese Popular Prints*. Leiden: E. J. Brill, 1996.

Maeda, Robert J. "*Chieh-hua*: Ruled-Line Painting in China." *Ars Orientalis* 10 (1975), 123–41.

———. "Spatial Enclosures: The Idea of Interior Space in Chinese Painting." *Oriental Art* 3 (Winter 1985–86), 370–91.

Mahoney, Michael S. "Drawing Mechanics." In Lefèvre, *Picturing Machines*, 281–306.

March, Benjamin. "Linear Perspective in Chinese Painting." *Eastern Art* (Philadelphia), 3(1931), 113–39.

Masuda Tsuna. *Kodô Zuroku; Illustrated Book on the Smelting of Copper*. With color woodcuts by Niwa Motokuni Tôkei. Trans. Zenryû Shirakawa. Edited with an introductory essay by Cyril Stanley Smith. Norwalk, CN: Burndy Library, 1983.

Matthies, Andrea L. "Medieval Treadwheels: Artists' Views of Building Construction." *TC* 33.3 (July 1992), 510–47.

Mayr, Otto. *Authority, Liberty and Automatic Machinery in Early Modern Europe*. Baltimore and London: The Johns Hopkins University Press, 1986.

McDermott, Joseph. "The Ascendance of the Imprint in China." In Brokaw and Chow, *Printing and Book Culture in Late Imperial China*, 55–104.

McDermott, Joseph. *A Social History of the Chinese Book: Books and Literati Culture in Late Imperial China*. Hong Kong: Hong Kong University Press, 2006.

McLaren, Anne E. *Chinese Popular Culture and Ming Chantefables*. Leiden: Brill, 1998.

Métailié, Georges. "The Representation of Plants: Engravings and Paintings." In Bray et al., *Graphics and Text*, 487–520.

Meyer-Fong, Tobie. "The Printed Word: Books, Publishing, Culture and Society in Late Imperial China." *Journal of Asian Studies* 66.3 (August 2007), 787–817.

Misa, Thomas J. *Leonardo to the Internet; Technology and Culture from the Renaissance to the Present.* Baltimore: Johns Hopkins University Press, 2004.

Moll-Murata, Christine, Song Jianze, and Hans Ulrich Vogel (eds.). *Chinese Handicraft Regulations of the Qing Dynasty*. Munich: Iudicium, 2005.

Mote, F[rederick] W. *Imperial China*. Cambridge, MA: Harvard University Press, 1999.

Mote, Frederick W. and Denis Twitchett (eds.). *The Cambridge History of China*. Vol. 7, "Ming Dynasty, 1368–1644, Part 1." Cambridge and New York: Cambridge University Press, 1988.

Murowchick, Robert E. *China: Ancient Culture, Modern Land*. Norman, OK: University of Oklahoma Press, 1994.

Murray, Julia K. "The Role of Art in Southern Sung Dynastic Revival." *Bulletin of Sung-Yuan Studies* 18 (1986), 41–59.

———. "Water under a Bridge: Further Thoughts on the *Qingming* Scroll." *Bulletin of Sung-Yuan Studies* 27 (1997), 99–107.

———. "Didactic Illustrations in Printed Books." In Brokaw and Chow, *Printing and Book Culture in Late Imperial China*, 417–50.

Needham, Joseph. *The Grand Titration: Science and Society in East and West*. Toronto: University of Toronto Press, 1969.

———. "Astronomy in Ancient and Medieval China." *Philosophical Transactions of the Royal Society of London* A. 276 (1974), 67–82.

———. *Science in Traditional China: a Comparative Perspective*. Cambridge, MA: Harvard University Press, 1981.

Needham, Joseph, Ho Ping-yü (Ho Peng Yoke), Lu Gwei-Djen and Wang Ling. "Chemistry and Chemical Technology: Military Technology, The Gunpowder Epic." In Joseph Needham (ed.). *Science and Civilisation in China (SCC)*, vol. 5 part 7. Cambridge: Cambridge University Press, 1965.

Needham, Joseph, Kenneth Girdwood Robinson and Ray Huang. "General Conclusions and Reflections." In *Science and Civilisation in China (SCC)*, ed. Joseph Needham, vol. 7, part 2. Cambridge: Cambridge University Press, 2004.

Needham, Joseph and Wang Ling. "History of Scientific Thought." In Joseph Needham (ed.). *Science and Civilisation in China (SCC)*, vol. 2. Cambridge: Cambridge University Press, 1956.

Needham, Joseph and Wang Ling. "Mathematics and the Sciences of the Heavens and the Earth." In Joseph Needham (ed.). *Science and Civilisation in China (SCC)*, vol. 3. Cambridge: Cambridge University Press, 1959.

Needham, Joseph and Wang Ling. "Physics and Physical Technology: Mechanical Engineering." In Joseph Needham (ed.). *Science and Civilisation in China (SCC)*, vol. 4, part 2. Cambridge: Cambridge University Press, 1965.

Needham, Joseph, Wang Ling and Derek J. de Solla Price. *Heavenly Clockwork; the Great Astronomical Clocks of Medieval China*. 2nd edition, with supplement by John H. Combridge. Cambridge: Cambridge University Press, 1986. (HC)

Needham, Joseph, Wang Ling and Lu Gwei-Djen. "Physics and Physical Technology: Civil Engineering and Nautics." In Joseph Needham (ed.). *Science and Civilisation in China (SCC)*, vol. 4, part 3. Cambridge: Cambridge University Press, 1971.

Needham, Joseph, Wang Ling and Kenneth Girdwood Robinson. "Physics and Physical Technology: Physics." In Joseph Needham (ed.). *Science and Civilisation in China (SCC)*, vol. 4, part 1. Cambridge: Cambridge University Press, 1962.

Needham, Joseph and Robin D. S. Yates. "Chemistry and Chemical Technology: Military Technology: Missiles and Sieges." In Joseph Needham (ed.). *Science and Civilisation in China (SCC)*, vol. 5, part 6. Cambridge: Cambridge University Press, 1965.

Nivison, David S. and Arthur F. Wright. *Confucianism in Action*. Stanford: Stanford University Press, 1959.

Nowotny, Helga (ed.). *Cultures of Techology and the Quest for Innovation*. New York: Berghahn Books, 2006.

Pacey, Arnold. *Technology in World Civilization; a Thousand-year History*. Cambridge, MA: MIT Press, 1991.

———. *The Maze of Ingenuity: Ideas and Idealism in the Development of Technology*. Cambridge, MA: MIT Press, 1992.

———. *Meaning in Technology*. Cambridge, MA: MIT Press, 2001.

Pagani, Catherine. *"Eastern Magnificence European Ingenuity": Clocks of Late Imperial China*. Ann Arbor: The University of Michigan Press, 2001.

Pan Guxi. "The Yuan and Ming Dynasties." In Steinhardt, *Chinese Traditional Architecture*, 199–259.

Pan Jixing 潘吉星. *Tian gong kai wu daodu* 天工開物導讀 [A guide to reading the *Tian gong kai wu*]. Chengdu: Ba Shu shushe, 1988.

———. *Tian gong kai wu jiao zhu ji yanjiu* 天工開物校注及研究 [Studies on an edited and annotated edition of the *Tian gong kai wu*]. Chengdu: Ba Shu shushe, 1989.

———. *Song Yingxing pingzhuan* 宋應星評傳 [A critical biography of Song Yingxing]. Nanjing: Nanjing daxue chubanshe, 1990.

Pelliot, Paul. "A Propos du *Keng Tche T'ou*." *Mémoires concernant l'Asie Orientale* 1 (1913), 65–122 and Pls. X–LXI.

Peterson, Willard J. "Calendar Reform Prior to the Arrival of Missionaries at the Ming Court." *Ming Studies* 21, (Spring 1986), 45–61.

———. "Western Natural Philosophy Published in Late Ming China." *Proceedings of the American Philosophical Society* 117.4 (August 1973), 295–322.

Peterson, Willard J., Andrew Plaks and Ying-shih Yü (eds.). *The Power of Culture: Studies in Chinese Cultural History*. Hong Kong: Chinese University Press, 1994.

Petroski, Henry. *The Pencil: A History of Design and Circumstances*. New York: Knopf, 1989.

Pfaffenberger, Bryan. "Social Anthropology of Technology." *Annual Review of Anthropology* 21 (1992), 491–516.

Pomeranz, Kenneth. *The Great Divergence: Europe, China and the Making of the Modern World Economy.* Princeton: Princeton University Press, 2000.

Porter, Tom. *How Architects Visualize*. New York: Van Nostrand, 1979.

Powers, Martin J. *Art and Political Expression in Early China*. New Haven: Yale University Press, 1991.

Qi qi tu shuo 奇器圖說 [Illustrations and explanations of wonderful machines]. *Siku quanshu* ed. (QQTS)

Rawski, Evelyn Sakakida. *Education and Popular Literacy in Ch'ing China*. Ann Arbor: University of Michigan Press, 1979.

———. "Economic and Social Foundations of Later Imperial Culture." In David Johnson, Andrew J. Nathan and Evelyn S. Rawski (eds.). *Popular Culture in Late Imperial China*, 3–33. Berkeley and Los Angeles: University of California Press, 1985.

Rawson, Jessica (ed.). *Mysteries of Ancient China; New Discoveries from the Early Dynasties*. New York: George Braziller, 1996.

Reed, Marcia and Paola Demattè (eds.). *China on Paper: European and Chinese Works from the Late Sixteenth to the Early Nineteenth Century*. Los Angeles: Getty Research Institute, 2007.

Reiter, Florian C. "Some Remarks on the Chinese Word *T'u, Chart, Plan, Design*." *Oriens* XXXII (1990), 308–27.

Ren Dayuan 任大援. "Wang Zheng: Xixue yu xin sixiang de chuanbozhe yihuo yige tianzhujiaotu?" 王徵：西學與新思想的傳播者抑或一个天主教徒? [Wang Zheng: A scientist, philosopher, and Catholic in Ming-dynasty China]. In Roman Malek (ed.), *Western Learning and Christianity in China: The Contribution and Impact of Johann Adam Schall von Bell, S.J. (1592–1666)*. vol. 1, 339–368. 2 vols. Sankt Augustin: China Zentrum and the Monumenta Serica Institute, 1998.

Reti, L. (ed.). *The Madrid Codices/Leonardo da Vinci*. New York: McGraw Hill, 1974.

——— (ed.). *The Unknown Leonardo*. London: Hutchinson, 1974.

Ronan, Colin A. *Science: Its History and Development among the World's Cultures*. New York: Facts on File, 1982.

Rossabi, Morris (ed.). *China among Equals: The Middle Kingdom and Its Neighbors, 10th to 14th Centuries*. Berkeley: University of California Press, 1983.

Rowe, William T. "Political, Social and Economic Factors Affecting the Transmission of Technical Knowledge in Early Modern China." In Schäfer, *Cultures of Knowledge*, 25–44.

Ruitenbeek, Klass. *Carpentry and Building in Late Imperial China: A Study of the Fifteenth-Century Carpenter's Manual* Lu Ban jing. Leiden: Brill, 1993.

Saban, Françoise. "L'industrie sucrière, le moulin a sucre et les relations sino-portugaises aux XVIe–XVIIIe siècles." *Annales, Histoire, Science Sociales* 49.4 (July–August 1994), 817–61.

Sahlins, Marshall. *Stone Age Economics*. Chicago: Aldine, 1972.

Saigusa Hiroto 三枝博音 (ed.). *Tenkô kaibutsu* 天工開物 [The *Tian gong kaiwu* of Song Yingxing]. Tokyo: Juichigumi shuppanbu, 1943.

Sakai, Tadao. "Confucianism and Popular Educational Works." In Wm. Theodore de Bary, *Self and Society in Ming Thought*, 331–62. New York: Columbia University Press, 1970.

SCC: Needham, Joseph et al. *Science and Civilization in China*. Cambridge: Cambridge University Press, 1954–.

Schäfer, Dagmar. *The Crafting of the 10,000 Things: Knowledge and Technology in Seventeenth-Century China*. Chicago: University of Chicago Press, 2011.

——— (ed.). *Cultures of Knowledge: Technology in Chinese History*. Leiden: Brill, 2012.

Schall, Adam. *Historica Relatio*. Ed. P. Henri Bernard; trans. P. Paul Bornet. Tientsin: Hautes Études, 1942.

Schirokauer, Conrad. *A Brief History of Chinese Civilization*. San Diego: Harcourt Brace Jovanovich, 1991.

Schlieder, Wolfgang, Pan Jixing and Sybille Girmond. *Chinesische Bambuspapier-herstellung: Ein Bilderalbum aus dem 18. Jahrhundert*. Berlin: Academie Verlag, 1993.

Screech, Timon. *The Western Scientific Gaze and Popular Imagery in Later Edo Japan: The Lens within the Heart*. Cambridge: Cambridge University Press, 1996.

Shaanxi sheng Qin yong kaogudui 陕西省秦俑考古隊and Qin Shihuang bingmayong bowuguan 秦始皇兵馬俑博物館 (eds.). *Qin ling er hao tong che ma* 秦陵二號銅車馬 [A bronze chariot and horses from the Necropolis of Qin]. Xi'an: Kaogu yu wenwu, 1983.

Shirley, John W. and F. David Hoeniger (eds.). *Science and the Arts in the Renaissance*. Washington, D.C.: Folger Shakespeare Library; London: Associated University Presses, 1985.

Sickman, Lawrence. "Chinese Painting before 1100." In Wai-Kam Ho et al., *Eight Dynasties of Chinese Painting*, xiii–xxiv. Cleveland: Cleveland Museum of Art, 1980.

Sickman, Lawrence and Alexander Soper. *The Art and Architecture of China*. Baltimore: Penguin Books, 1956.

Silbergeld, Jerome. *Chinese Painting Style; Media, Methods, and Principles of Form*. Seattle and London: University of Washington Press, 1982.

Sirén, Osvald. *Chinese Painting: Leading Masters and Principles*. Part 1 "The First Millennium." Vol. 1 "Early Chinese Painting." New York: The Ronald Press Company, 1956.

Sivin, Nathan. "Chinese Conceptions of Time," *The Earlham Review* 1966.1, 82–92.

———. "Science and Medicine in Chinese History." In Paul S. Ropp (ed.), *Heritage of China: Contemporary Perspectives on Chinese Civilization*, 164–96. Berkeley: University of California Press, 1990.

———. Review of Bray et al. *Graphics and Text in the Production of Technical Knowledge in China: The Warp and the Weft*. *China Review International* 15.4 (2008), 455–64.

Sivin, Nathan and Gari Ledyard. "Introduction to East Asian Cartography." In Harley and Woodward, *Cartography in the Traditional East and Southeast Asian Societies*, 23–31.

Smith, Paul Jakov and Richard von Glahn (eds.). *The Song-Yuan-Ming Transition in Chinese History*. Cambridge, MA: Harvard University Press, 2003.

Smith, Richard J. *China's Cultural Heritage: The Qing Dynasty 1644–1912*. 2nd ed. Boulder: Westview Press, 1994.

———. "Mapping China's World: Cultural Cartography in Late Imperial Times." In Wen-hsin Yeh, *Landscape, Culture and Power in Chinese Society*, 52–109. Berkeley: Institute of East Asian Studies, 1998.

Soper, Alexander Coburn (trans. and ed.). *Kuo Jo-Hsü's Experiences in Painting* (T'u-hua chien-wen chih): *An Eleventh Century History of Chinese Painting Together with the Chinese Text in Facsimile*. Washington, D.C.: American Council of Learned Societies, 1951.

Standaert, Nicolas (ed.). *Handbook of Christianity in China*. Vol. 1: 635–1800. Brill: Leiden, 2001.

———. "The Transmission of Renaissance Culture in Seventeenth-Century China." *Renaissance Studies* 17.3 (2003), 367–91.

Stearns, Peter N. et al. *World Civilizations: The Global Experience*. 3rd ed. New York: Longman, 2001.

Steinhardt, Nancy Shatzman. "Chinese Cartography and Calligraphy." *Oriental Art* 43.1 (1997), 10–20.

Steinhardt, Nancy S. (ed.). *Chinese Architecture*. New Haven and London: Yale University Press, 2002.

Struve, Lynn A. (ed.). *The Qing Formation in World-Historical Time*. Cambridge: Harvard University Asia Center, 2004.

Stunkel, Kenneth R. "Technology and Values in Traditional China and the West." *Comparative Civilizations Review* 23 (1990): 75–91; 24 (1991): 58–75.

Su Bai 宿白. *Baisha Song mu* 白沙宋墓 [The Song tombs at Baisha]. Beijing: Wenwu chubanshe, 1957.

Sullivan, Michael. "Some Possible Sources of European Influence on Late Ming and Early Ch'ing Painting." In *Proceedings of the International Symposium on Chinese Painting: National Palace Museum Republic of China 18th–24th June 1970*, 595–625. Taipei: National Palace Museum, 1972.

———. *The Arts of China*. Berkeley: University of California Press, 1973.

———. *Symbols of Eternity: The Art of Landscape Painting in China*. Stanford: Stanford University Press, 1979.

———. *Chinese Landscape Painting*. Vol. II. *The Sui and T'ang Dynasties*. Berkeley: University of California Press, 1980.

———. *The Meeting of Eastern and Western Art from the Sixteenth Century to the Present Day*. 2nd ed. Berkeley and Los Angeles: University of California Press, 1989.

Sun Dazhang, "The Qing Dynasty." In Steinhardt, *Chinese Architecture*, 261–343.

Sun, E-tu Zen and Shiou-Chuan Sun (trans.), *T'ien-kung k'ai-wu: Chinese Technology in the Seventeenth Century*. University Park and London: The Pennsylvania State University Press, 1966.

Thorp, Robert L. and Richard Ellis Vinograd. *Chinese Art and Culture*. New York: Harry N. Abrams, Inc., 2001.

Trousdale, William, "Architectural Landscapes Attributed to Chao Po-chü." *Ars Orientalis* 4 (1961), 285–313.

Tsien, Tsuen-hsuin. "Chemistry and Chemical Technology: Paper and Printing." In Joseph Needham (ed.). *Science and Civilisation in China (SCC)*, vol. 5, part 1. Cambridge: Cambridge University Press, 1985.

———. *Written on Bamboo and Silk: The Beginnings of Chinese Books and Inscriptions*. Chicago: University of Chicago Press, 1962; 2nd ed. 2004.

Tsuna, Masuda. *Kodô Zuroku* 鼓銅圖錄*: Illustrated Book on the Smelting of Copper*. With color woodcuts by Niwa Motokuni Tôkei. Trans. Zenryu Shirakawa. Norwalk, CT: Burndy Library, 1983.

Tu Lien-che. "Sung Ying-hsing." In *ECCP*, vol. 2, 690–91.

Twitchett, Denis. *Printing and Publishing in Medieval China*. New York: Frederic C. Beil, 1983.

Vogel, Hans-Ulrich. "History of Diagrams and Illustrations in Premodern China: Some Notes and Reflections." 1999. Electronic version downloadable from academia.edu (Hans Ulrich Vogel).

———. "Important Sources of the History of Premodern Chinese Salt Production Techniques." In Hua Jueming 華覺明 et al. (eds.). *Zhongguo keji dianji yanjiu* 中國科技典籍研究 [Study on ancient Chinese books and records of science and technology], 169–83. Zhengzhou: Daxiang chubanshe (Elephant Press), 1998.

———. "The Mining Industry in Traditional China; Intra-and Intercultural Comparisons." In Helga Nowotny (ed.), *Cultures of Technology and the Quest for Innovation*, 167–88. New York: Berghahn Books, 2006.

———. "Mechanical Knowledge in the Context of Pre-Modern Chinese Salt Industry." In Zhang and Renn, *Transformation and Transmission*, 99–126.

Volkov, Alexei. "Geometrical Diagrams in Traditional Chinese Mathematics." In Bray et al., *Graphics and Text*, 425–59.

Wagner, Donald B. "Iron Production in Three Ming Texts: *Tie ye zhi*, *Guangdong xinyu*, and *Tian gong kai wu*." Paper prepared for the Third International Symposium on Ancient Chinese Texts and Records on Science and Technology, Tübingen, March 31–April 3, 2003.

———. "Song Yingxing's Illustrations of Iron Production." In Bray et al., *Graphics and Text*, 615–32.

Waley, Arthur. *An Introduction to the Study of Chinese Painting*. New York: Charles Scribner's Son, 1923.

——— (trans.). *The Book of Songs*. New York: Grove Press, 1960.

Waley-Cohen, Joanna. "China and Western Technology in the Late Eighteenth Century." *American Historical Review* 98.5 (1993), 1525–44.

———. *The Sextants of Beijing: Global Currents in Chinese History*. New York: Norton, 1999.

Wang Bomin 王伯敏. *Zhongguo banhua shi* 中國版畫史 [A history of Chinese woodblock illustrations]. Shanghai: Renmin chubanshe, 1961.

Wang Chaosheng 王潮生. *Zhongguo gudai kengzhi tu* 中國古代耕織圖 [Farming and weaving pictures in ancient China]. Beijing (?): Zhongguo nongye chubanshe, 1995.

Wang Fang-yu. "Book Illustration in Late Ming and Early Qing China." In Edgren, *Chinese Rare Books*, 31–43.

Wang Qingzheng. "The Arts of Ming Woodblock-printed Images and Decorated Paper Albums." In Chu-Tsing Li and James C. Y. Watt (eds.), *The Chinese Scholar's Studio: Artistic Life in the Late Ming Period*, 56–61. New York: Thames and Hudson, 1987.

Wang Yuhu 王毓瑚 (ed.). *Wang Zhen Nongshu* 王禎農書 [Wang Zhen's *Agricultural Treatise*]. Beijing: Nongye chubanshe, 1981.

Wang Zheng 王徵. (*Xin zhi*) *Zhu qi tushuo* [新製]諸器圖說 [Diagrams and explanations of many machines (new printing)]. Beijing: Zhonghua shuju, 1985.

Watabe Takeshi 渡部武. "Chûgoku nôsho *Kôshikido* no ryûden to sono eikyô ni tsuite" 中國農書〈耕織圖〉の流傳とその影響について [Studies about the spread and influence of the *Pictures of Tilling and Weaving*]. *Bulletin of the Faculty of Letters,* Tokai University, 46 (1986), 1–37.

Watrous, James. *The Craft of Old-Master Drawings*. Madison: The University of Wisconsin Press, 1957.

Weller, Brant. "The European Discovery of External Cultures and Their Effect on European Expansion." *World History Bulletin* 24.1 (2008 Spring), 38–40.

White, Lynn, Jr. "The Flavor of Early Renaissance Technology." In B. E. Levy (ed.), *Developments in the Early Renaissance*, 36–57. Albany, NY: State University of New York Press, 1972.

Whitfield, Roderick. "Chang Tse-tuan's *Ch'ing-ming shang-ho t'u*." Ph.D. Dissertation. Princeton University, 1965.

———. "Material Culture in the Northern Song Dynasty–The World of Zhang Zeduan." In Kai-Yin Lo (ed.). *Bright as Silver, White as Snow: Chinese Ceramics from Late Tang to Yuan Dynasty*, 49–70. Hong Kong: Yong Ming Tang, 1998.

Wilkinson, Endymion. *Chinese History: A Manual*. Cambridge, MA: Harvard University Press, 2000.

Willmott, W. E. (ed.). *Economic Organization in Chinese Society*. Stanford: Stanford University Press, 1972.

Wood, Frances. *Chinese Illustration*. London: The British Library, 1985.

Wright, Arthur F. and Denis Twitchett (eds.). *Confucian Personalities*. Stanford: Stanford University Press, 1962.

Wright, M. T. "On the Lift Pump." *History of Technology* 18 (1996), 13–37.

Wu Hung. "The Origins of Chinese Painting (Paleolithic Period to Tang Dynasty)." In Yang, Barnhart et al. *Three Thousand Years*, 15–85.

Wu Hung. "Picturing or Diagramming the Universe." In Bray et al., *Graphics and Text*, 191–214.

Wu Jiming 吴繼明. *Zhongguo tuxue shi* 中國圖學史 [A history of Chinese drawing]. Wuchang: Huazhong ligong daxue, 1988.

Wu, K. T. "Illustrations in Sung Printing." *The Quarterly Journal of the Library of Congress* 28 (July 1971), 173–95.

Wu, Nelson I. "Tung Ch'i-ch'ang (1555–1636): Apathy in Government and Fervor in Art." In Arthur F. Wright and Denis Twitchett (eds.), *Confucian Personalities*, 260–93. Stanford: Stanford University Press, 1962.

Xiao Zhenshi 蕭振士. *Zhongguo gudai jianzhu jishu shi* 中國古代建築技術史 [A history of building technology in China]. 3 vols. Taipei: Po-yuan, 1993.

Xu Dixin and Wu Chengming (eds.). *Chinese Capitalism, 1522–1840*. New York: St. Martin's Press, 2000.

Yabu'uchi Kiyoshi 藪内清. *Tenkô kaibutsu* 天工開物 [(Translation of the) *Tiangong kaiwu*]. Tokyo: Heibonsha, 1969.

Yang, Lien-sheng. "Review of YABU'UCHI Kiyoshi, *Tenkô kaibutsu no kenkyû* (Tokyo: Kôsei, 1953)." *Harvard Journal of Asiatic Studies* 17 (1954), 307–16.

———. *Studies in Chinese Institutional History*. Cambridge, MA: Harvard University Press, 1963.

Yang Xin. "Approaches to Chinese Painting." In Yang, Barnhart et al., *Three Thousand Years*, 1–4.

Yang Xin, Richard M. Barnhart, et al. *Three Thousand Years of Chinese Painting*. New Haven and Beijing: Yale University Press and Foreign Languages Press, 1997.

Yee, Cordell D. K. "Cartography in China: Chinese Cartography Among the Arts." In Harley and Woodward, *A History of Cartography*, vol. 2, book 2, 128–69.

———. "Reinterpreting Traditional Chinese Geographical Maps." In Harley and Woodward, *A History of Cartography*, vol. 2, book 2, 35–67.

———. "Taking the World Measure: Chinese Maps between Observation and Text." In Harley and Woodward, *A History of Cartography*, vol. 2, book 2, 96–127.

Yetts, W. Perceval. "A Chinese Treatise on Architecture." *Bulletin of the the School of Oriental and African Studies* 4 (1927), 473–92.

Yonezawa Yoshio 米沢嘉圃. "Chûgoku kaiga ni okeru shomin–toku ni nôminga ni tsuite" 中國繪畫における庶民ー特に農民畫について [Common people in Chinese paintings, especially peasant paintings]. *Tôyô bunka* 2 (1950), 22–64.

Yoshida Tora. *Salt Production Techniques in Ancient China: The "Aobo Tu."* Trans. and rev. Hans Ulrich Vogel. Leiden: E. J. Brill, 1993.

You Zhanhong 游戰洪. "Lun junqi zeli" 論軍器則例 [A study of regulations and precedents on weapons and military equipment]. In Moll-Murata et al., *Chinese Handicraft Regulations of the Qing Dynasty*, 307–24.

Yu Weichao (ed.). *A Journey into China's Antiquity*. 4 vols. Beijing: Morning Glory Publishers, 1997.

Zhang Baichun 張柏春. "Wang Zheng *Xin zhi Zhuqi tushuo* bianxi" 王徵新制諸器圖說辨析 [An inquiry into Wang Zheng's *Diagrams and Explanations of Various Newly Built Machines*]. *Zhongguo keji shiliao* 17.1 (1996), 88–91.

Zhang Baichun. "Archimedian Mechanical Knowledge in 17th Century China." Paper presented at the International Congress of History of Science, Mexico City, 2001, 16pp.

Zhang Baichun and Jürgen Renn (eds.). *Transformation and Transmission: Chinese Mechanical Knowledge and the Jesuit Intervention*. Berlin: Max Planck Institute for the History of Science, 2006.

Zhang Baichun and Tian Miao. "Wang Zheng and the Transmission of Western Mechanical Knowledge to China." In Zhang and Renn, *Transformation and Transmission*, 75–88.

Zhang Baichun 張柏春 and Tian Miao 田淼. "Zhongguo gudai jixie yu qiwu de tuxiang biaoda" 中國古代機械與器物的圖像表達 [The visual presentation of ancient Chinese machines and devices]. *Gugong bowuyuan yuankan* 故宮博物院院刊 3 (2006), 81–97.

Zhang Yinlin 張蔭麟. "Zhongguo lishi shang zhi qiqi ji qi zuozhe" 中國歷史上之奇器及其作者 [Inventions and inventors in Chinese History]. In Zhang Yinlin , *Zhang Yinlin wenji* 張蔭麟文集 [The collected works of Zhang Yinlin], 64–85. Taipei and Hong Kong: Zhonghua congshu, 1956.

Zheng Wei 鄭為. "*Zhakou panche tujuan*" 閘口盤車圖卷 [Transport carts at the mill]. *Wenwu* (February 1966), 17–25.

Zhu Qi Tushuo [新製]諸器圖說. See Wang Zheng. (ZQTS)

Zhu Renfu 朱仁夫. *Zhongguo gudai shufa shi* 中國古代書法史 [The history of calligraphy in ancient China]. Beijing: Peking University Press, 1992.

Index

Page numbers in italics refer to illustrations.

www.ingramcontent.com/pod-product-compliance
Ingram Content Group UK Ltd.
Pitfield, Milton Keynes, MK11 3LW, UK
UKHW021835100926
8556IPUK00004B/353

9 789888 208159